Lecture Notes in Mathematics

Edited by A. Dold and B. Eckmann

1280

Erhard Neher

Jordan Triple Systems by the Grid Approach

Springer-Verlag
Berlin Heidelberg New York London Paris Tokyo

Author

Erhard Neher
Department of Mathematics, University of Ottawa
585 King Edward, Ottawa, Ontario, Canada K1N 6N5

Mathematics Subject Classification (1980): 14J26, 17C10, 17C20, 17C40, 17C65, 46K15, 46L10, 46L35

ISBN 3-540-18362-0 Springer-Verlag Berlin Heidelberg New York
ISBN 0-387-18362-0 Springer-Verlag New York Berlin Heidelberg

Printed in Germany

Printing and binding: Druckhaus Beltz, Hemsbach/Bergstr.
2146/3140-543210

TABLE OF CONTENTS

INTRODUCTION

Some background and definitions

In these notes we study Jordan triple systems. A prominent class of examples of Jordan triple systems are associative algebras. To consider an associative algebra A with a bilinear product $(x,y) \to xy$ as a Jordan triple system means to forget the bilinear product of A and its unit element and instead work just with the triple product $(x,y) \to P(x)y = xyx$ and its linearization $(x,y,z) \to \{xyz\} = xyz + zyx$. There are good reasons for doing this, some of which are indicated below. Of course, the associative law has to be rephrased in terms of the triple product, leading to the general definition due to K. Meyberg: A Jordan triple system consists of a module V over a commutative associative ring k and a quadratic map $P: V \to End_k V$ (the triple product) satisfying the following identities in all scalar extensions:

(1) $L(x,y)P(x) = P(x)L(y,x)$

(2) $L(P(x)y,y) = L(x,P(y)x)$

(3) $P(P(x)y) = P(x)P(y)P(x)$,

where the linear map $L(x,y) \in End_k V$ is defined by $L(x,y)z = \{xyz\} = P(x+z)y - P(x)y - P(z)y$.

Jordan triple systems received their name because they are generalizations of Jordan algebras (which in turn are generalizations of associative algebras): although this is not the usual definition, a Jordan algebra can be defined as a Jordan triple system which contains a unit element, i.e. an element e with $P(e) = Id$. Over the past 20 years the theory of Jordan algebras has seen major advances mainly due to the work of N. Jacobson, K. McCrimmon and the Russian School, notably E.I. Zel'manov. An exposition of this has been given by N. Jacobson in his book [19] and his lecture notes [20] [21] and by K. McCrimmon in [41].

Besides Jordan algebras another class of Jordan triple systems, which has been thoroughly investigated in recent years, is the class of Jordan pairs. A Jordan pair is a Jordan triple system which has a polarization: V is the direct sum of two submodules, $V = V^+ \oplus V^-$, satisfying

$$P(V^\varepsilon)V^\varepsilon = 0 = P(V^+,V^-)V, \quad P(V^\varepsilon)V^{-\varepsilon} \subset V^\varepsilon.$$

The theory of Jordan pairs was proposed by K. Meyberg and pursued and described by O. Loos in his lecture notes [31].

One of the driving forces of the development of the Jordan pair and

Jordan triple theory was the increasing number of applications found in other areas of mathematics showing that Jordan triple systems are more than just generalizations of associative and Jordan algebras.

The most important of these applications was found by M. Koecher, the category of circled bounded symmetric domains in $\mathbb{C}^n$ is equivalent to a certain category of Jordan triple systems (namely finite-dimensional positive hermitian Hilbert triples in the terminology of these notes, see IV §2). An exposition of this fundamental theorem, from different points of view, was given by M. Koecher in his notes [28], O. Loos in [32] and I. Satake in [52]. This theorem has been generalized by W. Kaup and his collaborators to infinite dimensions, using certain infinite-dimensional Banach Jordan triple systems (see IV §3). An exposition of this part of Jordan theory can be found in H. Upmeier's notes [55]. Banach Jordan algebras are studied in [11] and [17].

There are many more applications of Jordan structures known - so many that we can only mention some areas of applications without going into details: Lie algebras, algebraic and Lie groups, symmetric spaces, in particular symmetric R-spaces, Siegel domains, cones, various types of geometries and mathematical physics. Some of these applications have been described in the surveys [6, 37, 16].

The grid approach motivated by examples

In these notes we present a new method in Jordan theory: the grid approach. For a first understanding of grids it might be helpful to keep in mind the theory of split semisimple Lie algebras in characteristic 0. Roughly speaking, covering grids in Jordan triple systems play the same rôle as Chevalley bases for split semisimple Lie algebras, a spanning system of elements closed under the product. We will make this more precise in the course of this introduction starting with two instructive examples. But first we need to introduce some general notions.

Grids will be special families of tripotents. An element e in a Jordan triple system V over k is called a <u>tripotent</u> if $P(e)e = e$, in which case V has a <u>Peirce decomposition</u> with respect to e:

$$V = V_2(e) \oplus V_1(e) \oplus V_0(e),$$

with $V_i(e) = \{v \in V;\ L(e,e)v = iv\}$ if $1/2 \in k$, for a general k see I §1.3. We can define <u>generalized Peirce spaces</u> with respect to an arbitrary family E of tripotents:

$$V_I(E) = \bigcap_{e \in E} V_{i(e)}(e),\quad I = (i(e))_{e \in E} \in \{0,1,2\}^E$$

In general, V will not be the sum of the Peirce spaces $V_I(E)$ nor will it be determined by E. Some information can only be expected for the <u>cover of</u> E which is

$$C_V(E) = \oplus \{V_I(E);\ E \cap V_I(E) \neq \emptyset\}.$$

We say E <u>covers</u> V if $C_V(E) = V$.

<u>Example 1</u> (rectangular matrix system): For an associative k-algebra D with involution $d \to \bar{d}$ the $(p\times q)$-matrices over D form a Jordan triple system $\mathrm{Mat}(p,q;D,^-)$ over k with quadratic representation $P(x)y = x\bar{y}^t x$. A Chevalley-type D-basis of $V = \mathrm{Mat}(p,q;D,^-)$ is given by the matrix units E_{ij}:

$$R(p,q) := \{E_{ij};\ 1\leq i\leq p, 1\leq j\leq q\},$$

is a concrete realization of what has been called a rectangular grid. It is straightforward to check that $R(p,q)$ has the following two decisive properties:

(G) $R(p,q)$ is a family of tripotents which is closed under triple products (the "grid property"), i.e. for $i \neq k$, $l \neq j$ one has

$$\{E_{ij}\ E_{ij}\ E_{kj}\} = E_{kj} = \{E_{kl}\ E_{kl}\ E_{kj}\},\ i\neq k,\ l\neq j,$$
$$\{E_{ij}\ E_{kj}\ E_{kl}\} = E_{il},$$

whereas all other types of products vanish,

(C) $R(p,q)$ covers $V = \mathrm{Mat}(p,q;D,^-)$:

$$V = \oplus\, DE_{ij}$$

where DE_{ij} is a Peirce space of V relative to $R(p,q)$:

$$DE_{ij} = V_2(E_{ij}) \subset V_1(E_{ik}) \cap V_1(E_{lj}) \cap V_0(E_{lk})$$

It is easily seen that in this example the whole Jordan triple structure, i.e. the module and the triple product, is determined by

(i) the covering rectangular grid $R(p,q)$ and

(ii) the coordinate system $(D,^-)$.

<u>Example 2</u> (hermitian matrix system): Let $(D,^-)$ be as in Example 1 and π be a second involution of D commuting with $^-$. Then the π-hermitian $(p\times p)$-matrices $x=x^{\pi t}$ form a Jordan triple system $H_p(D,\pi,^-)$ with triple product $P(x)y = x\bar{y}^t x$ (which is a Jordan algebra iff $\pi = {}^-$). In this case a Chevalley-type (D,π)-basis (use $\mathrm{Fix}\,\pi$ on the diagonal) is formed by the hermitian matrix units $H_{ii} = E_{ii}$, $H_{ij} = H_{ji} = E_{ij} + E_{ji}$, $i \neq j$:

$$H(p) = \{H_{ij};\ 1\leq i\leq j\leq p\}$$

is a concrete realization of a hermitian grid. It has similar properties to $R(p,q)$:

(G) $H(p)$ is a family of tripotents closed under triple products in the following sense: For i,j,k ≠ one has

$\{H_{ij}\ H_{ij}\ H_{ii}\} = 2H_{ii}$, $\{H_{ii}\ H_{ii}\ H_{ij}\} = H_{ij}$

$\{H_{ij}\ H_{ij}\ H_{ik}\} = H_{ik}$,

$P(H_{ij})H_{ii} = H_{jj}$

$\{H_{im}\ H_{mn}\ H_{nj}\} = H_{ij}$ (m arbitrary) ,

$\{H_{im}\ H_{mn}\ H_{ni}\} = 2H_{ii}$, (m,n arbitrary)

whereas all other types of products vanish.

(C) $H(p)$ covers $V = H_p(D,\pi,^-)$:

$V = \oplus_i\ (\text{Fix}\ \pi)\ H_{ii} \oplus (\oplus_{i<j} DH_{ij})$

where for i,j,k,m ≠:

$(\text{Fix}\ \pi)H_{ii} = V_2(H_{ii}) = V_2(H_{ij}) \cap V_0(H_{jj}) \cap V_0(H_{jk})$

$DH_{ij} = V_2(H_{ij}) \cap V_1(H_{ii}) \cap V_1(H_{jj}) \subset V_1(H_{jk}) \cap V_0(H_{km})$.

Analogously to the rectangular matrix case the Jordan triple $H_p(D,\pi,^-)$ is determined by

(i) the covering hermitian grid $H(p)$ and

(ii) the coordinate system $(D,\pi,^-)$.

These two examples suggest the following program: Develop a theory for grids in general such that $R(p,q)$ and $H(p)$ become examples and such that interesting classes of Jordan triple systems allow the

GRID APPROACH (finite-dimensional version):

Look at a Jordan triple system as an object determined by:

(i) a finite covering grid (carrying all the combinatorial information), and

(ii) a coordinate system.

We want to point out that this philosophy appears in different areas in mathematics, for example in the theory of split semisimple Lie algebras or algebraic groups or in the theory of W* - algebras.

Although we will find that important classes of Jordan triple systems are accessible to the finite-dimensional version of the grid approach, the Jordan triple systems appearing in some applications of Jordan theory require an extended version which we want to motivate by the following.

Example 3 (Hilbert-Schmidt operators). The Hilbert-Schmidt operators from E to F, where E and F are Hilbert spaces over $\mathbb{K} = \mathbb{R}$, $\mathbb{C}$ or $\mathbb{H}$, form a real Jordan triple system $L_2(E,F;\mathbb{K})$ with triple product $P(x)y = xy^*x$ (y^* = adjoint of y). With respect to Hilbert space bases $(e_j)_{j\in J}$ of E and $(e_i)_{i\in I}$ of F we can identify $L_2(E,F;\mathbb{K})$ with matrices

$x = (x_{ij})_{i\in I, j\in J}$ over $\mathbb{K}$ such that $\sum_{i,j}|x_{ij}|^2 < \infty$. Then the product becomes $P(x)y = x\bar{y}^t x$ ($\bar{}$ = canonical involution of $\mathbb{K}$), and the whole triple system behaves like an infinite rectangular matrix system. Indeed, we again have a rectangular grid

$$R(I,J) = \{e_i \otimes e_j^* \; ; \; i\in I, j\in J\},$$

satisfying the multiplication rules (G) of Example 1. However, a new phenomenon appears: $R(I,J)$ does not cover $V = L_2(E,F;\mathbb{K})$, we only have that

($\bar{C}$) the cover of $R(I,J)$ is dense in V with respect to the Hilbert-Schmidt operator norm: $V = \overline{C_V(R(I,J))}$.

Nevertheless, one can recover the structure of $L_2(E,F;\mathbb{K})$ from the cover of $R(I,J)$. Thus, here we need a topology as a third ingredient (besides a grid and a coordinate system) in order to completely determine the structure of $L_2(E,F;\mathbb{K})$. This example suggests the

<u>GRID APPROACH (topological version):</u>

Look at a Jordan triple system as an object determined by

(i) a grid of arbitrary size,

(ii) a coordinate system, and

(iii) a topology with respect to which the cover of the grid is dense.

We see that both versions of the grid approach lead to the following four fundamental questions which will be answered in these notes:

- What is the general definition of a grid so that $R(p,q)$, $H(p)$ and $R(I,J)$ are examples of grids? - Chapter I
- Can grids be classified? - Chapter II
- What does the cover of a grid look like? - Chapter III
- Which classes of Jordan triple systems allow the grid approach? - Chapter I § 5,6 and Chapter IV.

A short description of the contents can be found below ("the new results") or at the beginning of each chapter and section.

The origins of grids

One special grid has long been known to Jordan algebraists: In our terminology Jacobson's Coordinatization Theorem states that a Jordan algebra covered by a hermitian grid $H(p)$, $3 > p$, is a hermitian matrix algebra, a fundamental result in the classification theory of Jordan algebras. Since semi-primitive Jordan algebras with capacity are always covered by a hermitian grid, it was however not necessary to develop a general theory for grids in Jordan algebras.

The first hint about grids in Jordan triple systems can be found in K. McCrimmon's review [38] of O. Loos' notes [31]. To explain this, let us recall the procedure of [31] for classifying Jordan pairs: Choose a maximal tripotent e so that $V = V_2(e) \oplus V_1(e)$. If $V_1(e) = 0$ then V is a mutation of a Jordan algebra hence is known modulo Jordan algebra theory. Otherwise V is the standard imbedding of an alternative polarized triple system (called alternative pair in [31]) which is defined on $V_1(e)$. In view of the "complicated and asymmetric identities of alternative pairs" McCrimmon remarked that it may be possible to "analyse the Jordan pair directly using 'collinear' rather than merely 'orthogonal' family of tripotents", [38] p. 689.

Subsequently ([40]) he considered "compatible" families of tripotents and then defined special classes of grids. However, he did not develop a theory for grids in general nor did he give a classification-free proof that covering grids exist for certain classes of semisimple Jordan triple systems. Nevertheless the merits of grids became clear in the paper [42] by K. McCrimmon and K. Meyberg, where the covers of finite rectangular, symplectic and hermitian grids were coordinatized.

Independently of [40] and [42] (actually before their publication) and unaware of McCrimmon's remark in [38] the author was led to grids by studying weight space decompositions of compact Jordan triple systems with respect to the structure algebra: The weight spaces turn out to be exactly the generalized Peirce spaces of a suitable grid. Details will appear elsewhere.

The new results

Since grids are a new concept in Jordan theory, most of the results in these notes are new. In case a result has been previously proven we have indicated this to the best of our knowledge. We apologize in advance for any omissions.

The results can be summarized according to the various themes of the chapters:

Chapter I ("What is the general definition of a grid so that $\mathcal{R}(p,q)$, $\mathcal{H}(p)$ and $\mathcal{R}(I,J)$ are examples of grids ?") - We develop a theory for families of Peirce compatible tripotents (called cogs), leading to the fundamental notion of a grid. Roughly speaking a grid is a cog which is multiplicatively closed up to association. For example, $\mathcal{R}(p,q) \subset Mat(p',q';D,^{-})$, $p \leq p'$, $q \leq q'$, is a grid, but also

$$\mathcal{R}(p,q,(\zeta_{ij})) = (\zeta_{ij} E_{ij};\ 1 \leq i \leq p,\ 1 \leq j \leq q)$$

for any choice of scalars $\zeta_{ij} \in D$ satisfying $\zeta_{ij}\bar{\zeta}_{ij} = 1$.

Chapter II ("Can grids be classified?") - Indeed they can be classified modulo "association". We show that every grid is a unique disjoint union of connected grids and that every connected grid is associated to one of seven standard grids. For example, $\mathcal{R}(p,q,(\zeta_{ij}))$ is associated to $\mathcal{R}(p,q)$. The seven standard grids fall into 5 types of grids of arbitrary size (for example $\mathcal{R}(I,J)$ is one type) and 2 exceptional grids consisting of 16 resp. 27 tripotents. This last grid, called Albert grid, naturally carries a geometry, which is the same as the geometry of the 27 lines on a cubic surface.

Chapter III ("What does the cover of a grid look like?") - We generalize the coordinatization theorems of McCrimmon and Meyberg to grids of arbitrary size and prove new coordinatization theorems for the remaining four types of grids.

Chapter IV ("Which classes of Jordan triple systems allow the grid approach?") - The results of chapters I - III give, among others, a classification of a new class of Jordan triple systems, namely the ones covered by a grid. This class is distinct from the class of Jordan triple systems recently classified by E. Zel'manov ([56]). In the intersection of both classes lie the simple Jordan triple systems covered by a grid. These are identified in IV §1. In the last two sections we give examples of how the topological version of the grid approach works. We classify Hilbert triples over $\mathbb{R}$ and $\mathbb{C}$, generalizing results of Kaup, and we present the theory of atomic JBW*-triples in a new way which is independent of the elaborate theory of JB-algebras.

The advantages of grids

Grids provide a more direct and natural approach to Jordan triple systems than presently existing ones, thus justifying McCrimmon's remark as quoted above. Grids make it superfluous to leave the category of Jordan triple systems (or Jordan pairs) and study alternative triple systems (or alternative pairs) first in order to gain information about Jordan triple systems.

It is also no longer necessary to reduce triple theory to algebra theory as has been customary until now. In fact the theory of grids as presented in these notes is independent of the theory of Jordan algebras. Indeed, parts of Jordan algebra theory (e.g. classification) are special cases of our results.

Since grids can have arbitrary size, systems of infinite rank are treated at the same time as the finite rank case. This is not only a technical improvement from the algebraic point of view, but also allows a more direct application of the algebraic theory in a functional analytic setting (see for example IV§§2,3).

Besides these fundamental advantages we would like to point out two more innovations: The coordinatization theorems proven by McCrimmon and Meyberg in [42] and supplemented here in Chapter III immediately give information about unital bimodules of Jordan triple systems covered by a grid. Details have not been included here, however they are easily deduced along the lines of [42] §6.

And, at last we give a confirmation (via the Albert grid) of the conjecture that there is a connection between the 27 lines on a cubic surface and the 27 dimensional exceptional Jordan structures.

Concluding remarks

The notes are self-contained up to some elementary facts concerning Peirce decompositions (to be found in [31] § 5 or [44]) and the results of the papers [40] and [42]. However all that is needed is put together in I §1, including the necessary definitions.

These notes contain the proofs of the results of the survey [47]. These results were also announced at the Oberwolfach meeting on Jordan algebras in 1982 and in an improved form again in 1985.

Major parts of the research for these notes were done at the University of Virginia, Charlottesville, in the academic year 1980/81 while the author held a Forschungsstipendium der Deutschen Forschungsgemeinschaft (Research fellowship of the German Research Council). The notes were completed at the University of Ottawa with the support of a NSERC operating grant.

Besides DFG and NSERC the author wants to thank all mathematicians whose interest and encouragement were of great help during his research on grids, in particular thanks are due to J.Dorfmeister, J.Faulkner, G.Horn, W.Kaup, M.Koecher, K.McCrimmon, K.Meyberg and M.Racine.

Ottawa, Winter 1986 E. Neher

CHAPTER I. SPECIAL FAMILIES OF COMPATIBLE TRIPOTENTS

This first chapter is the foundation for the whole notes. After a review of the basic definitions, results and examples for Jordan triple systems in §1 we consider in the following sections special families of compatible tripotents, i.e. tripotents whose Peirce projections commute. We derive the fundamental properties of these families (§2-§5) and prove existence theorems (§§ 5,6).

§1 Basic definitions, known results, examples

In this section we recall some of the basic facts about Jordan triple systems which are needed in the following. We also give the examples which will play a fundamental role throughout these notes.

1.1. The results of this section are mostly contained in [30], [31] or [44] which are the standard texts on Jordan triple systems. To fix our notation and assumptions we begin with the definition:

A Jordan triple system V is a module V over an arbitrary ring k of scalars together with a quadratic map $P: V \to \operatorname{End} V$ such that the following identities hold in all scalar extensions:

$$(1.1) \qquad L(x,y)P(x) = P(x)L(y,x)$$

$$(1.2) \qquad L(P(x)y,y) = L(x,P(y)x)$$

$$(1.3) \qquad P(P(x)y) = P(x)P(y)P(x)$$

where $L(x,y)z = P(x,z)y = P(x+z)y - P(x)y - P(z)y$. We use the notation

$$\{xyz\} = L(x,y)z = P(x,z)y = \{zyx\}$$

We refer to P resp. $\{\ldots\}$ as the triple product.

A homomorphism $\phi: V \to V'$ of Jordan triple systems over k is a k-linear map such that $\phi P(x)y = P(\phi x)\phi y$ for all $x,y \in V$. Isomorphisms and automorphisms are defined in the usual way.

A subsystem (resp. an ideal) of a Jordan triple system V is a submodule W of V such that $P(W)W \subset W$ (resp. $P(V)W + P(W)V + \{VVW\} \subset W$).

The defining identities for Jordan triple systems imply of course more identities (eg. by linearization, see [31]§2), of which we list in particular the following

$$(1.4)\qquad L(x,y)P(x) = P(x,P(x)y) = P(x)L(y,x)$$

$$(1.5)\qquad P(x,z)L(y,x) + P(x)L(y,z) = P(x,\{xyz\}) + P(z,P(x)y) = L(x,y)P(x,z) + L(z,y)P(x)$$

$$(1.6)\qquad L(x,\{yxz\}) + P(x)P(y,z) = L(x,z)L(x,y) + L(P(x)y,z) = L(x,y)L(x,z) + L(P(x)z,y)$$

$$(1.7)\qquad L(\{xyz\},y) = L(z,P(y)x) + L(x,P(y)z)$$

$$(1.8)\qquad L(x,\{yxz\}) = L(P(x)y,z) + L(P(x)z,y)$$

$$(1.9)\qquad L(x,y)L(z,y) = P(x,z)P(y) + L(x,P(y)z)$$

$$(1.10)\qquad P(x,z)L(y,x) = P(P(x)y,z) + L(z,y)P(x)$$

$$(1.11)\qquad L(x,y)P(x,z) = P(P(x)y,z) + P(x)L(y,z)$$

$$(1.12)\qquad L(x,y)P(z) + P(z)L(y,x) = P(z,\{xyz\})$$

$$(1.13)\qquad L(x,y)L(x,z) = L(P(x)y,z) + P(x)P(y,z)$$

$$(1.14)\qquad [L(x,y),L(u,v)] = L(\{xyu\},v) - L(u,\{yxv\}),$$

where as usual $[A,B] = AB - BA$.

The map

$$B(x,y) = \mathrm{Id} - L(x,y) + P(x)P(y) \in \mathrm{End}\ V$$

is called the Bergman operator (because of reasons explained in [28], [32]). It is used to define the Jacobson radical

$$(1.15)\qquad \mathrm{Rad}\ V = \{x \in V;\ B(x,y) \text{ is invertible for all } y \in V\},$$

see [31]§4 or [44]XIII. If Rad $V = 0$ one calls V semisimple. The Bergman operator satisfies the identity

$$(1.16)\qquad P(B(x,y)z) = B(x,y)P(z)B(y,x),$$

so $B(x,y)$ is an automorphism of V if it is invertible with inverse $B(y,x)$. A special example of this fact is Theorem 1.13.

1.2. In this subsection we present the basic examples of Jordan triple systems.

Example 1.1. Let J be a quadratic Jordan algebra over k with quadratic representation U (see [20], [21], [31] or [44] for a definition). By forgetting the squaring (or the identity element if J has one) and putting $P = U$ one obtains a Jordan triple system, denoted by $V(J)$ and called the Jordan triple system associated to J. For an abstract characterization of such Jordan triple systems see [48].

Example 1.2. Let $V = (V^+,V^-)$ be a Jordan pair over k and put $V = V^+ \oplus V^-$, $P(x^+ \oplus x^-)(y^+ \oplus y^-) = P(x^+)y^- \oplus P(x^-)y^+$. Then V is a Jordan triple system called the Jordan triple system of a Jordan pair. The

Jordan triple systems arising in this way are exactly the <u>polarized Jordan triple systems</u>, i.e. Jordan triple systems $V = V^+ \oplus V^-$ where the P-operator satisfies ($\varepsilon = \pm$)

$$P(V^\varepsilon)V^\varepsilon = 0,\ P(V^\varepsilon)V^{-\varepsilon} \subset V^\varepsilon ,$$
$$L(V^\varepsilon,V^\varepsilon)V^{-\varepsilon} = P(V^\varepsilon,V^{-\varepsilon})V^\varepsilon = 0.$$

In this way we will identify Jordan pairs = polarized Jordan triple systems.

<u>Example 1.3.</u> (<u>rectangular matrix system</u>) This class of examples consists of two subclasses:

a) Let V be the 1×2 matrices with entries in a unital alternative algebra D over k, which has a k-linear involution $d \to \bar{d}$, and define the quadratic representation by $P(x)y = x(\bar{y}^t x)$. We denote this Jordan triple system by Mat(1,2;D) or Mat(1,2;D,$^-$).

b)(i) Before we define the second subclass we make the following remark: For any unital algebra D over k with an involution $d \to \bar{d}$ we may look at the $p \times q$ matrices Mat(p,q;D), $p + q \geq 3$, together with the product $P(x)y = x(\bar{y}^t x)$. It was shown in [34] that this defines a Jordan triple system iff D is alternative and even associative if $p + q \geq 4$. We will generalize this example to matrices of arbitrary size.

(ii) Let I and J be arbitrary index sets and D an associative algebra over k. We define a <u>matrix of type</u> $I \times J$ <u>over</u> D as a family $x = (x_{ij})_{(i,j) \in I \times J}$ of elements of D such that for every $i \in I$ the <u>i-th row</u> $(x_{ij})_{j \in J}$ and for every $j \in J$ the <u>j-th column</u> $(x_{ij})_{i \in I}$ contain only a finite number of non-zero elements. Assuming that $d \to \bar{d}$ is a k-linear involution of D we can define a Jordan triple system on the space of all matrices of type $I \times J$ over D by putting as above $P(x)y = x\bar{y}^t x$. (Note that $x\bar{y}^t x$ makes sense). This triple system is denoted by Mat(I,J;D) or Mat(I,J;D,$^-$) if we want to exhibit the involution. In case $\#I = p < \infty$ and $\#J = q < \infty$ we also write Mat(p,q;D) for Mat(I,J;D).

(iii) A D-basis of Mat(I,J;D) is given by all <u>rectangular matrix units</u> $E_{ij} = (x_{kl})$ with $x_{kl} = \delta_{ki}\delta_{lj}$:

$$R := R(I,J) := \{E_{ij};\ i \in I,\ j \in J\}$$

We have a decomposition

$$\mathrm{Mat}(I,J;D) = \bigoplus_{i \in I} \bigoplus_{j \in J} DE_{ij}$$

and the product is given by the following rules which are stated in such a way that they also hold for the example a):

$P(aE_{ij})bE_{ij} = a(\bar{b}a)E_{ij}$,
$\{aE_{ij}\ bE_{ij}\ cE_{kj}\} = (c\bar{b})aE_{kj}$, $\{aE_{ij}\ bE_{ij}\ cE_{il}\} = a(\bar{b}c)E_{il}$,
$\{aE_{ij}\ bE_{kj}\ cE_{kl}\} = a(\bar{b}c)E_{il}$,

all other products (modulo symmetry in the outer variables) vanish.

(iv) That Mat(I,J;D) actually is a Jordan triple system may be best seen in the following more general context: Let A be an associative algebra over k with an involution $a \to a^*$ and define $P(a)b = ab^*a$. Then the identities (1.1) - (1.3) are easily checked in $K \otimes A$ for any extension $K \supset k$. Hence we have a Jordan triple system; call it $V(A,*)$. Now every subsystem W of $V(A,*)$ is again a Jordan triple system. For $A = \mathrm{Mat}(I \dot\cup J, I \dot\cup J; D)$, $a^* = \bar{a}^t$, we can imbed Mat(I,J;D) into A in a canonical way so that it becomes a subsystem of A.

Example 1.4. (symplectic matrix system) Let I be an arbitrary index set and define the alternating matrices of type $I \times I$ as those skew-symmetric matrices $x = -x^t$ of type $I \times I$ with diagonal elements $x_{ii} = 0$. We assume that the entries of x belong to an extension $K \supset k$, i.e. a unital commutative and associative algebra over k. Moreover, let $a \to \bar{a}$ be a k-linear automorphism of K with period 2. Then the alternating matrices form a subsystem of $\mathrm{Mat}(I,I;K,{}^-)$ with product $P(x)y = -x\bar{y}x$, which we denote by $A(I;K,{}^-)$ or $A(I;K)$ for short or $A(p;K)$ in case $\#I = p$ and which we call the symplectic (or alternating in case ${}^- = \mathrm{Id}$) matrix system. We point out that the involution ${}^-$ is only needed to define the product, but not the underlying module of $A(I;K,{}^-)$.

The symplectic matrix units are the matrices $F_{ij} = E_{ij} - E_{ji} = -F_{ji}$ for $i,j \in I$, $i \neq j$. To write down a K-basis for $A(I;K)$ we have to use a total order $\leqslant$ on I:

$$S = S(I) = \{F_{ij};\ i<j\}$$

The triple product of $A(I;K)$ is completely determined by the following rules (for $i,j,k,l \neq$):

$P(aF_{ij})bF_{ij} = a\bar{b}aF_{ij}$, $\{aF_{ij}\ bF_{ij}\ cF_{ik}\} = a\bar{b}cF_{ik}$,

$\{aF_{ij}\ bF_{kj}\ cF_{kl}\} = a\bar{b}cF_{il}$,

all other products (modulo symmetry) vanish.

Finally, since $A(1;K) = 0$, $A(2;K) \cong K$ and $A(3;K) \cong \mathrm{Mat}(1,3;K)$ (see [40](0.11)) we always assume $\#I \geqslant 4$ for this example.

Example 1.5. (hermitian matrix system) Here we start with a unital alternative algebra D over k with a nuclear k-linear involution $a \to a^\pi$ (i.e. all norms aa^π are in the nucleus $N(D) = \{n \in D;\ (nd_1)d_2 = n(d_1d_2)$ for all $d_1,d_2 \in D\}$) and an ample subspace D_0 (i.e. a subspace of π-symmetric elements in the nucleus of D containing the unit 1 of D and closed under $aD_0a^\pi \subset D_0$ for all $a \in D$). Also, let I be an index set with $\#I \geqslant 2$. The underlying module of this example is formed by the $I \times I$ hermitian matrices $(x^\pi = x^t \in \mathrm{Mat}(I,I;D))$ over D whose diagonal elements lie in D_0. To define the triple product we moreover assume that we have another involution $a \to \bar{a}$ of D commuting with the given involution π and leaving D_0 invariant ($a^\pi = \bar{a}$ allowed).

For #I $\leqslant$ 3 it is well-known that the hermitian matrices over D form a Jordan algebra J (see e.g. [20]). Let U be its quadratic representation and define a new quadratic representation by $P(x)y = U(x)\bar{y}^t$. That this so-called mutation actually defines a Jordan triple system follows from [44]10 Theorem 2. For #I $\geqslant$ 4 we assume in addition that D is associative and define the quadratic representation for the hermitian matrices by $P(x)y = x\bar{y}^t x$, i.e. we consider them as a subsystem of $Mat(I,I;D,^-)$.

In both cases the resulting triple system is called a <u>hermitian matrix system</u> and denoted by $H_I(D,D_0,\pi,^-)$ or $H_p(D,D_0,\pi,^-)$ if #I = p. It is denoted by $H_I(D,D_0,^-)$ in case $\bar{a} = a^\pi$. We remark that $H_I(D,D_0,^-)$ is actually the Jordan triple system associated to the obvious Jordan algebra. Often the choice of D_0 is clear (for example, if $\frac{1}{2} \in k$ then $D_0 = \mathrm{Fix}\,\pi$ is the only possible choice for D_0), then $H_I(D,D_0,\pi,^-) = H_I(D,\pi,^-)$ and $H_I(D,^-,^-) = H_I(D,^-)$. In particular for D = k the only choices are $D_0 = k$, $\pi = {}^- = Id$ and we put $H_I(k,k,Id,Id) = H_I(k)$ or $H_p(k)$ for #I = p. In general, $H_I(D,D_0,\pi,^-)$ is spanned by elements of type

$$a[ij] = aE_{ij} + a^\pi E_{ji} \quad (a \in D,\ i \neq j), \quad a_0[ii] = a_0 E_{ii} \quad (a_0 \in D_0)$$

with products for distinct i,j,k,l

$$P(a_0[ii])b_0[ii] = a_0(\bar{b}_0 a_0)[ii]$$
$$P(a[ij])b[ij] = a(\bar{b}a)[ij]$$
$$P(a[ij])b_0[jj] = a(\bar{b}_0 a^\pi)[ii]$$
$$\{a[ij]\ b[ij]\ c[ik]\} = a(\bar{b}c)[ik]$$
$$\{a[ij]\ b[kj]\ c[ki]\} = (a(\bar{b}c) + (c^\pi \bar{b}^\pi)a^\pi)[ii] \quad (k = i \text{ allowed})$$
$$\{a[ij]\ b[kj]\ c[kl]\} = a(\bar{b}c)[il]$$
$$(k = j \text{ or } i = j \text{ or } k = i = j \text{ or } i = j,\ k = l \text{ allowed})$$

whereas all other products between these generators vanish. Finally, in analogy to the symplectic matrix units we define the <u>hermitian matrix units</u> H_{ij} as the matrices $H_{ij} = 1[ij] = H_{ji}$ (i = j allowed)

Note that

$$\mathcal{H}(I) = \{H_{ij}, i,j \in I\}$$

is a (D,D_0)-basis of $H_I(D,D_0,\pi,^-)$.

<u>Example 1.6.</u> (<u>quadratic form triple</u>) Here the ingredients are the following: K is an extension over k, i.e. a unital commutative associative algebra over k, with a k-linear automorphism $\kappa: K \to K: a \to \bar{a}$ of period 2, V is a module over K, $q: V \to K$ is a quadratic form with bilinearization $q(x,y) = q(x+y)-q(x)-q(y)$ and finally $S: V \to V$ is a k-linear map with $S^2 = Id$, $S(cv) = \bar{c}S(v)$ for $c \in K$, $v \in V$ and $q(Sv) =$

$\overline{q(v)}$. It is then straightforward to check that

$$P(x)y = q(x,Sy)x - q(x)Sy$$

satisfies (1.1) - (1.3). Since for any extension $R \supset k$ the product on $R \otimes V$ is of the same type, we obtain a Jordan triple system over k denoted by $[K,{}^-,V,q,S]$. Note that the product is K-linear in x and κ-linear in y. There are 2 special cases of particular interest:

a) For the first one we assume that V has a K-basis $\{e_i^\varepsilon;\ \varepsilon = \pm,\ i \in I\}$ for some index set I and that q is given by

$$q(e_i^\varepsilon) = 0 = q(e_i^\varepsilon, e_j^\mu) \text{ for } i \neq j \text{ and } q(e_i^+, e_i^-) = -1.$$

So V is free over K and a hyperbolic space relative to q. (To stop the readers' astonishment about our choice of the basis satisfying $q(e_i^+, e_i^-) = -1$ instead of the more usual $q(e_i^+, e_i^-) = 1$, we remark that with our setting this special case fits more naturally into the second special case to be considered below.) We define a k-linear map $S: V \to V$ by

$$S(ce_i^\varepsilon) = -\bar{c}e_i^{-\varepsilon},\ c \in K,\ i \in I,\ \varepsilon = \pm.$$

It is then easily checked that S has the properties required above. So V carries the structure of a Jordan triple system, which will be called an <u>even-dimensional quadratic form triple</u> (even in the case $\#I = \infty$) and denoted by $[K,{}^-,2I]$ or $[K,{}^-,2m]$ for $\#I = m$.

Relative to the K-basis

$$\mathcal{Q}_e(I) = \{e_i^\varepsilon;\ \varepsilon = \pm,\ i \in I\}$$

the product of V is given by the following formulas ($a,b,c \in K$)

$$P(ae_i^\varepsilon)be_i^\varepsilon = a^2\bar{b}e_i^\varepsilon ,$$

$$\{ae_i^\varepsilon,\ be_i^\varepsilon,\ ce_j^\mu\} = a\bar{b}ce_j^\mu,\ i \neq j,$$

$$\{ae_i^+,\ be_j^\varepsilon,\ ce_i^-\} = -a\bar{b}ce_j^{-\varepsilon},\ i \neq j,$$

whereas all other products between the basis vectors e_i^ε vanish.

b) The second special case is an extension of a): We take a Jordan triple system $[K,{}^-,X,q_X,S_X]$ and a Jordan triple system $[K,{}^-,2I]$ and glue them together

$$V = X \oplus K^{2I} \quad , K^{2I} = \bigoplus_{i \in I}(Ke_i^+ \oplus Ke_i^-)$$

$$q: = q_X \oplus q_{2I} \quad \text{(orthogonal sum)},$$

$$S: = S_X \oplus S_{2I}$$

where q_{2I} and S_{2I} are as in a). It is then straightforward to see that the properties required for a Jordan triple system $[K,{}^-,V,q,S]$ are fulfilled; it is denoted by $[K,{}^-,X,q_X,S_X] \oplus [K,{}^-,2I]$ or $[K,{}^-,X \oplus 2I]$ for short and called an <u>odd-dimensional quadratic form triple</u>, since it behaves like $[K,{}^-,2I]$ with one dimension added.

Clearly, $[K,{}^-,X,q_X,S_X]$ and $[K,{}^-,2I]$ are subsystems. The only non-zero products with at least one factor in each subsystem are ($x,y \in X$, $a,b \in K$)

$$P(x)ae_i^{\varepsilon} = q(x)\bar{a}e_i^{-\varepsilon},$$
$$\{x,y,ae_i^{\varepsilon}\} = q(x,Sy)ae_i^{\varepsilon}$$
$$\{ae_i^{\varepsilon}, be_i^{\varepsilon}, x\} = a\bar{b}x$$
$$\{ae_i^{+}, x, be_i^{-}\} = abSx.$$

These rules in particular show, that this example is not a direct triple system sum of the subsystems $[K,^-,X,q_X,S_X]$ and $[K,^-,2I]$.

We remark that later the following special case will occur: X contains an element e_0 satisfying $q(e_0) = 1$ and $Se_0 = e_0$. Then $P(e_0)e_0 = e_0$, so e_0 is an example of a so-called tripotent to be defined in the next subsection. Also, e_0 operates relative to the other tripotents as follows:

$$\{e_0e_0e_i^{\varepsilon}\} = 2e_i^{\varepsilon},\ P(e_0)e_i^{\varepsilon} = e_i^{-\varepsilon}\ ,$$
$$\{e_i^{+}e_0e_i^{-}\} = e_0,\ \{e_i^{\varepsilon}e_i^{\varepsilon}e_0\} = e_0.$$

In particular, we have this situation for $X = Ke_0$, $q(e_0) = 1$, $S(ce_0) = \bar{c}e_0$, in which case the triple system is denoted by $[K,^-,1+2I]$ and we have the natural K-basis

$$\mathcal{Q}_0(I) = \{e_0\} \cup \{e_i^{\varepsilon};\ i \in I,\ \varepsilon = \pm\}.$$

<u>1.3.</u> An element e of a Jordan triple system V is called a <u>tripotent</u> if $P(e)e = e$. Every such element determines a so-called <u>Peirce decomposition</u>

$$V = V_2(e) \oplus V_1(e) \oplus V_0(e),$$

which is induced by the decomposition

$$Id = E_2 \oplus E_1 \oplus E_0$$

where the maps $E_i = E_i(e)$, $i = 2,1,0$ are orthogonal projections with image $V_i(e)$, called <u>Peirce projection operators</u> and explicitely given by

$$E_2(e) = P(e)^2,\ E_0(e) = B(e,e) \text{ and}$$
$$E_1(e) = L(e,e)-2P(e)^2. \tag{1.17}$$

The Peirce spaces $V_i(e)$ have the following elemental characterization

$$V_2(e) = \text{im } P(e) = \text{im } P(e)^2$$
$$V_1(e) = \{x \in V;\ L(e,e)x = x\}, \tag{1.18}$$
$$V_0(e) = \{x \in V;\ P(e)x = 0 = L(e,e)x\}.$$

Moreover,

$$V_j(e) \subset \{x \in V;\ L(e,e)x = jx\},\ j = 0,1,2 \tag{1.19}$$

and we have equality if $\frac{1}{2} \in k$.

There are the following <u>Peirce multiplication rules</u> where $V_i = V_i(e)$ (= 0 if $i \neq 2,1,0$):

(1.20) $P(V_i)V_j \subset V_{2i-j}$,

(1.21) $\{V_iV_jV_k\} \subset V_{i-j+k}$,

(1.22) $\{V_2V_0V\} = 0 = \{V_0V_2V\}$.

Note in particular, every V_i is a subsystem. The Peirce space $V_2(e)$ comes along with an automorphism of order ≤ 2:

(1.23) $x \to P(e)x =: \bar{x}$ is an involution of V_2.

We put $V_2^{\pm}(e) = \{x \in V;\ P(e)x = \pm x\}$ and have

$$V_2(e) = V_2^+(e) \oplus V_2^-(e), \text{ if } \tfrac{1}{2} \in k.$$

Moreover, one knows (see e.g. [42])

(1.24) $P(e)\{x_1y_1e\} = \{y_1x_1e\}$, $\{x_0y_0z_1\} = \{x_0\{y_0z_1e\}e\}$

(1.25) $L(x_2,e) = L(e,\bar{x}_2)$

(1.26) $L(x_2,y_1) = L(e,\{\bar{x}_2ey_1\})$, $L(y_1,x_2) = L(\{y_1e\bar{x}_2\},e)$

(1.27) $L(x_0,y_1) = L(\{x_0y_1e\},e)$, $L(y_1,x_0) = L(e,\{ey_1x_0\})$

(1.28) $P(x_1)y_1 = \{\{x_1y_1e\}ex_1\} - \{ey_1P(x_1)e\}$

For every tripotent e the map $B(e,2e)$ is an automorphism of V of order ≤ 2, called the <u>Peirce reflection of</u> e, which operates as

$$x_2 + x_1 + x_0 \to x_2 - x_1 + x_0.$$

We consider some

<u>Examples 1.7.</u> a) $V = V(J)$ for a Jordan algebra J. Tripotents of V(J) can be constructed using idempotents, i.e. elements $c \in J$ such that $c^2 = c$. Indeed, $\pm c$ for an idempotent c of J is a tripotent of V(J), i.e. $c^3 = c$ holds in J, and more generally, $e = e_1-e_2$ is a tripotent, if e_1,e_2 are orthogonal idempotents of J. If $\frac{1}{2} \in k$ then all tripotents of V(J) arise in this way: $e_1 = \frac{1}{2}(e+e^2)$ and $e_2 = \frac{1}{2}(e-e^2)$ are orthogonal idempotents of J with $e = e_1-e_2$. The Peirce spaces of $e = e_1-e_2$ are given by

$$V_2(e) = J_{11} \oplus J_{12} \oplus J_{22},\quad V_1(e) = J_{13} \oplus J_{23},\quad V_0(e) = J_{33}$$

where J_{ij} are the Peirce spaces of J relative to the orthogonal system $(e_1,e_2,1-e_1-e_2)$. Note that in this case $P(e)|J_{11} \oplus J_{22} = \mathrm{Id}$ and $P(e)|J_{12} = -\mathrm{Id}$.

If the tripotent e of V(J) splits in the form $e = e_1-e_2$ as described above, then $e^2 = e_1+e_2$ is an idempotent in J. This is not true in general as the following example ([40]1.18) shows: Let $J = k[x]/K$ for $k[x]$ the polynomial ring in one indeterminate and K the Jordan ideal spanned by $x - x^3$, $2x^2-2x^4$ and x^i-x^j for $i \equiv j \bmod 4$. Then J is spanned by $1, x, z = x^4-x^2$ and satisfies the relations $x^2 = 1+z$, $x^3 = x$, $x^4 = 1$, $2z = 0$ but $z \neq 0$ if $\frac{1}{2} \notin k$. However, for any tripotent $e \in V(J)$ the element e^4 is always an idempotent of J, since $(e^4)^2 = U(e)^3e^2 = U(e)e^2 = e^4$ ([40]1.17).

b) Let $V = V^+ \oplus V^-$ be a polarized Jordan triple system. Here $e = e^+ \oplus e^-$ is a tripotent of V iff (e^+,e^-) is an idempotent of the Jordan pair $\mathcal{V} = (V^+,V^-)$. In this case the Peirce space $V_i(e)$ is again polarized, namely the sum of the two components of the Peirce space $\mathcal{V}_i(e^+,e^-)$.

c) rectangular matrix system Mat(I,J;D): Examples of tripotents are the matrix units E_{ij}. Their Peirce spaces are given by

$$V_2(E_{ij}) = DE_{ij},\ V_1(E_{ij}) = \sum_{\ell \neq i} DE_{\ell j} \oplus \sum_{k \neq j} DE_{ik},$$

$$V_0(E_{ij}) = \sum_{\{k,\ell\} \cap \{i,j\} = \emptyset} DE_{k\ell}$$

In this case the involution $P(E_{ij})$ on $V_2(E_{ij})$ coincides with the involution $^-$ on D, hence is $\neq$Id in general.

Another example of a tripotent is $E_p = E_{11}+E_{22}+\ldots+E_{pp} \in \mathrm{Mat}(p,q;D)$ assuming $p \leq q$ with Peirce spaces

$$V_2(E_p) = \sum_{1 \leq i,j \leq p} DE_{ij},\ V_1(E_p) = \sum_{1 \leq i \leq p} \ \sum_{p<j \leq q} DE_{ij}$$

and $V_0(E_p) = 0$.

Note that $V_1(E_p) = 0 \leftrightarrow p=q$.

d) symplectic matrix system A(I;K): Here the basic examples of tripotents are the symplectic matrix units F_{ij}.

e) hermitian matrix system $H_I(D,D_0,\pi,{}^-)$: The basic examples of tripotents are the hermitian matrix units H_{ij}.

f) quadratic forms: The basis elements e_i^ε in a quadratic form triple are tripotents and the same holds for e_0 in the odd-dimensional case.

After so many tripotents we need to point out that not every Jordan triple system contains non-zero tripotents. As a prominent example take the reals with product $P(x)y = -xyx$. However, there are reasonable classes of Jordan triple systems which have a lot of non-zero tripotents, see §6 and IV §§2,3.

1.4. The philosophy of tripotents in the study of Jordan triple systems is to use them for breaking up the whole system in smaller pieces which are easier to handle. As can be expected the Peirce decomposition relative to a single tripotent described in 1.3 does not suffice for this in general, one has to use several of them. But arbitrary tripotents will not help, at least one wants to have some relation between the various Peirce spaces.

Considering the case of two tripotents a natural condition is that they induce a simultaneous Peirce decomposition, a condition introduced and studied in [40]: Two tripotents e and f are called compatible, if their Peirce projections commute:

$$[E_i(e),E_j(f)] = 0 \text{ for } i,j = 0,1,2$$

There is the following elemental characterization of compatible tripotents.

Compatibility Criterion 1.8. ([40]1.6-1.9). The following are equivalent for tripotents e,f:

a) e and f are compatible,

b) $\{eef\}$ lies in $V_2(f)$,

c) $f = f_2+f_1+f_0$ for elements $f_i \in V_i(e) \cap V_2(f)$.

In this case, $\{eef\}$ is symmetric under the involution $P(f)$ of $V_2(f)$:

$$P(f)\{eef\} = \{eef\}.$$

A special case: e and f compatible with $f_2 = 0$ in c). Then f_1 and f_0 are orthogonal tripotents (see 1.5.). The same holds, if $f_1 = 0$ or $f_0 = 0$.

By definition two compatible tripotents have a simultaneous Peirce decomposition which is written as

$$V = \bigoplus_{i,j=0,1,2} V_{(ij)}$$

where

$$V_{(ij)} = V_i(e) \cap V_j(f).$$

We remark that we write $V_{(ij)}$ with parentheses (instead of V_{ij}) since V_{ij} usually denotes Peirce spaces of two orthogonal tripotents (see Theorem 1.10). Contrary to the situation there the Peirce spaces $V_{(ij)}$ depend on the order of i and j: $V_{(ij)} \neq V_{(ji)}$ in general.

In general, a family E of tripotents is called compatible if each two $e,f \in E$ are compatible.

For an arbitrary family E of tripotents and for $I = (i_e)_{e\in E}$ with al $i_e = 0,1$ or 2 we define the I-Peirce space of E as

$$V_I(E) = V_I = \bigcap_{e\in E} V_{i_e}(e)$$

The Peirce spaces of E inherit the multiplication rules from (1.20), (1.21):

(1.29) $$P(V_I)V_J \subset V_{2I-J}, \quad \{V_IV_JV_K\} \subset V_{I-J+K}$$

where the meaning of 2I-J or I-J+K is the obvious one. In particular, every single Peirce space V_I and

$$PS_V(E) := \oplus_I V_I(E) \quad \text{(Peirce sum of } E \text{ in } V)$$

are subsystems of V. Contrary to the situation of a single tripotent, $PS_V(E)$ in general is a proper subsystem (see e.g. Example 6.5.). For a finite E, however

$$V = PS_V(E) \leftrightarrow E \text{ is compatible}$$

For a general E it will sometimes be of advantage to group together the Peirce spaces in a way that reminds of a single tripotent:

$$V_2(E) = \sum_{2\in I} V_I(E)$$
$$V_1(E) = \sum_{2\notin I,\ (0)\neq I} V_I(E)$$
$$V_0(E) = V_{(0)}(E) = \cap_{e\in E} V_0(e)$$

Then obviously

$$PS_V(E) = V_2(E) \oplus V_1(E) \oplus V_0(E).$$

We note that in general $V_2(E)$ and $V_1(E)$ are not subsystems of V. Their multiplication rules, which will not be needed in the following, are given in [40]2.2.

The following subspace of $V_2(E)$ will become important later:

$$C_V(E) = \oplus_{V_I \cap E \neq \emptyset} V_I \qquad (\underline{\text{cover of } E \text{ in } V})$$

Again, the cover of E in V is not a subsystem in general. Conditions under which it is a subsystem are derived in Theorem 4.14. If every non-zero Peirce space $V_I(E)$ contains a tripotent of E, we call E <u>Peirce-dense</u>, thus

$$E \text{ Peirce-dense} \leftrightarrow C_V(E) = PS_V(E) < V$$

In particular this is the case if $C_V(E) = V$. We then say E <u>covers</u> V.

For a finite compatible E we have a decomposition of V as

$$V = V_2(E) \oplus V_1(E) \oplus V_0(E)$$

which suggests the following generalization of the Peirce reflection of a single tripotent, defined in <u>1.3.</u> We put

$$S_E(x_2+x_1+x_0) = x_2-x_1+x_0, \quad |E| < \infty$$

where $x_i \in V_i(E)$ and call this map the <u>Peirce reflection relative to</u> E. We point out that in general S_E is not an automorphism of V in contrast to the case of a single tripotent. Conditions under which S_E actually is an automorphism of V can be found in [40] §4.

<u>1.5.</u> It will be enough for our purposes to consider the following special case of compatibility: Two tripotents e,f are called <u>Peirce-compatible</u> if one lies in a single Peirce space of the other, e.g. $f \in V_i(e)$, i=2,1 or 0. That Peirce-compatible tripotents actually are compatible follows from the Compatibility Criterion 1.8.c.

Special cases of Peirce-compatible tripotents e,f are:

e,f are orthogonal (written as $e \perp f$): $e \in V_0(f)$ and $f \in V_0(e)$
e,f are collinear (written as $e \top f$) : $e \in V_1(f)$ and $f \in V_1(e)$
e governs f (written as $e \vdash f$ or $f \dashv e$): $e \in V_1(f)$ and $f \in V_2(e)$.

These 3 relations are fundamental for the following. We therefore present some known properties and examples. First we recall:

Lemma 1.9.([31]5.11) *Let e,f be two tripotents. Then*

$$e \in V_0(f) \leftrightarrow f \in V_0(e)$$

In this case we have:

a) $V_2(f) \subset V_0(e)$ *and* $V_2(e) \subset V_0(f)$,

b) $e+f$ *is a tripotent.*

By definition, an orthogonal system is an ordered set of pairwise orthogonal tripotents. In case the system is finite it is usually denoted by $(e_1,\dots,e_r)$.

Theorem 1.10. ([31]5.14) *Let* $(e_1,\dots,e_r)$ *be an orthogonal system.*

a) *There is a decomposition*

(1.30) $$V = \bigoplus_{0 \leq i \leq j \leq r} V_{ij}$$

of V into subsystems where

$$V_{00} = \cap_i V_0(e_i), \quad V_{ii} = V_2(e_i), \quad 1 \leq i \leq r,$$
$$V_{0j} = V_{j0} = V_1(e_j) \cap \bigcap_{j \neq i} V_0(e_i), \quad 1 \leq j \leq r,$$
$$V_{ij} = V_{ji} = V_1(e_i) \cap V_1(e_j) \quad , \quad 1 \leq i < j \leq r$$

b) *For* $I \subset \{1,\dots,r\}$ *put* $e_I = \sum_{i \in I} e_i$. *Then* e_I *is a tripotent with Peirce spaces*

(1.31) $$V_2(e_I) = \sum_{i,j \in I} V_{ij}, \quad V_1(e_I) = \sum_{i \in I} \sum_{j \notin I} V_{ij} \text{ and}$$
$$V_0(e_I) = \sum_{i,j \notin I} V_{ij}.$$

c) *The following composition rules are satisfied:*

(1.32) $$\{V_{ij}V_{jm}V_{mn}\} \subset V_{in},$$
$$P(V_{ij})V_{ij} \subset V_{ij},$$
$$P(V_{ij})V_{ii} \subset V_{jj}$$

whereas all other products which cannot be written in the form (1.32) *are zero.*

We remark that for an arbitrary orthogonal system $(e_j)_{j \in J}$ one can define the Peirce spaces V_{ij} as in the theorem above. Then (1.30) need not be true in general, however the multiplication rules (1.32) are still true.

As an example for this theorem we consider the rectangular matrices Mat(p,q;D), p≤q. Two tripotents $E_{ij}, E_{k\ell}$ are orthogonal iff $\{i,j\} \cap \{k,\ell\} = \emptyset$. In particular $(E_{11},\dots,E_{rr})$, r≤p, is an orthogonal system whose Peirce spaces are

$$V_{00} = \sum_{r<i\leq p} \sum_{r<j\leq q} DE_{ij},$$
$$V_{ii} = DE_{ii},$$
$$V_{0j} = \sum_{r<k\leq q} DE_{jk} + \sum_{r<k\leq p} DE_{kj}$$
$$V_{ij} = DE_{ij} + DE_{ji},$$

which for r=2, p=3, p<q may be visualized as

11	12	01
12	22	02
01	02	00

Note that in the general case $V_{00} = 0 \leftrightarrow r=p$ and $V_{0j} = 0 \leftrightarrow r=p=q$.

Now we turn to collinear tripotents and look first at two basic examples:

In Mat(p,q;D) two tripotents E_{ij}, $E_{k\ell}$ are collinear iff either i=k, j≠ℓ or i≠k, j=ℓ, i.e. iff they lie in the same row or column (this description justifies the name "collinear"). The Peirce decomposition of the collinear tripotents E_{11}, E_{12} is given by

$$\text{Mat}(p,q;D) = V_{(21)} \oplus V_{(12)} \oplus V_{(11)} \oplus V_{(10)} \oplus V_{(01)} \oplus V_{(00)}$$

where

$$V_{(21)} = DE_{11},\ V_{(12)} = DE_{12},\ V_{(11)} = \sum_{k\geq 3} DE_{1k}$$
$$V_{(10)} = \sum_{j\geq 2} DE_{j1},\ V_{(01)} = \sum_{j\geq 2} DE_{j2} \quad \text{and}$$
$$V_{(00)} = \sum_{j\geq 2} \sum_{k\geq 3} DE_{jk}$$

which may be visualized as

(21)	(12)	(11)
(10)	(01)	(00)

In the hermitian matrix system $H_p(k)$ two different tripotents $H_{ij} = E_{ij}+E_{ji}$ $(i < j)$ and H_{kl} are collinear iff they have exactly one index in common. The two collinear tripotents H_{12} and H_{13} of $H_4(D,D_0,\pi,^-)$ have a Peirce decomposition $V = \oplus V_{(ij)}$ where all possible Peirce spaces $V_{(ij)}$, $i,j \in \{0,1,2\}$ occur ($H_{ii}=E_{ii}$):

$$V_{(22)} = D_0H_{11},\ V_{(21)} = DH_{12},\ V_{(20)} = D_0H_{22}$$
$$V_{(02)} = D_0H_{33},\ V_{(12)} = DH_{13},\ V_{(11)} = DH_{23} + DH_{14}$$
$$V_{(10)} = DH_{24},\ V_{(01)} = DH_{34},\ V_{(00)} = D_0H_{44}.$$

These two examples show the two possibilities for the Peirce spaces $V_{(ij)}$ relative to two collinear tripotents e,f: Either $V_{(20)} = V_{(02)} = V_{(22)} = 0$, in which case e,f are called <u>rigid</u> or <u>rigidly imbedded</u>, or $V_{(22)}$, $V_{(20)}$ and $V_{(02)}$ are isomorphic non-zero k-submodules, as follows from

<u>Lemma 1.11.</u> ([40]Lemma 3.4). <u>The Peirce spaces relative to collinear tripotents</u> e,f <u>satisfy</u>

a) $V_{(20)} = P(e)V_{(22)}$, $V_{(02)} = P(f)V_{(22)}$,
$V_{(22)} = P(e)V_{(20)} = P(f)V_{(02)}$,

b) $V_{(10)} = L(e,f)V_{(01)}$, $V_{(01)} = L(f,e)V_{(10)}$.

As a corollary we obtain:

(1.32) Two collinear tripotents e,f are rigid iff $V_{(22)} = 0$ iff $V_2(e) \subset V_1(f)$ iff $V_2(f) \subset V_1(e)$.

The following criterion which we will need later is a supplement to [40] Proposition 3.7:

<u>Criterion 1.12.</u> <u>Let</u> $0 \neq e$ <u>be a tripotent of</u> V <u>and assume</u> $V_2(e)$ <u>is a domain</u> (i.e. $P(x)|V_2(e)$ <u>is injective for all</u> $0 \neq x \in V_2(e)$). <u>Then for any nonzero tripotent</u> $f \in V_1(e)$ <u>we either have</u> $e \dashv f$ <u>or</u> $e \top h$. <u>In the latter case</u> (e,f) <u>are automatically rigid</u>.

<u>Proof.</u> It was shown in [40] Proposition 3.7 that $P(f)e = 0$ implies that (e,f) are rigid collinear. If $P(f)e \neq 0$ then also $x = P(f)^2e \neq 0$ since otherwise $0 = P(f)^3e = P(f)e$. Because the Peirce projections of f leave $V_2(e)$ invariant we have

$$V_2(e) = V_2(e) \cap V_2(f) \oplus V_2(e) \cap V_1(f) \oplus V_2(e) \cap V_0(f)$$

with $x \in V_2(e) \cap V_2(f)$. But $P(x)$ annihilates $V_1(f)$ and $V_0(f)$, therefore $V_2(e) \cap V_i(f) = 0$ for $i = 0,1$ and $V_2(e) \subset V_2(f)$, thus $e \dashv f$. ⊞

We note that any tripotent $e \neq 0$ with $V_2(e)$ a domain is necessarily minimal in the sense of <u>§5.1.</u>

An important feature of collinear tripotents e,f is that they induce an automorphism $T_{e,f}$ of order 2 exchanging each other. It is therefore called the exchange automorphism (between e and f).

Theorem 1.13.([42]1.1) If e,f are collinear tripotents, then

$$T_{e,f} = B(e+f,\ e+f)$$

is an automorphism of V of order 2 exchanging e and f and reducing on the various Peirce spaces to

$L(f,e)$ on $V_{(21)}$, $L(e,f)$ on $V_{(12)}$

$L(e,f)L(f,e) - Id = P(e,f)^2 - Id$ on $V_{(11)}$

$-L(f,e)$ on $V_{(10)}$, $-L(e,f)$ on $V_{(01)}$

$P(f)P(e)$ on $V_{(20)}$, $P(e)P(f)$ on $V_{(02)}$ and

Id on $V_{(00)} \oplus V_{(22)}$.

We will come across families of pairwise collinear tripotents, which will be called collinear systems.

§2 Elementary configurations: quadrangle, triangle and diamond

In this section we look at three configurations of tripotents which will become building blocks for the families of tripotents to be considered later.

Throughout, V is a Jordan triple system over a ring k as defined in 1.1. The basic relations between two tripotents e,f of V have been defined in 1.5: $e \perp f$, $e \top f$ and $e \vdash f$. We will also put together these signs in an obvious meaning, e.g. $e_1 \dashv e_2 \top e_3$ means $e_1 \dashv e_2$ and $e_2 \top e_3$.

2.1. In this subsection we review the theory of quadrangles which was developped in [40]: By definition, a quadrangle (of tripotents) is an ordered quadruple (e_1,e_2,e_3,e_4) of tripotents satisfying

(2.1) $$e_i \top e_{i+1}\ ,\ e_i \perp e_{i+2}\ ,\ \{e_i\ e_{i+1}\ e_{i+2}\} = e_{i+3}$$
(indices mod 4)

The name "quadrangle" is motivated by the standard example of a quadrangle in Mat(p,q;D) - see below. Note that all products in a quadrangle are known:

(2.2) $$\begin{aligned} &P(e_i)e_j = \delta_{ij}e_i \\ &\{e_i\ e_i\ e_{i+1}\} = e_{i+1},\ \{e_i\ e_i\ e_{i-1}\} = e_{i-1} \\ &\{e_i\ e_{i+1}\ e_{i+2}\} = e_{i+3}, \end{aligned}$$

and all other products are zero or can be brought into the form (2.2) by using the symmetry in the outer two variables. To decide whether a given quadruple of tripotents is a quadrangle one does not need to check all the properties (2.1) because of the following

Quadrangle Criterion 2.1.([40]3.2) Let e_1,e_2,e_3 be tripotents satisfying

$$e_1 \top e_2 \top e_3 \quad \text{and} \quad e_1 \perp e_3.$$

Then $e_4 = \{e_1\ e_2\ e_3\}$ is a tripotent such that (e_1,e_2,e_3,e_4) is a quadrangle.

Since a quadrangle consists of Peirce-compatible tripotents, it induces a simultaneous Peirce decomposition. The properties of this decomposition are summarized in the

Quadrangle Decomposition Theorem 2.2.([40]3.3) Let $\mathcal{E} = (e_1,e_2,e_3,e_4)$ be a quadrangle. Then the multiplication operators satisfy

(2.3) $$\begin{aligned} &L(e_i,e_{i+1}) = L(e_{i+3},e_{i+2}) \qquad \text{(indices mod 4)} \\ &P(e_i)P(e_{i+1}) = P(e_{i+3})P(e_{i+2}) \end{aligned}$$

$$L(e_1,e_1) - L(e_2,e_2) + L(e_3,e_3) - L(e_4,e_4) = 0.$$

The Peirce decomposition relative to E *is* $V = V_2(E) \oplus V_1(E) \oplus V_0(E)$ *where*

$$\begin{aligned} V_2(E) &= (V_{(2200)} + V_{(0022)} + V_{(2002)} + V_{(0220)}) \\ &\quad + (V_{(2101)} + V_{(1210)} + V_{(0121)} + V_{(1012)}) \\ V_1(E) &= V_{(1111)} + V_{(1001)} + V_{(0110)} + V_{(1100)} + V_{(0011)} \\ V_0(E) &= V_{(0000)}, \end{aligned} \tag{2.4}$$

while all other Peirce spaces vanish.

Examples of quadrangles are $(E_{ir},E_{is},E_{js},E_{jr})$ in Mat(p,q;D), p,q>2. For $(E_{11},E_{12},E_{22},E_{21})$ the quadrangle decomposition is (in an obvious meaning)

(2101)	(1210)	(1100)	
(1012)	(0121)	(0011)	
(1001)	(0110)	(0000)	} p-2
		q-2	

Notice the Peirce spaces $V_{(2200)}$, $V_{(0022)}$, $V_{(2002)}$, $V_{(0220)}$ and $V_{(1111)}$ vanish. Some of the other Peirce spaces might also be zero depending on p and q. The situation is different for the quadrangle $(H_{13},H_{14},H_{24},H_{23})$ of $H_p(D,D_0,\pi,{}^-)$, p>4, where the decomposition becomes

(2200)	(1111)	(2101)	(1210)	(1100)	
	(0022)	(1012)	(0121)	(0011)	
		(2002)	(1111)	(1001)	
	*		(0220)	(0110)	
				(0000)	} p-4
				p-4	

<u>2.2.</u> Before we proceed to triangles and diamonds it is useful to make a short digression and to introduce another relation between two tripotents e, f of V. We say that e and f are *associated* ($e \approx f$) iff $V_j(e) = V_j(f)$ for j = 0,1,2. Obviously, $\approx$ is an equivalence relation. Note that $-e \approx e$. For the next theorem which characterizes $\approx$ we recall that an element x of a subsystem W of V is called *invertible in* W iff $P(x)|W$ is bijective.

Theorem 2.3. *Let* e, f *be tripotents of* V. *Then*

$$e \approx f \leftrightarrow V_2(e) = V_2(f) \leftrightarrow e \in V_2(f) \text{ and } f \in V_2(e)$$
$$\leftrightarrow e \text{ is invertible in } V_2(f) \leftrightarrow f \text{ is invertible in } V_2(e).$$

In this case we have

(2.5) $L(e,e) = L(f,f)$.

Conversely, (2.5) *implies* $e \approx f$ *if* $\frac{1}{2} \in k$.

Proof. Clearly, $e \approx f \to V_2(e) = V_2(f) \to e \in V_2(f)$, $f \in V_2(e)$. To prove the converse we note that for every tripotent $e \in V_2(f)$ we have $V_2(e) = P(e)V = P(e)(V_2(f) \oplus V_1(f) \oplus V_0(f)) \subset V_2(f)$ by (1.18) and (1.20). Hence $e \in V_2(f)$ and $f \in V_2(e)$ imply $V_2(e) = V_2(f)$. Under this assumption we get $V_0(e) \subset V_0(f)$ from (1.18), (1.20), (1.22) and thus by symmetry $V_0(e) = V_0(f)$. But then the Peirce projection $E_1(e) = \mathrm{Id} - E_2(e) - E_0(e) = \mathrm{Id} - E_2(f) - E_0(f) = E_1(f)$, and we have $V_j(e) = V_j(f)$ for $j = 0,1,2$, i.e. $e \approx f$.

If $e \approx f$ then $P(e)|V_2(f) = P(e)|V_2(e)$ is invertible by (1.23). Conversely, if e is invertible in $V_2(f)$ then $V_2(e) = P(e)V = P(e)V_2(f)$ (since $e \in V_2(f)$) $= V_2(f)$ and $e \approx f$ by what we already proved. The last statement is a consequence of (1.19). ∎

Examples 2.4. a) Let k be a field, V a vector space over k and $q: V \to k$ a quadratic form. Then

$$P(x)y = q(x,y)x - q(x)y$$

defines a Jordan triple system on V. Indeed, this is a special case of Example 1.6 (put K=k, $^{-}$=Id and S=Id). It is easily seen:

(2.6) Let $0 \neq e \in V$. Then e is a tripotent $\leftrightarrow$ $q(e)=1$.

For a non-zero tripotent e one computes $P(e)y = q(e,y)e-y$, i.e. $-P(e)$ is the usual reflection relative to e. As a consequence,

(2.7) $V_2(e) = V$ for every non-zero tripotent e,

and each two non-zero tripotents are associated.

b) Let V = V(J) for a Jordan algebra J. As shown in Example 1.7.a) for any tripotent $e \in V(J)$ the element e^4 is an idempotent of J. Clearly, $e^4 \in J_2(e)$, and $e^3 = e$ implies $e = U(e^3)e^3 = e^9 = U(e^4)e$, so $e \in J_2(e^4)$ whence $e \approx e^4$ for any tripotent $e \in V(J)$.

If e is of the form $e = e_1 - e_2$ for orthogonal idempotents e_1, e_2, then $e \approx e_1 + e_2$. We also note that "association" becomes "equality" for idempotents: Since an idempotent c is the unit element of $J_2(c)$, we have

$$c \approx d \leftrightarrow c = d \quad \text{for idempotents } c, d \in J$$

c) In $V = \mathrm{Mat}(I,J;D,{}^{-})$ an element aE_{ij} is a tripotent iff $a\bar{a}a = a$. We have $aE_{ij} \approx E_{ij}$ iff a is invertible with inverse $a^{-1} = \bar{a}$, i.e. a is a unitary element of $(D,{}^{-})$.

<u>2.3.</u> The first new configuration we consider is a <u>triangle</u>, which by definition is an ordered triple $(e_0;e_1,e_2)$ of tripotents satisfying

(2.8) $\quad e_1 \dashv e_0 \vdash e_2, \quad e_1 \perp e_2$ and

(2.9) $\quad P(e_0)e_1 = e_2 \ , \ P(e_0)e_2 = e_1 \ , \ P(e_1,e_2)e_0 = e_0.$

The name "triangle" is motivated by the standard example (H_{ij},H_{ii},H_{jj}), $i \neq j$, in an hermitian matrix system (see Examples 1.5, 1.7.e). Another example is given by the K-basis (e_0,e_1^+,e_1^-) of the Jordan triple system $[K,{}^{-},3]$ considered in Example 1.6.b.

As for quadrangles there is also a Triangle Criterion and a Triangle Decomposition Theorem.

<u>Triangle Criterion 2.5.</u> a) <u>Assume</u> $e_1 \dashv e_0$. <u>Then</u> $e_2 = P(e_0)e_1$ <u>is a tripotent such that</u> (e_0,e_1,e_2) <u>is a triangle</u>.

b) <u>Assume</u> $c_1 \dashv e_0 \vdash c_2 \perp c_1$. <u>Then</u> $c_2 \approx P(e_0)c_1$ <u>and</u> $e_0 \approx c_1 + c_2$.

<u>Proof</u>. a) From (1.23) follows that e_2 is a tripotent and (1.20) implies $e_2 \in V_2(e_0) \cap V_0(e_1)$, in particular $e_1 \perp e_2$. Because $\{e_2\ e_2\ e_0\} = \{e_0\ P(e_0)e_1\ P(e_0)e_1\} =$ (by (1.4)) $\{e_0\ e_1\ P(e_0)^2e_1\} = \{e_0\ e_1\ e_1\} = e_0$ we also have $e_2 \dashv e_0$. Clearly, $P(e_0)e_2 = e_1$ and $\{e_1\ e_0\ e_2\} = \{e_1\ e_0\ P(e_0)e_1\} =$ (by (1.12)) $\{e_0\ e_1\{e_1\ e_0\ e_0\}\} - P(e_0)\{e_0\ e_1\ e_1\} = -P(e_0)e_0 + 2\{e_0\ e_1\ e_1\} = e_0$.

b) We know that $V_2(c_i) \subset V_2(e_0)$. By (1.20) we have $P(e_0)c_2 \in V_2(c_1)$, whence $c_2 \in P(e_0)V_2(c_1) = V_2(P(e_0)c_1)$. Since $P(e_0)c_1 \in V_2(c_2)$ we have $c_2 \approx P(e_0)c_1$ by Theorem 2.3. Obviously, $c_1 + c_2 \in V_2(e_0)$. Because $e_0 \in V_1(c_i)$ it follows $e_0 \in V_2(c_1 + c_2)$ by (1.31), so $e_0 \approx c_1 + c_2$. ∎

<u>Triangle Decomposition Theorem 2.6.</u> <u>Let</u> $\mathcal{E} = (e_0,e_1,e_2)$ <u>be a triangle. Denote by</u> $V_{(ijk)}$ <u>the Peirce spaces relative to</u> $\mathcal{E}$ <u>and by</u> V_{ij} <u>the usual Peirce spaces relative to the orthogonal system</u> (e_1,e_2). <u>Then</u>

$$V_{11} = V_{(220)},\ V_{12} = V_{(211)},\ V_{22} = V_{(202)}$$
$$V_{01} = V_{(110)},\ V_{02} = V_{(101)},\ V_{00} = V_{(000)},$$

i.e. <u>the Peirce decomposition of</u> V <u>relative to</u> $\mathcal{E}$ <u>and relative to</u> (e_1,e_2) <u>coincide. The multiplication operators satisfy</u>

(2.10) $\quad L(e_0,e_0) = L(e_1,e_1) + L(e_2,e_2).$

Proof. By the Triangle Criterion 2.5.b) we know $e_0 \approx e_1 + e_2$ and from Theorem 1.10.b) we get $V_2(e_1+e_2) = V_{11} \oplus V_{12} \oplus V_{22}$, $V_1(e_1+e_2) = V_{10} \oplus V_{20}$, $V_0(e_1+e_2) = V_{00}$. With the definition of the Peirce spaces V_{ij} the first claim now easily follows, and because of (2.5) we have $L(e_0,e_0) = L(e_1+e_2, e_1+e_2) = L(e_1,e_1) + L(e_2,e_2)$. ⊞

2.4. The next configuration we consider is the diamond, which by definition is an ordered quadruple of tripotents (e_0,e_1,e_2,e_3) such that

(2.11) $\quad e_1 \vdash e_0 \dashv e_3$, $e_0 \perp e_2$, $e_1 \top e_2 \top e_3 \top e_1$ and

(2.12) $\quad \{e_0\ e_1\ e_2\} = e_3.$

The name "diamond" is motivated by the standard example $(H_{11},H_{12},H_{23},H_{13})$ of a diamond in $H_3(k)$ - see below.

Further multiplication rules between elements of a diamond will be established in the proof of the

Diamond Criterion 2.7. a) *If e_0, e_1, e_2 are tripotents with $e_0 \dashv e_1 \top e_2 \perp e_0$, then $(e_0,e_1,e_2,\{e_0\ e_1\ e_2\})$ is a diamond.*
b) *If e_0, e_1, e_3 are tripotents with $e_1 \vdash e_0 \dashv e_3 \top e_1$, then $(e_0, e_1,\{e_1\ e_0\ e_3\},e_3)$ is also a diamond.*

Proof. a) Since $e_0 + e_2$ is a tripotent with $e_1 \in V_1(e_0) \cap V_1(e_2) \subset V_2(e_0 + e_2)$ and $e_3 := \{e_0\ e_1\ e_2\} = P(e_0 + e_2)e_1$ we get that e_3 is a tripotent too. Using (1.14) we show

$$\{e_3\ e_0\ e_1\} = \{e_1\ e_0\{e_0\ e_1\ e_2\}\} = \{\{e_1\ e_0\ e_0\}\ e_1\ e_2\} - \{e_0\{e_0\ e_1\ e_1\}e_2\} + \{e_0\ e_1\{e_1\ e_0\ e_2\}\} = e_2 - 0 + 0 = e_2 \text{ because } e_0 \perp e_2,$$

$$\{e_2\ e_3\ e_0\} = \{e_2\{e_2\ e_1\ e_0\}e_0\} = \{\{e_1\ e_2\ e_2\}e_0\ e_0\} - \{e_1\ e_2\{e_2\ e_0\ e_0\}\} + \{e_2\ e_0\{e_1\ e_2\ e_0\}\} = e_1 - 0 + 0 = e_1,$$

$$\{e_1\ e_2\ e_3\} = \{\{e_1\ e_2\ e_0\}e_1\ e_2\} - \{e_0\{e_2\ e_1\ e_1\}e_2\} + \{e_0\ e_1\{e_1\ e_2\ e_2\}\} = 0 - 0 + 2e_0 = 2e_0.$$

From the linearized form of (1.7) we get

$L(\{e_1e_2e_3\},e_0) + L(\{e_1e_0e_3\},e_2) = L(e_1,\{e_2e_3e_0\}) + L(e_3,\{e_2e_1e_0\})$ thus

(*) $\quad 2L(e_0,e_0) + L(e_2,e_2) = L(e_1,e_1) + L(e_3,e_3)$

and hence $\{e_3e_3e_0\} = 2e_0$, $\{e_3e_3e_1\} = e_1$ and $\{e_3e_3e_2\} = e_2$. Since $e_3 \in V_1(e_0) \cap V_1(e_1) \cap V_1(e_2)$ we get $e_1 \top e_3 \top e_2$, and $e_3 \vdash e_0$ if $\frac{1}{2} \in k$. To show $e_0 \in V_2(e_3)$ in general, we have to argue more:

$P(e_3)e_0 = P(P(e_0+e_2)e_1)e_0 =$ (by (1.3)) $P(e_0+e_2)P(e_1)P(e_0+e_2)e_0 =$ (because $e_0 \perp e_2$) $P(e_0+e_2)P(e_1)e_0 = P(e_2)P(e_1)e_0$ since $e_0 \perp P(e_1)e_0$ by the triangle criterion. We have $P(e_1)e_0$, $P(e_2)P(e_1)e_0 \in V_2(e_2)$ and therefore $P(e_3)^2e_0 = P(e_0+e_2)P(e_1)P(e_0+e_2)P(e_2)P(e_1)e_0 = P(e_0+e_2)P(e_1)P(e_2)^2P(e_1)e_0 = P(e_0+e_2)P(e_1)^2e_0 = P(e_0+e_2)e_0 = e_0$, i.e. $e_0 \in V_2(e_3)$.

b) We first want to show that $e_2 = \{e_1e_0e_3\}$ is a tripotent: We specialize [31](JP21) and obtain
$P(\{e_1e_0e_3\}) = P(e_1)P(e_0)P(e_3) + P(e_3)P(e_0)P(e_1) + L(e_1,e_0)P(e_3)L(e_0,e_1)$, applying to $e_2 = \{e_1e_0e_3\} \in V_1(e_1) \cap V_1(e_3)$ gives

$$P(e_2)e_2 = L(e_1,e_0)P(e_3)L(e_0,e_1)e_2 ,$$

whence e_2 is a tripotent as soon as

$$(**) \qquad L(e_0,e_1)e_2 = e_3$$

To show (**) we use (1.14): $\{e_0e_1\{e_1e_0e_3\}\} = \{\{e_0e_1e_1\}e_0e_3\} - \{e_1\{e_1e_0e_0\}e_3\} + \{e_1e_0\{e_0e_1e_3\}\}$. Because $\{e_0e_1e_3\} \in V_{2-1+2}(e_3) = 0$ this last sum equals e_3, i.e. (**) holds. By construction $e_2 \in V_0(e_0) \cap V_1(e_1)$. Also, $\{e_0e_3e_2\} = \{e_0e_3\{e_3e_0e_1\}\} = \{\{e_0e_3e_3\}e_0e_1\} - \{e_3\{e_3e_0e_0\}e_1\} + \{e_3e_0\{e_0e_3e_1\}\} = 2e_1 - e_1 + 0 = e_1$ and so by (1.12) $\{e_2e_2e_1\} = P(e_1,\{e_3e_0e_1\})e_2 = L(e_3,e_0)P(e_1)e_2 + P(e_1)L(e_0,e_3)e_2 = P(e_1)e_1 = e_1$, hence $e_1 \top e_2 \perp e_0$ and the claim follows from a) and (**). ∎

We collect the additional multiplication rules established in the proof of the Diamond Criterion.

<u>Corollary 2.8.</u> <u>Let</u> (e_0,e_1,e_2,e_3) <u>be a diamond. Then</u>

(2.13) $\{e_0e_1e_2\} = e_3$, $\{e_3e_0e_1\} = e_2$, $\{e_2e_3e_0\} = e_1$, <u>but</u> $\{e_1e_2e_3\} = 2e_0$,

(2.14) <u>all products between three different factors including</u> e_0 <u>vanish or are (up to symmetry in the outer two variables) the products of</u> (2.13),

(2.15) $2L(e_0,e_0) + L(e_2,e_2) = L(e_1,e_1) + L(e_3,e_3)$,

(2.16) $P(e_3)e_0 = P(e_2)P(e_1)e_0$, $P(e_1)e_0 = P(e_2)P(e_3)e_0$.

<u>Proof.</u> Except (2.14) all formulas were established in the previous proof (note that the second formula of (2.16) follows by applying $P(e_2)$ to the first). Regarding (2.14) we have $L(e_0,e_2) = 0 = L(e_2,e_0)$ and so the only remaining products are $\{e_0e_1e_2\} = e_3$, $\{e_0e_1e_3\} \in V_{2-1+2}(e_3) = 0$, $\{e_0e_3e_1\} \in V_{2-1+2}(e_1) = 0$, $\{e_0e_3e_2\} = e_1$ and $\{e_1e_0e_3\} = e_2$. ∎

<u>Corollary 2.9.</u> <u>Let</u> (e_0,e_1,e_2,e_3) <u>be a diamond. Then</u> $(P(e_1)e_0,e_1,e_3,e_2)$ <u>is a diamond too.</u>

<u>Proof.</u> Because $e_0 \dashv e_1$ we know $P(e_1)e_0 \dashv e_1$ from the Triangle Criterion

2.5. Further, $e_1 \top e_3$ and $P(e_1)e_0 \in V_0(e_3)$, i.e. $e_3 \perp P(e_1)e_0$. Hence, from the Diamond Criterion 2.7. a) we get
$(P(e_1)e_0, e_1, e_3, \{P(e_1)e_0, e_1, e_3\})$ is a diamond, but $\{e_3\, e_1\, P(e_1)e_0\} =$
$-P(e_1)\{e_1 e_3 e_0\} + \{e_1 e_0 \{e_3 e_1 e_1\}\} = \{e_1 e_0 e_3\} = e_2$. ■

Up to now we do not know all products between tripotents of a diamond, e.g. what is $\{e_2 e_3 e_1\}$? To determine this, is our next goal. We first look at an example constructed with hermitian matrices (see Examples 1.5, 1.7.e): It is easily checked that $(H_{ii}, H_{ij}, H_{jk}, H_{ki})$ for $i,j,k \neq$ is a diamond.

It is an important fact that this is no "real" example: We will see below that each diamond can be imbedded in a family of tripotents which behave like the usual hermitian matrix units. To this end we recall the following special case of a definition from [42]§5 (for the general case see II §1).

A <u>hermitian grid</u> (<u>of size</u> 3) is a family $(h_{k\ell} = h_{\ell k},\ 1 \leq k \leq \ell \leq 3)$ of tripotents which satisfies for $i,j,k \neq$:

(2.17) $\quad (h_{ii},\ h_{ij},\ h_{jk},\ h_{ki})$ is a diamond,

(2.18) $\quad P(h_{ij})h_{ii} = h_{jj}$.

Clearly, the usual hermitian matrix units $(H_{k\ell};\ 1 \leq k \leq \ell \leq 3)$ form an example of an hermitian grid. In general, we have enough information to determine all possible products between the $h_{k\ell}$:

<u>Lemma 2.10.</u> <u>Let</u> $(h_{k\ell};\ 1 \leq k \leq \ell \leq 3)$ <u>be a hermitian grid. Then for</u> $i,j,k \neq$:

(2.19) $\quad h_{ii} \perp h_{jj},\ h_{ii} \perp h_{jk},\ h_{ii} \dashv h_{ij},\ h_{ij} \top h_{ik}$

(2.20) $\quad P(h_{ij})h_{ii} = h_{jj}$

(2.21) $\quad \{h_{ii} h_{ij} h_{jm}\} = h_{im}$ ($i \neq m$, <u>but</u> $j = m$ <u>allowed</u>)

(2.22) $\quad \{h_{ji} h_{ii} h_{ik}\} = h_{jk}$

(2.23) $\quad \{h_{ij} h_{jk} h_{ki}\} = 2h_{ii}$

(2.24) $\quad \{h_{k\ell} h_{mn} h_{pq}\} = 0$, <u>if the indices cannot be linked</u>.

<u>Proof.</u> By (2.17) we know $h_{ii} \dashv h_{ki} \perp h_{jj}$ whence $h_{ii} \perp h_{jj}$. The other relations of (2.19) follow immediately from the definition of a diamond. (2.20) = (2.18), (2.21), $j = m$, follows from (2.9), because (h_{ij}, h_{ii}, h_{jj}) is a triangle, and (2.21), $j \neq m$, is a consequence of (2.12). (2.22) and (2.23) follow from (2.13). Finally, (2.24) reflects the fact that $h_{k\ell} \perp h_{mn}$ whenever $\{k,\ell\} \cap \{m,n\} = \emptyset$. ■

Theorem 2.11. *Let* (e_0,e_1,e_2,e_3) *be a diamond. Put*

$$h_{11} = e_0, \quad h_{22} = P(e_1)e_0, \quad h_{33} = P(e_3)e_0,$$
$$h_{12} = e_1, \quad h_{23} = e_2, \quad h_{13} = e_3$$

Then $(h_{ij})_{1 \le i,j \le 3}$ *is a hermitian grid.*

Proof. By (2.13), (e_0,e_1,e_2,e_3) is a diamond iff (e_0,e_3,e_2,e_1) is a diamond. Hence (2.17) holds iff $(h_{11},h_{12},h_{23},h_{31})$, $(h_{22},h_{12},h_{13},h_{23})$ and $(h_{33},h_{13},h_{12},h_{23})$ are diamonds. For the first family this holds by definition and for the second and third by Corollary 2.9.

Because $h_{ii} \dashv h_{ij} \vdash h_{jj}$ (2.18) holds iff $P(h_{12})h_{11} = h_{22}$, $P(h_{13})h_{11} = h_{33}$ and $P(h_{23})h_{22} = h_{33}$. The first two equations are valid by definition and the third by (2.16). ∎

As for quadrangles and triangles there is also for diamonds a

Diamond Decomposition Theorem 2.12. *Let* (e_0,e_1,e_2,e_3) *be a diamond and define the family* $(h_{k\ell};\ 1 \le k \le \ell \le 3)$ *as in Theorem* 2.11. *Then the Peirce decompositions relative to* $\mathcal{D} = (e_0,e_1,e_2,e_3)$ *and relative to* $\mathcal{O} = (h_{11},h_{22},h_{33})$ *coincide. More precisely, let* $V_{(ijk\ell)}$ *be the Peirce spaces relative to* $\mathcal{D}$ *and* V_{ij} *the Peirce spaces relative to* $\mathcal{O}$, *then*

$$V_{11} = V_{(2202)}, \quad V_{22} = V_{(0220)}, \quad V_{33} = V_{(0022)}$$
$$V_{12} = V_{(1211)}, \quad V_{13} = V_{(1112)}, \quad V_{23} = V_{(0121)}$$
$$V_{01} = V_{(1101)}, \quad V_{02} = V_{(0110)}, \quad V_{03} = V_{(0011)}$$
$$V_{00} = V_{(0000)},$$

and all other Peirce spaces vanish.

Proof. It is enough to prove $V_{mn} \subset V_{(ijk\ell)} = V_i(e_0) \cap V_j(e_1) \cap V_k(e_2) \cap V_\ell(e_3)$. This is clear for $e_0 = h_{11}$ and follows for $e_1 = h_{12}$, $e_2 = h_{23}$ and $e_3 = h_{13}$ by noticing that (h_{ij},h_{ii},h_{jj}) is a triangle and that therefore each Peirce space of h_{ij} is contained in a Peirce space of (h_{ii},h_{jj}) by the Triangle Decomposition Theorem 2.6. The precise identification is straightforward. ∎

We use the results derived so far to characterize collinear tripotents in a Jordan triple system $V = V(J)$ where J is a Jordan algebra. An important example of a Jordan algebra is formed by hermitian matrices - see Example 1.5: The triple system $H_p(D,D_0,Id)$ is actually the triple system of a Jordan algebra with the unit matrix as unit element. The tripotents $1[12]$ and $1[23]$ are collinear, and we will show that the general situation resembles this example:

Theorem 2.13. *Let (e,f) be collinear non-zero tripotents in a Jordan algebra J. Then*

$h_{11} = U(e)f^2$, $h_{22} = U(e)^2 f^2$, $h_{33} = (U(f)e^2)^2$

$h_{12} = e$, $h_{23} = f$, $h_{13} = e\circ f$ $(:= \{e1f\})$

is a hermitian grid.

Proof. We put $c = U(e)f^2$ and show that c is a tripotent in J such that $e \vdash c \perp f$. Indeed, by compatibility the projections onto the Peirce-2-spaces of e and f commute, i.e. $[U(f)^2, U(e)^2] = 0$, and therefore $c^3 = U(U(e)f^2)U(e)f^2 = U(e)U(f^2)U(e)^2 f^2 = U(e)^3 U(f)^2 f^2 = U(e)f^2 = c$. The Peirce multiplication rules relative to f immediately show $c \in J_0(f)$, whence $c \perp f$. It is also clear that $c \in U(e)J = J_2(e)$. Thus $c \dashv e$ if we can prove $e \in J_1(c)$, i.e. $\{cce\} = e$. We have $\{cce\} = U(e,U(e)f^2)U(e)f^2$ = (by(1.4)) $V(e,f^2)U(e)^2 f^2 = V(e,f^2)U(e^2)f^2$ = (by(1.12)) $- U(e^2)\{f^2ef^2\} + \{\{ef^2e^2\}f^2e^2\}$. Because $\{f^2ef^2\} = 2U(f)^2e = 0$ the first summand vanishes. As to the second we show $\{ef^2e^2\}$ = (using $V(x,y) + V(y,x) = V(x\circ y)$ in a Jordan algebra) $- \{f^2e\,e^2\} + (e\circ f^2)\circ e^2 = - f^2\circ e^3 + e\circ e^2$ (since $e\circ f^2 = \{eff\} = e$) $= -e+2e = e$. Applying this formula twice in the second summand of $\{cce\}$ gives $\{cce\} = e$.

We can now apply the Diamond Criterion 2.7.a and obtain that $(c,e,f,\{cef\})$ is a diamond. We compute $\{cef\} = \{f,e,U(e)f^2\} = -U(e)\{eff^2\} + \{\{fee\}f^2e\} = -U(e)(e\circ f^3) + \{ff^2e\} = -U(e)(e\circ f) + f\circ e = f\circ e$ because $U(e)(e\circ f) = e\circ(U(e)f) = 0$. Finally, the result follows from Theorem 2.11 with the exception of the formula for h_{33}, but we have $h_{33} = U(h_{23})h_{22} = U(f)U(e)^2f^2 = U(U(f)e^2)1 = (U(f)e^2)^2$. ∎

Remark 2.14. We point out that the elements h_{ii} need not be idempotents. However, it is straightforward to check that h_{33} is indeed an idempotent and that h_{11} and h_{22} are at least strict tripotents ($h_{ii} = h_{ii}^3$ and $h_{ii}^2 = h_{ii}^4$). If one assumes that e and f are strict tripotents, then all h_{ii} are idempotents ([40]1.19).

Corollary 2.15. *In a Jordan algebra a pair of non-zero tripotents can never be collinear, if one of them is an idempotent.*

Proof. If $f^2 = f$, then $h_{11} = 0 = h_{22}$, contradiction. ∎

Corollary 2.16. *In a Jordan algebra two non-zero collinear tripotents can never be rigid-imbedded.*

§3 Cogs

While we considered small families of tripotents in the last section we start our investigations of certain types of large families of Peirce-compatible tripotents called cogs in this section. We retain the general setting of §2.

3.1. Recall 1.4: A finite family of compatible tripotents in a Jordan triple system V gives rise to a simultaneous Peirce decomposition of V. Although this is a useful property of compatible families they are too general to say something about their internal structure. This deficiency makes them unsuitable for many purposes, e.g. for coordinatization. We will therefore impose certain internal conditions on compatible families. These conditions will be restrictive enough to allow a classification of these compatible families whereas they are still weak enough to show the existence of such families in many interesting classes of examples.

As defined in 1.5 two tripotents are called Peirce-compatible if one lies in a single Peirce space of the other. Peirce-compatible families are still too general for our purposes. The reason is that the definition of Peirce-compatibility is not symmetric: If $e \in V_j(f)$, it does not follow in general that also f lies in some $V_k(e)$.

This observation motivates our first condition. We require "symmetry": $e \in V_j(f)$ and $f \in V_k(e)$. Our second condition can be thought of as a "minimality-condition": By definition, two associated tripotents have the same Peirce spaces, hence we obtain the same generalized Peirce spaces if we throw away associated tripotents. Now to the precise definition: A cog is an arbitrary family E of non-zero tripotents such that for each pair $(e,f) \in E \times E$ the following holds:

(3.1) e lies in a Peirce space of f, and

(3.2) e and f are not associated, unless e = f.

An alternative description of cogs is given in

Lemma 3.1. *Let E be a family of non-zero tripotents. Then E is a cog iff for each pair $(e,f) \in E \times E$ with $e \neq f$ we have*

(3.3) $e \top f$ *or* $e \perp f$ *or* $e \vdash f$ *or* $e \dashv f$.

Proof. This is immediate using Lemma 1.9 and Theorem 2.3. ∎

We note that (3.3) motivated the choice of the name "cog": It should remind the reader of the three possible relations in such a family: collinear, orthogonal and governing. (This is a suggestion of K. McCrimmon). Examples of cogs are the families $\mathcal{R}$, $\mathcal{S}$, $\mathcal{H}$ defined in Examples 1.3..., and any subfamily of these.

Later we will need the following results:

Lemma 3.2. Let c, e, f be tripotents such that $c \approx e$, $c \in V_k(f)$ and $e \in V_\ell(f)$. Then $k = \ell$.

Proof. If $k = 0$ then $V_2(e) = V_2(c) \subset V_0(f)$ by Lemma 1.9.a), whence $\ell = 0$ and, by symmetry $k = 0 \leftrightarrow \ell = 0$. If $k = 2$ then $V_2(e) = V_2(c) \subset V_2(f)$, therefore $k = 2 \leftrightarrow \ell = 2$. Since $k, \ell \in \{0,1,2\}$ the assertion follows. ∎

Corollary 3.3. Let E be a cog and $e \in E$. Assume c is a tripotent which is associated to e and lies in a Peirce space relative to E. Then for every $f \in E$, $f \neq e$, the pairs (e,f) and (c,f) fulfill the same relation ($\top$, $\perp$, $\vdash$ or $\dashv$). In particular, e and c lie in the same Peirce space of E.

Proof. Since $f \in E$ lies in some $V_j(e)$ by (3.1) we get $f \in V_j(c)$ because $e \approx c$. If $e \in V_\ell(f)$ Lemma 3.2 implies $c \in V_\ell(f)$, whence the corollary. ∎

3.2. Since the possible relations between two tripotents in a cog are known, our next object is to list all possible relations between three tripotents in a cog. First we note the following restriction, valid for arbitrary tripotents:

No-Tower-Lemma 3.4. Three non-zero tripotents d, e, f cannot satisfy the relation $d \dashv e \dashv f \vdash d$. In particular, if d, e, f belong to a cog, they cannot form a tower: $d \dashv e \dashv f$ is impossible.

Proof. Whenever $d \dashv e \dashv f$ we have $V_2(d) \subset V_2(e) \subset V_2(f)$, whence $d \dashv f$ if d, e, f belong to a cog.

To contradict $d \dashv e \dashv f \vdash d$ we put $c = P(e)d$. Then (e,d,c) is a triangle by the Triangle Criterion 2.5 and $L(e,e) = L(d,d) + L(c,c)$ by (2.10), whence $L(c,c)f = 0$. Moreover, $P(c)f = P(e)P(d)P(e)f = 0$ and thus $c \perp f$. But $c \in V_2(e) \subset V_2(f)$, contradiction. ∎

<u>Local-Structure-Theorem 3.5.</u> <u>Let</u> e_1, e_2, e_3 <u>be distinct elements of a cog. Then only the following relations can occur:</u>

Case-No.	1	2	3	4	5	6	7	8	9	10	11
$e_1:e_2$	$\perp$	$\top$	$\top$	$\perp$	$\top$	$\perp$	$\perp$	$\top$	$\dashv$	$\perp$	$\top$
$e_2:e_3$	$\perp$	$\top$	$\perp$	$\top$	$\perp$	$\top$	$\perp$	$\top$	$\top$	$\dashv$	$\perp$
$e_3:e_1$	$\perp$	$\perp$	$\top$	$\top$	$\perp$	$\perp$	$\top$	$\top$	$\perp$	$\top$	$\dashv$

Case-No.	12	13	14	15	16	17	18	19	20	21	22
$e_1:e_2$	$\vdash$	$\top$	$\perp$	$\dashv$	$\perp$	$\perp$	$\vdash$	$\perp$	$\perp$	$\dashv$	$\perp$
$e_2:e_3$	$\perp$	$\vdash$	$\top$	$\perp$	$\dashv$	$\perp$	$\perp$	$\vdash$	$\perp$	$\vdash$	$\dashv$
$e_3:e_1$	$\top$	$\perp$	$\vdash$	$\perp$	$\perp$	$\dashv$	$\perp$	$\perp$	$\vdash$	$\perp$	$\vdash$

Case-No.	23	24	25	26	27	28	29
$e_1:e_2$	$\vdash$	$\dashv$	$\top$	$\vdash$	$\vdash$	$\top$	$\dashv$
$e_2:e_3$	$\perp$	$\vdash$	$\dashv$	$\top$	$\dashv$	$\vdash$	$\top$
$e_3:e_1$	$\dashv$	$\top$	$\vdash$	$\dashv$	$\top$	$\dashv$	$\vdash$

Note that the cases between two |-signs only differ by a permutation.

<u>Proof</u>. The first 8 cases are the only ones where only $\perp$ and $\top$ occur. Assume now that exactly one of the relations is $\dashv$, say $e_1 \dashv e_2$. Then $e_2 \top e_3 \top e_1$ is impossible: By the Triangle Criterion 2.5 we know that $(e_2, e_1, \overline{e}_1 = P(e_2)e_1)$ is a triangle and by (2.10) we have $L(e_2,e_2) = L(e_1,e_1) + L(\overline{e}_1,\overline{e}_1)$. Since $e_3 \in V_1(e_2) \cap V_1(e_1)$ we obtain $e_3 \in V_0(\overline{e}_1)$. But e_2, $e_1 \in V_1(e_3)$ shows $\overline{e}_1 = P(e_2)e_1 \in V_1(e_3)$, which is a contradiction. Also $e_1 \dashv e_2 \perp e_3$ implies $e_1 \perp e_3$. Thus the possibilities with $e_1 \dashv e_2$ are $e_2 \top e_3 \perp e_1$ (Case 9 and by permutation 10-11) or $e_2 \perp e_3 \perp e_1$ (Cases 15-17). By interchanging e_1 and e_2 the cases 9-11 resp. 15-17 go over to 12-14 resp. 18-10.

Finally, assume that at least two signs of type $\dashv$ or $\vdash$ occur. Since a cog does not contain a tower, we must have (up to permutation of the indices) $e_1 \dashv e_2 \vdash e_3$ or $e_1 \vdash e_2 \dashv e_3$. The same reason shows that the remaining sign must be $\perp$ or $\top$. Because $e_1 \vdash e_2 \dashv e_3 \perp e_1$ is impossible $(e_2 \in V_2(e_1) \cap V_2(e_3))$, we are left with the cases 21-29. ∎

Using the examples of Jordan triple systems given in <u>1.2.</u> it is easy to show that all 29 cases can occur.

3.3. The deficiency of cogs is that they are not multiplicatively closed. Whenever e_1, e_2 belong to cog E we do not know $P(e_1)e_2 \in E$, not even that $P(e_1)e_2$ or $\{e_1 e_2 e_3\}$ is a tripotent. A first answer to this question is given in the following theorem which for later use is stated somewhat more general than presently needed.

Theorem 3.6.   Let E be a cog in V and e_1, e_2, $e_3 \in E$ be pairwise distinct lying in the Peirce spaces V_I resp. V_J resp. V_K relative to E.
  a) We have $V_{2I-J} = 0$, in particular $P(V_I)V_J = 0 = P(e_1)e_2$, unless $e_1 \vdash e_2$, in which case $P(e_1)e_2$ is a non-zero tripotent in V_{2I-J}.
  b) In the cases 1, 3-7, 10-12, 14-20, 22, 23, 28 and 29 of the Local-Structure-Theorem 3.5 we have $\{V_I V_J V_K\} = 0$, in particular $\{e_1 e_2 e_3\} = 0$.
  c) In the cases 2, 9, 13, 21 and 27 $\{e_1 e_2 e_3\}$ is a non-zero tripotent in V_{I-J+K}.

Proof. a) If $e_1 \perp e_2$, $e_1 \top e_2$ or $e_1 \dashv e_2$ we have $V_J \subset V_j(e_1)$ with $j \in \{0,1\}$, thus the component of e_1 in $2I-J$ is $4-j \geq 3$, which forces V_{2I-J} to vanish. Then also $P(V_I)V_J = 0$ because $P(V_I)V_J \subset V_{2I-J}$ by (1.29). The remaining case is $e_1 \vdash e_2$. Here $P(e_1)e_2$ is a non-zero tripotent since $0 \neq e_2 \in V_2(e_1)$ and $P(e_1)|V_2(e_1)$ is an automorphism.
  b) If $e_1 \perp e_2$ then $\{V_I V_J V_K\} \subset \{V_2(e_1) V_0(e_1) V\} = 0$ by (1.22) and by symmetry $\{V_I V_J V_K\} = 0$ for $e_1 \perp e_3$. This settles all cases except 28 and 29. But in these cases we have $\{V_I V_J V_K\} \subset V_{2-1+2}(e_1) = 0$ resp. $\{V_I V_J V_K\} \subset V_{2-1+2}(e_3) = 0$.
  c) If $e_1 \perp e_3$ and $e_2 \in V_1(e_1) \cap V_1(e_3)$, then $\{e_1 e_2 e_3\}$ is a non-zero tripotent because $e_2 \in V_2(e_1+e_3)$ by Theorem 1.10 and $\{e_1 e_2 e_3\} = P(e_1+e_3)e_2$. This remark settles all cases except 27. But here (e_2, e_1, c, e_3) is a diamond by Theorem 2.7.b).   ∎

The cases not covered by Theorem 3.6 are the cases 8 and 24-26. We can handle case 8 under additional assumptions:

Lemma 3.7.   (The case 8, part I). Let (e_1, e_2, e_3) be a collinear system. Then
  a) $\{e_1 e_2 e_3\} = 0 \leftrightarrow \{e_2 e_3 e_1\} = 0 \leftrightarrow \{e_3 e_1 e_2\} = 0$.
  b) If one of the pairs (e_i, e_j) is rigid-collinear (see (1.32)), then $\{e_1 e_2 e_3\} = 0$.
  c) If there exists a tripotent $c \in V$ satisfying for $i, j, k \neq$:

1) $e_i \vdash c$ and

2) $e_j \vdash c \dashv e_k$ or $e_j \perp c \perp e_k$,

then $\{e_1e_2e_3\} = 0$.

Proof. a) For $i,j,k \neq$ we have $\{e_ie_je_k\} \in V_2(e_i) \cap V_0(e_j) \cap V_2(e_k)$ and $P(e_i)\{e_ie_je_k\} = \{e_ie_ke_j\}$ by (1.12). Therefore $\{e_1e_2e_3\} = 0 \leftrightarrow 0 = P(e_1)\{e_1e_2e_3\} = \{e_1e_3e_2\} \leftrightarrow 0 = P(e_2)\{e_1e_3e_2\} = \{e_3e_1e_2\}$.

b) This follows from a) since $\{e_ie_je_k\} \in V_2(e_i) \cap V_0(e_j) = 0$.

c) Assume $e_j \vdash c \dashv e_k$. Then $e_i \vdash d = P(e_i)c \in V_0(e_j) \cap V_0(e_k)$. Hence there always exists a tripotent $d \in V$ satisfying $e_i \vdash d$ and $e_j \perp d \perp e_k$. But then $\{e_je_ie_k\} \in V_{-1}(d) = 0$, and therefore $\{e_1e_2e_3\} = 0$ by a). ∎

Lemma 3.8. (The case 8, part II) Let (e_1,e_2,e_3) be a collinear system which belongs to a cog E. If there exists $c \in E$ with $c \dashv e_i$, then the assumptions of Lemma 3.7.c) are fulfilled (and thus $\{e_1e_2e_3\} = 0$) or (c,e_i,e_j,e_k) (for suitable j,k) fulfill the relations of a diamond:

$$e_i \vdash c \dashv e_k,\ c \perp e_j,\quad e_1 \top e_2 \top e_3 \top e_1.$$

Proof. For simpler notation assume $c \dashv e_1$. Applying the Local-Structure-Theorem to (e_1,c,e_2) shows that only $c \perp e_2$ (Case 12) or $c \dashv e_2$ (Case 27) is possible. Since the same holds for e_3 we get the following possibilities: $e_2 \perp c \perp e_3$, $e_2 \perp c \dashv e_3$, $e_2 \vdash c \perp e_3$ or $e_2 \vdash c \dashv e_3$. Applying 3.7.c) to the first and last case gives $\{e_1e_2e_3\} = 0$, and in the remaining ones we have the relations of a diamond. ∎

We remark that in case (c,e_i,e_j,e_k) is actually a diamond (i.e. if $\{ce_ie_j\} = e_k$) we have $\{e_ie_je_k\} = 2c$ by (2.13).

We now turn to the cases 24-26. The product $\{e_1e_2e_3\}$ in these cases can be handled simultaneously in the setting of case 24, for $\{e_1e_2e_3\}$ in case 25 resp. 26 becomes $\{e_3e_1e_2\}$ resp. $\{e_2e_3e_1\}$ in case 24.

Lemma 3.9. (The cases 24-26). Assume the tripotents $e_i \in V$ satisfy $e_1 \dashv e_2 \vdash e_3 \top e_1$. Then

a) $\{e_1e_2e_3\} = 0 \leftrightarrow \{e_2e_3e_1\} = 0 \leftrightarrow \{e_3e_1e_2\} = 0$

b) Let $e_1 \in V_I = V_I(E)$, $e_2 \in V_J$ and $e_3 \in V_K$. Then $\{V_IV_JV_K\} = 0 \leftrightarrow \{V_JV_KV_I\} = 0 \leftrightarrow \{V_KV_IV_J\} = 0$

c) If (e_1,e_3) is rigid imbedded, then $\{V_IV_JV_K\} = 0$, in particular $\{e_1e_2e_3\} = 0$.

Proof. Since the assumptions are symmetric in e_1 and e_3, we only have to show the first equivalence in a) and b). By (1.24) we have $P(e_1)\{e_1e_2e_3\} = \{e_1e_3e_2\}$ and $P(e_1)\{e_1e_3e_2\} = \{e_1e_2e_3\}$ which implies a). For b) we use (1.12) to obtain $P(e_1)\{V_IV_JV_K\} + \{V_JV_IP(e_1)V_K\} = \{e_1V_K\{V_JV_Ie_1\}\}$. But $e_3 \in V_K$ and $e_3 \top e_1$ implies $V_K \subset V_1(e_1)$, hence $P(e_1)V_K = 0$. Because $\{V_JV_Ie_1\} \subset V_I$ by (1.29) we get $P(e_1)\{V_IV_JV_K\} \subset \{V_IV_KV_J\}$. Similarly, $P(e_1)\{V_IV_KV_J\} \subset \{V_IV_JV_K\}$. But $\{V_IV_JV_K\} \oplus \{V_IV_KV_J\} \subset V_2(e_1)$ and $P(e_1)|V_2(e_1)$ is an automorphism. Thus $P(e_1)$ interchanges $\{V_IV_JV_K\}$ and $\{V_IV_KV_J\}$, which implies b).
c) We have $\{V_IV_JV_K\} \subset V_2(e_1) \cap V_2(e_3) = 0$ by rigid-collinearity. ∎

We remark that the situation of Lemma 3.9. occurs e.g. in $[K,{}^-,2m+1]$, $m > 2$: In the notation of Example 1.6 we have for $j \neq k$ $e_j^\varepsilon \dashv e_0 \vdash e_k^\delta \top e_j^\varepsilon$. Because $(e_j^\varepsilon, e_k^\delta)$ is rigid imbedded, we get $\{e_1e_2e_3\} = 0$. On the other side it also occurs in the Jordan triple system of hermitian matrices $H_p(k)$, $p \geq 4$: In the notation of Example 1.7.e we have $H_{13} \dashv H_{12}+H_{34} \vdash H_{14} \top H_{13}$, but $\{H_{13}, H_{12}+H_{34}, H_{14}\} = 2H_{11}$. It is therefore clear that one cannot say something about the cases 24-26 in general.

3.4. An <u>ortho-collinear system</u> is a family of non-zero tripotents where two distinct elements are either orthogonal or collinear. Such families were already introduced in [40]. Obviously, an ortho-collinear system is a special case of a cog (its nick name is "oc"). On the other hand every cog is the disjoint union of two oc's. To see this we put for a cog E

(3.4) $\quad E^{(2)} = \{e \in E \;;\; \text{there exists } f \in E \text{ with } e \vdash f\}$,

(3.5) $\quad E^{(1)} = E \setminus E^{(2)}$.

Lemma 3.10. *$E^{(1)}$ and $E^{(2)}$ are ortho-collinear systems such that* $E = E^{(1)} \dot\cup E^{(2)}$.

Proof. This is clear for $E^{(1)}$ and follows for $E^{(2)}$ from the No-Tower-Lemma 3.4. ∎

Another concept we shall need is that of connectedness: Two tripotents e, f of a cog E are called <u>connected</u> if there exists a finite sequence $(f_1,\dots,f_n) \subset E$, $n \geq 1$, such that

(3.6) $\quad f_1 = e,\ f_n = f$ and $f_i \not\perp f_{i+1}$ for $1 \leq i \leq n-1$,

where $f_i \not\perp f_{i+1}$ of course means that f_i is not orthogonal to f_{i+1}. We call $(f_1,\dots,f_n)$ a <u>connecting chain of length</u> n. Clearly, connectedness is an equivalence relation for E. The corresponding equivalence classes are called connected components, and E is called connected if each pair of tripotents of E is connected.

Theorem 3.11. Let E be a connected ortho-collinear system. Then for each pair $(e,f) \subset E$ there exists an automorphism ϕ of V such that

a) $\phi e = f$,

b) ϕ permutes the Peirce spaces of V relative to E.

Proof. By assumption there exists a connecting chain $(e = f_1, f_2, \ldots, f_n = f) \subset E$ such that $f_i \top f_{i+1}$ for $1 \leq i \leq n-1$. Thus the theorem follows from

Lemma 3.12. Let E be a cog and let $e,f \in E$ such that $e \top f$. Then the automorphism $T_{e,f}$, which by Theorem 1.13 interchanges e and f, permutes the Peirce spaces of V relative to E. In particular, if $e \in V_I(E)$ and $f \in V_J(E)$, then $T_{e,f} V_I(E) = V_J(E)$.

Proof. The general case follows if we can prove the assertion for Peirce spaces $V_{(ijk)}$ relative to the family $(e,f,h) \subset E$. Keeping in mind that e and f belong to Peirce spaces of h, the form of $T_{e,f}$ specified in Theorem 1.13 shows $T_{e,f} V_{(ijk)} \subset V_{(ji\ell)}$ for a unique ℓ, e.g. for $i = j = 1$ we get $T_{e,f} V_{(11k)} \subset V_{(11k)}$ because $L(e,f)L(f,e)V_k(h) \subset V_k(h)$. Since $T_{e,f}$ is of order 2 we have $T_{e,f} V_{(ijk)} = V_{(ji\ell)}$. The last assertion follows from $T_{e,f} e = f$. ∎

Later on we will need the following result which is analogous to Lemma 3.12:

Lemma 3.13. Let E be a cog and assume for e, $f_i \in E$ that $f_1 \dashv e \vdash f_2 \perp f_1$. Then the involutorical automorphism $\phi = P(e)|V_2(e)$ of $V_2(e)$ satisfies

a) $\phi V_2(f_1) = V_2(f_2)$,

b) ϕ permutes the Peirce spaces of E which are contained in $V_2(e)$; in particular, if $f_1 \in V_I(E)$ and $f_2 \in V_J(E)$ then $\phi V_I(E) = V_J(E)$.

Proof. That ϕ is an involutorical automorphism was already stated in (1.23).

a) By the Triangle Criterion 2.5.b we know $f_2 \approx P(e)f_1$, hence $\phi V_2(f_1) = V_2(P(e)f_1) = V_2(f_2)$.

b) For every $h \in E$ we have $P(e)V_k(h) \subset V_\ell(h)$ and $P(e^2)V_k(h) \subset V_k(h)$.

This implies that every $V_I(E) \subset V_2(e)$ is mapped by $P(e)$ onto another Peirce space $V_J(E) \subset V_2(e)$. The last assertion now follows from Corollary 3.3 and $f_2 \approx P(e)f_1$. ⊞

§4 Closed cogs and grids

In this section we study cogs which are up to association multiplicatively closed relative to the elementary configurations of §2.

Throughout, E always denotes a cog in a Jordan triple system V. We will use the abbreviation $c \approx E$, if c is a tripotent which is associated to a tripotent of E.

4.1. The starting point of our considerations is Theorem 3.6: In several cases $P(e_1)e_2$ or $\{e_1 e_2 e_3\}$ is a non-zero tripotent which in general does not belong to E or is at least associated to some tripotent of E. We will require this as an additional condition for the cogs considered in this section. But before we give the precise definition we want to minimize the requirements. For this we need the following

Lemma 4.1. *a) If* $e_1 \dashv e_2 \top e_3 \perp e_1$ *and* $P(e_2)e_1 \approx c$, *then* $e_2 \vdash c \dashv e_3$ *and* $\{e_1\ e_2\ e_3\} \approx \{e_2\ c\ e_3\}$.

b) If $c_1 \vdash c_2 \dashv c_3 \top c_1$ *and* $P(c_1)c_2 \approx e$, *then* $e \dashv c_1 \top c_3 \perp e$ *and* $\{c_1\ c_2\ c_3\} \approx \{e\ c_1\ c_3\}$.

Proof. a) By the Diamond Criterion 2.7 $(e_1,e_2,e_3,\{e_1\ e_2\ e_3\})$ is a diamond, and by Theorem 2.11 the corresponding family of 3×3 hermitian matrix units is given by

$$h_{11} = e_1,\ h_{22} = P(e_2)e_1,\ h_{33} = P(\{e_1e_2e_3\})e_1 = P(e_3)P(e_2)e_1,$$
$$h_{12} = e_2,\ h_{23} = e_3,\qquad h_{13} = \{e_1e_2e_3\}.$$

Because $h_{22} \approx c$ we get $e_2 \vdash c \dashv e_3$ and therefore $(c,e_2,\{e_2\ c\ e_3\},e_3)$ is also a diamond whose corresponding family of 3×3 hermitian matrix units is given by

$$\tilde{h}_{11} = c,\quad \tilde{h}_{22} = P(e_2)c,\quad \tilde{h}_{33} = P(e_3)c$$
$$\tilde{h}_{12} = e_2,\ \tilde{h}_{23} = \{e_2\ c\ e_3\},\ \tilde{h}_{13} = e_3\ .$$

Now $\tilde{h}_{11} \approx h_{22}$ by assumption, and applying the automorphism $P(e_2)|V_2(e_2)$ resp. $P(e_3)|V_2(e_3)$ shows $\tilde{h}_{22} \approx h_{11}$ resp. $\tilde{h}_{33} \approx h_{33}$. Thus the Peirce decompositions relative to (h_{11},h_{22},h_{33}) and to $(\tilde{h}_{11},\tilde{h}_{22},\tilde{h}_{33})$ coincide which implies $V_2(\tilde{h}_{23}) = V_2(\tilde{h}_{22}+\tilde{h}_{33}) = V_2(h_{11}+h_{33}) = V_2(h_{13})$, i.e. $\{e_2ce_3\} \approx \{e_1e_2e_3\}$.

b) follows from a): We have $e \dashv c_1 \top c_3 \perp e$ and $P(c_1)e \approx P(c_1)^2c_2 = c_2$, hence $\{ec_1c_3\} \approx \{c_1c_2c_3\}$. ⊞

The minimization of the requirements mentioned above is contained in

Corollary 4.2. *Let* $e_i \in E$ *and assume* $P(e_1)e_2 \approx E$, *whenever* $e_1 \vdash e_2$.

<u>Then the condition</u>

i) <u>If</u> $e_1 \vdash e_2 \dashv e_3 \top e_1$, <u>then</u> $\{e_1e_2e_3\} \approx E$

<u>is equivalent to</u>

ii) <u>If</u> $e_1 \dashv e_2 \top e_3 \perp e_1$, <u>then</u> $\{e_1e_2e_3\} \approx E$.

<u>Proof</u>. Assume $e_1 \dashv e_2 \top e_3 \perp e_1$ and let $P(e_2)e_1 \approx c \in E$. By Lemma 4.1.a we have $\{e_1e_2e_3\} \approx \{e_2ce_3\}$ and $e_2 \vdash c \dashv e_3 \top e_2$, so by i) $\{e_1ce_3\} \approx E$, whence $\{e_1e_2e_3\} \approx E$.

Conversely, let $e_1 \vdash e_2 \dashv e_3 \top e_1$ and $P(e_1)e_2 \approx e \in E$. By Lemma 4.1.b and ii) we have $e \dashv e_1 \top e_3 \perp e$ and $\{e_1e_2e_3\} \approx \{ee_1e_3\} \approx E$. ∎

We call E <u>closed</u>, if for $e_1,e_2,e_3 \in E$ the following three conditions hold:

(4.1) If $e_1 \vdash e_2$, then $P(e_1)e_2 \approx E$.

(4.2) If $e_1 \vdash e_2 \dashv e_3 \top e_1$, then $\{e_1e_2e_3\} \approx E$.

(4.3) If $e_1 \top e_2 \top e_3 \perp e_1$, then $\{e_1e_2e_3\} \approx E$.

We have the following characterization of closed cogs:

<u>Theorem 4.3.</u> <u>Let</u> e_i <u>be distinct elements of the cog</u> E. <u>Then</u> E <u>is closed iff</u> $P(e_1)e_2$ <u>and</u> $\{e_1e_2e_3\}$ <u>are tripotents which are zero or associated to some tripotent in</u> E, <u>except in the following cases for</u> $\{e_1e_2e_3\}$:

1) (e_1,e_2,e_3) <u>is a collinear system</u>,

2) <u>after a change of indices we have</u> $e_1 \dashv e_2 \vdash e_3 \top e_1$.

<u>Proof</u>. Let E be closed. By Theorem 3.6 either $P(e_1)e_2 = 0$ or $e_1 \vdash e_2$ in which case $P(e_1)e_2 \approx E$ by (4.1). Now note that the exceptions 1) and 2) of the theorem are just the cases 8 and 24-26 of the Local-Structure-Theorem 3.5. Hence, by Theorem 3.6, we have to show $\{e_1e_2e_3\} \approx E$ in the cases 2,9,13,21 and 27. The case 2 follows from (4.3), the cases 9,13 and 27 from (4.1), (4.2) together with Corollary 4.2 and in case 21 we have $e_1+e_3 \approx e_2$ by the Triangle Criterion 2.5.b), $\{e_1e_2e_3\} = P(e_1+e_3)e_2 \approx e_2$ by the criterion (4.4) proven below. The converse direction is clear. ∎

(4.4) <u>If</u> $c \approx d$, <u>then</u> $P(c)d \approx d$.

<u>Proof</u> of (4.4): Because $P(c)$ is an automorphism of $V_2(c)$ and d is invertible in $V_2(c)$ by Theorem 2.3, also $P(c)d$ is invertible, hence $P(c)d \approx c \approx d$. ∎

4.2. In this subsection we study the ortho-collinear subfamilies of a cog E introduced in 3.4. Recall

(4.5) $E^{(2)} = \{e \in E\,;\ \text{there exists } f \in E \text{ with } e \vdash f\}$,

(4.6) $E^{(1)} = E \setminus E^{(2)}$.

Lemma 4.4. Assume E is closed and let $e,f \in E$ such that $e \top f$. Then for $i = 1,2$

$$e \in E^{(i)} \leftrightarrow f \in E^{(i)}.$$

Proof. It is enough to show $e \in E^{(2)} \to f \in E^{(2)}$. So we may assume there exists $c \in E$ such that $c \dashv e$. Then the Local-Structure-Theorem 3.5 shows either $c \dashv f$, in which case $f \in E^{(2)}$, or $c \perp f$, in which case $(c,e,f,\{cef\})$ is a diamond and so $f \vdash P(e)c \approx E$, whence $f \in E^{(2)}$ too. ∎

Theorem 4.5. Let E be connected and closed. Then

a) $E^{(2)}$ is connected and closed,

b) $E^{(1)}$ is closed.

c) Assume $E^{(2)} \neq \emptyset$. Then

(4.7) $E^{(1)} = \{f \in E\,;\ \text{there exists } e \in E^{(2)} \text{ with } e \vdash f\}$.

Proof. a) and b): We first prove connectedness of $E^{(2)}$: For $e,f \in E^{(2)}$ we choose a connecting chain $(e = f_1,\dots,f_n = f) \subset E$ of shortest length and claim all $f_i \in E^{(2)}$. Indeed, assume $f_{i-1} \in E^{(2)}$, $1 < i < n$. Then $f_{i-1} \dashv f_i$ contradicts the No-Tower-Lemma 3.4., so $f_{i-1} \vdash f_i$ or $f_{i-1} \top f_i$. Also $f_{i-1} \perp f_{i+1}$ since otherwise the chain could be shortened. Going through the various cases of the Local-Structure-Theorem 3.5 for $f_{i+1} \perp f_{i-1} \vdash f_i$ shows $f_i \perp f_{i+1}$, a contradiction. Hence $f_{i-1} \top f_i$ and $f_i \in E^{(2)}$ by Lemma 4.4.

To prove closedness of $E^{(i)}$, $i = 1,2$ we only have to verify (4.3) for $e_i \in E^{(i)}$. We know $\{e_1e_2e_3\} \approx e_4 \in E$ and $e_1 \top e_4$ by Corollary 3.3, thus $e_4 \in E^{(i)}$ by Lemma 4.4.

c) Let $f \in E^{(1)}$. Then f can be connected with the elements of $E^{(2)}$, and under all such chains we choose one of shortest length, say $(f=f_1,f_2,\dots,f_n)$. Then $f_1,\dots,f_{n-1} \in E^{(1)}$, $f_n \in E^{(2)}$ and $f_{n-1} \dashv f_n$ by Lemma 4.4. If $n \geq 3$ we have $f_{n-2} \top f_{n-1}$ and the Local-Structure-Theorem shows $f_n \vdash f_{n-2}$, hence the length could be shortened. So $n=2$, $f = f_1 \dashv f_2 \in E^{(2)}$. The other direction is clear. ∎

For the connected closed cog $H = \{H_{ij}\,;\ 1 \leq i \leq j \leq p\} \subset H_p(k)$ one gets

$$H^{(1)} = \{H_{ii}\,;\ 1 \leq i \leq p\},$$
$$H^{(2)} = \{H_{ij}\,;\ 1 \leq i < j \leq p\}.$$

In particular, $H^{(1)}$ is an orthogonal system, hence totally disconnected.

Our next aim is to minimize the connecting chains between the various pieces of E.

Lemma 4.6. Let E be an ortho-collinear system which is connected and closed. Then each two tripotents of E can be connected by a chain of length ≤ 3.

Proof. Let $(f_1,\ldots,f_n) \subset E$ be a connecting chain of shortest length and assume $n \geq 4$. Then (f_2,f_3,f_4) is also of shortest length, hence $f_2 \top f_3 \top f_4 \perp f_2$ and $\{f_2 f_3 f_4\} \approx g \in E$ by closedness. Because $f_3 \perp f_1 \perp f_4$ we get $g \in V_1(f_1)$ and thus (f_1,g,f_4) is a connecting chain, contradiction. ∎

Theorem 4.7. Let E be connected and closed.

a) If $f \in E^{(1)}$ and $e \in E^{(2)}$, then $f \dashv e$ or there exists $h \in E^{(2)}$ such that $f \dashv h \top e \perp f$.

b) If $f_1, f_2 \in E^{(1)}$, $f_1 \perp f_2$ and $E^{(2)} \neq \emptyset$, then there exists $e \in E^{(2)}$ such that $f_1 \dashv e \vdash f_2$.

Proof. a) By Lemma 4.4. we either have $f \dashv e$ or $f \perp e$. Assume the latter. By (4.7) we may choose $g_1 \in E^{(2)}$ with $f \dashv g_1$. If $g_1 \top e$ we are done. Otherwise $g_1 \perp e$ and by Lemma 4.6 there exists $g_2 \in E^{(2)}$ such that $g_1 \top g_2 \top e$. If $f \dashv g_2$ we are done again, so we can assume $f \perp g_2$. But then $(f,g_1,g_2,\{f g_1 g_2\})$ is a diamond with $\{f g_1 g_2\} \in V_1(e)$. Let $\{f g_1 g_2\} \approx h \in E^{(2)}$. Then $f \dashv h \top e$.

b) We choose $e_1 \in E^{(2)}$ with $f_1 \dashv e_1$. In case $e_1 \vdash f_2$ we are done, otherwise $e_1 \perp f_2$ and by a) there exists $e_2 \in E^{(2)}$ such that $e_1 \top e_2 \vdash f_2$. Again we are done if $f_1 \dashv e_2$. So we may assume $f_1 \perp e_2$, whence $(f_1,e_1,e_2,\{f e_1 e_2\})$ is a diamond. Let $\{f e_1 e_2\} \approx e \in E^{(2)}$. Then $f_1 \dashv e \vdash f_2$. ∎

Corollary 4.8. Let E be connected and closed.

a) Then each two elements of E are connected by a chain of length ≤ 3.

b) For $f,g \in E^{(i)}$ denote by $V_I(E)$ resp. $V_J(E)$ the Peirce spaces of E containing f resp. g, i.e., we have $f \in V_I(E) \subset V_2(f)$ and $g \in V_J(E) \subset V_2(g)$. Then there exists an isomorphism

$$\phi: V_2(f) \to V_2(g) \text{ with } \phi V_I(E) = V_J(E).$$

Proof. a) If $E^{(2)} = \emptyset$ then E is ortho-collinear and the claim follows from Lemma 4.6. The same argument applies to $e_1, e_2 \in E^{(2)}$. The cases $e \in E^{(2)}$, $f \in E^{(1)}$ and $f_1, f_2 \in E^{(1)}$ with $f_1 \perp f_2$ are covered by Theorem 4.7. But the remaining possibility is $f_1, f_2 \in E^{(1)}$ such that $f_1 \top f_2$.

b) For i=2 the claim follows from Theorem 4.5.a and Theorem 3.11 if one notices that $g = \phi f \in \phi V_I(E)$ which is a Peirce space of E. In the same way we are done if f and g lie in a connected component of $E^{(1)}$. If not, then $f \perp g$ and Theorem 4.7.b) tells us that there is an $e \in E^{(2)}$ satisfying $f \dashv e \vdash g$. Now the existence of ϕ follows from Lemma 3.13.∎

So far we investigated the ortho-collinear systems $E^{(i)}$ of a cog E. We now look at a certain subfamily of $E^{(1)}$ which later will be used to distinguish between various types of closed cogs.

Theorem 4.9. Let E be a closed cog with $e \in E$ and assume $E_e = \{f \in E ; f \dashv e\} \neq \emptyset$. Then E_e is a closed ortho-collinear system such that:

a) The rank of E_e is 2, more precisely, for every $f_1 \in E_e$ there exist a unique $f_2 \in E_e$ with $f_1 \perp f_2$.

b) If $E_e = \{f_1, f_2\}$, then for every $h \in E \setminus \{e, f_1, f_2\}$ we have

$V_1(f_1) \cap V_1(f_2) \subset V_2(e) \cap V_k(h)$ for some $k \in \{0,1\}$,

$V_2(f_i) \subset V_2(e) \cap V_j(h)$ for some $j \in \{0,2\}$;

in particular, $V_1(f_1) \cap V_1(f_2)$ and $V_2(f_i)$ are Peirce spaces relative to E.

c) If $\# E_e > 3$, then E_e is connected. Moreover $E \cap V_1(e) = \emptyset$, and $\{e\} \cup E_e$ is a connected component of E.

Proof. Clearly, $E_e \subset E^{(1)}$ is ortho-collinear by Lemma 3.10, and is closed because $E^{(1)}$ is.

a) For every $f_1 \dashv e$ we know that $(e, f_1, P(e)f_1)$ is a triangle and $P(e)f_1 \approx f_2 \in E$ by (4.1). Applying Corollary 3.3 shows $f_1 \perp f_2$ and $f_2 \in E_e$. For any other $g \in E_e$ with $f_1 \perp g$ we can apply the Triangle Criterion 2.5.b to obtain $g \approx P(e)f_1 \approx f_2$, and so $g = f_2$. This shows a).

b) Since $(e, f_1, P(e)f_1)$ is a triangle with $f_2 \approx P(e)f_1$ we obtain from Theorem 2.6 that

$$V_2(e) = V_2(f_1) \oplus V_1(f_1) \cap V_1(f_1) \oplus V_2(f_2).$$

By the No-Tower-Lemma and our assumption every $h \in E \setminus \{e, f_1, f_2\}$ satisfies $e \perp h$ or $e \top h$. In the first case $V_2(e) \subset V_0(h)$ and we are done. So we consider $e \top h$, which implies $h \in E^{(2)}$ by Lemma 4.4. Since

$$V_1(e) = V_1(f_1) \cap V_0(f_2) \oplus V_0(f_1) \cap V_1(f_2)$$

we may assume $h \vdash f_1$. Then $(f_1, e, \{ef_1h\}, h)$ is a diamond and the corresponding hermitian grid (h_{ij}) of size 3 satisfies $h_{11} = f_1$, $h_{22} \approx f_2$ and $h_{13} = h$. Because $h_{13} \approx h_{11} + h_{33}$ we get $V_1(f_1) \cap V_1(f_2) \subset V_1(h)$, $V_2(f_1) \subset V_2(h)$ and $V_2(f_2) \subset V_0(h)$.

c) As in b) let $(f_1, f_2) \subset E_e$ be an orthogonal pair and let g be another element of E_e. Then $g \in V_1(f_1) \cap V_1(f_2) \in E^{(1)}$, which shows connectedness of E_e. Assume there exists $h \in E \cap V_1(e)$. As in b) we then get a hermitian grid (h_{ij}) of size 3 with $h = h_{13} \in E^{(2)}$ and $g \in V_1(h)$ which contradicts $g \in E^{(1)}$. The last assertion is obvious. ∎

We note that both possibilities for E_e considered in Theorem 4.9 actually occur: For b) one may take the family $H \subset H_p(k)$ of hermitian matrix units, for c) the canonical cog $\{e_0\} \cup \{e_j^\varepsilon;\ 1 \le j \le m,\ \varepsilon = \pm\}$ in $[K, \bar{\ }, 2m+1]$.

4.3. In this subsection we will show that every cog can be imbedded in a closed cog. We will need

Lemma 4.10. *Let E be a cog and assume the tripotent c is given in one of the following four ways using $e_i \in E$:*

a) $c = P(e_1)e_2$ *for* $e_1 \vdash e_2$,

b) $c = \{e_1e_2e_3\}$ *for* $e_1 \vdash e_2 \dashv e_3 \top e_1$,

c) $c = \{e_1e_2e_3\}$ *for* $e_1 \top e_2 \top e_3 \perp e_1$,

d) $c = \{e_1e_2e_3\}$ *for* $e_1 \dashv e_2 \top e_3 \perp e_1$.

Then c lies in a Peirce space relative to E, and the Peirce spaces relative to E and $E \cup \{c\}$ coincide. In particular, every $e \in E$ lies in a Peirce space of c.

Proof. Since e_i lies in a Peirce space relative to every $e \in E$, the same holds for c by (1.20), (1.21). To prove the converse we note that by the results of §2 in each of the cases a)-d) the Peirce decomposition of (e_1, e_2) resp. (e_1, e_2, e_3) is the same as the Peirce decomposition of (e_1, e_2, c) resp. (e_1, e_2, e_3, c). Because every $e \in E$ lies in a Peirce space of (e_1, e_2) resp. (e_1, e_2, e_3) the claim follows. ∎

Theorem 4.11. *For every cog E there exists a closed cog E_c such that*

1) $E \subset E_c$,

2) *every tripotent of E_c is connected to a tripotent of E and*

3) *the Peirce spaces of V relative to E and E_c coincide.*

We call a closed cog with the properties 1) - 3) of Theorem 4.11 a closure of E. It is not unique. Note that whenever E is connected every closure of E is connected too.

Proof. I. In the first step we construct a cog E_1 such that E is closed in E_1 (i.e. all tripotents constructed in Lemma 4.10 are associated to a tripotent of E_1) and 1) - 3) hold for E_1 in place of E_c.

Let X be the set of tripotents occuring in the cases a)-c) of Lemma 4.10, but which are not associated to a tripotent of E. Let Y be a system of representatives for the equivalence relation $\approx$ on X. Then E is closed in $E_1 = E \cup Y$, 1) - 3) hold for E_1, so in particular E_1 is a cog: Every $c \in E_1$ lies in a Peirce space relative to E whence also relative to E_1.

II. Now we define successively $E_0 = E$, $E_{n+1} = (E_n)_1$ for $(E_{\dots})_1$ constructed in I. and put $E_c = \bigcup_{n>0} E_n$. Since a finite subset of E_c lies in some cog E_n all claims of the theorem are easily established. ∎

4.4. As it was pointed out at the end of §3.3 one cannot say anything about the cases 24-26 of the Local-Structure-Theorem 3.5 in general: For tripotents e_i in a closed cog satisfying $e_1 \dashv e_2 \vdash e_3 \top e_1$ we could have $\{e_1e_2e_3\} = 0$ or $\{e_1e_2e_3\} = 2c$ for another tripotent c. For the families which we will classify in Chap. II and use for coordinatization in Chap. III we require the first possibility. For these families we will also be able to handle the troublesome case 8 of the Local-Structure-Theorem 3.5. We define:

A grid E is a closed cog satisfying the following two rigidity conditions:

(4.8) If $(e_1,e_2,e_3) \subset E$ is a collinear system and $\{e_1e_2e_3\} \neq 0$, then there exists $c \in E$ such that $e_1 \vdash c \dashv e_3$ and $c \perp e_2$.

(4.9) If $e_1 \dashv e_2 \vdash e_3 \top e_1$, then $\{e_1e_2e_3\} = 0$.

We note that the notion grid was already used in [40] and [42], however in a more restrictive meaning. Every grid in [40] is a grid in our sense but not conversely. The reason for calling (4.8) and (4.9) rigidity conditions lies in

Theorem 4.12. Let E be a closed cog and assume that every collinear pair in $E^{(1)}$ is rigid-imbedded. Then E is a grid.

Proof. Let $(e_1,e_2,e_3) \subset E$ be a collinear system. Then either

$(e_1,e_2,e_3) \subset E^{(1)}$, in which case $\{e_1e_2e_3\} = 0$ by Lemma 3.7.b, or $(e_1,e_2,e_3) \subset E^{(2)}$. By Theorem 4.5.c there exists $c \in E^{(1)}$ such that $c \dashv e_1$. Now there are the following possibilities for (e_2,c,e_3): α) $e_2 \perp c \perp e_3$, β) $e_2 \vdash c \dashv e_3$, γ) $e_2 \perp c \dashv e_3$ or δ) $e_2 \vdash c \perp e_3$. In the first two cases $\{e_1e_2e_3\} = 0$ by Lemma 3.7.c), the last two cases fall under the exceptions of (4.8): This is clear for γ), and for δ) we choose $d \in E^{(1)}$ such that $d \approx P(e_1)c$. Then $e_1 \vdash d \dashv e_3$ and $d \perp e_2$. This proves (4.8). (4.9) follows from Lemma 3.9. ▣

Of course, now the question arises: When are collinear pairs rigid-imbedded? One sufficient condition for this was given in the Collinearity Criterion 1.12. Another one will be established in the next section (Corollary 5.4.).

Using Lemma 3.9 the following sharpening of Theorem 4.3 is straightforward to prove:

Theorem 4.13. Let E be a grid and $e_1,e_2,e_3 \in E$ be pairwise distinct. Then $P(e_1)e_2$ and $\{e_1e_2e_3\}$ are zero or associated to some tripotent in E, except when there exists $c \in E$ such that (c,e_1,e_2,e_3) fulfills

$$e_1 \vdash c \dashv e_3, \; c \perp e_2 \text{ and } e_1 \top e_2 \top e_3 \top e_1.$$

The span of a grid E is in general not a subsystem (although we will see that it is associated to a grid $\tilde{E}$ where this is the case) nor does this hold for the cover

$$C_V(E) = \oplus_{E \cap V_I \neq \emptyset} V_I$$

We need extra conditions, determined in

Theorem 4.14. The cover of a closed cog E is a subsystem of V iff the following condition for $e_i \in E$ with $e_1 \in V_I$, $e_2 \in V_J$ and $e_3 \in V_K$ holds:

$$\text{(4.10)} \quad \begin{array}{ll} \text{If} & \text{i) } e_1 \top e_2 \top e_3 \top e_1 \text{ and } \{f \in E\,;\, f \dashv e_1\} = \emptyset \\ \text{or} & \text{ii) } e_1 \dashv e_2 \vdash e_3 \top e_1, \\ \text{then} & \{V_IV_JV_K\} = 0. \end{array}$$

In this case, E is already a grid.

Proof. Certainly (4.10) implies (4.9), but also (4.8) using Lemma 3.8. Clearly, $C_V(I)$ is a subsystem iff all submodules $P(V_I)V_J$ and $\{V_IV_JV_K\}$ belong to $C_V(I)$. From (1.29) we recall

$$P(V_I)V_J \subset V_{2I-J}, \quad \{V_IV_JV_K\} \subset V_{I-J+K}.$$

Now assume that $C_V(E)$ is a subsystem and let (e_1,e_2,e_3) fulfill the relations of i) resp. ii). Then in both cases $\{V_I V_J V_K\} \subset V_{I-J+K} \subset V_2(e_1) \cap V_2(e_3)$. If $\{V_I V_J V_K\} \neq 0$ there exists $c \in E \cap V_{I-J+K}$ and it follows $c \dashv e_1$ which in both cases gives a contradiction. (For case ii) use e.g. the No-Tower-Lemma 3.4.).

Conversely, assume (4.10). Then it is to show that all submodules $P(V_I)V_J$ resp. $\{V_I V_J V_K\}$ vanish or V_{2I-J} resp. V_{I-J+K} contains a tripotent of E. Since the cases $I = J$ or $J = K$ are obvious, we may assume (e_1,e_2,e_3) pairwise distinct.

If $P(V_I)V_J \neq 0$ then $e_1 \vdash e_2$ by Theorem 3.6.a and $0 \neq P(e_1)e_2$ is a tripotent in V_{2I-J}, which, by closedness of E is associated to some $e \in E$. Then $e \in V_{2I-J}$ by Corollary 3.3. A similar argument, using Theorem 3.6.b) and c) handles all the cases of the Local-Structure-Theorem, except 8 and 24-26.

Let us first look at the case 8, in which (e_1,e_2,e_3) is a collinear system. By assumption i) we may assume $e_1 \in E^{(2)}$ and thus every $e_i \in E^{(2)}$. Moreover, Theorem 4.9 implies $\{f \in E ; f \dashv e_1\} = \{f_1,f_2\}$ with $f_1 \perp f_2$. Since $e_2 \in E^{(2)}$ we either have $f_j \dashv e_2$ or $f_j \perp e_2$. In case $f_1 \dashv e_2 \vdash f_2$ we get $e_1 \approx f_1 + f_2 \approx e_2$ by the Triangle Criterion 2.5, in case $f_1 \perp e_2 \perp f_2$ we would have $e_1 \approx f_1 + f_2 \perp e_2$. Therefore we can assume $f_1 \perp e_2 \vdash f_2$ and $f_i \perp e_3 \vdash f_j$ for $(i,j) = (1,2)$ or $(2,1)$.

Assume the first, i.e. $f_1 \perp e_3 \vdash f_2 \dashv e_2 \perp f_1$. We know $\{f \in E ; f \dashv e_2\} = \{f_2, g\}$ with $e_1 \perp g \perp e_3$ (the latter because $f_2 \dashv e_3$). Therefore $V_I + V_K \subset V_0(g)$. Clearly $V_J \subset V_1(g)$, whence $\{V_I V_J V_K\} \subset V_{-1}(g) = 0$.

To finish the case 8 it remains to consider the following subcase: $f_1 \dashv e_3 \perp f_2 \dashv e_2 \perp f_1$. Then the component of f_1 in $I - J + K$ is $1 - 0 + 1 = 2$, so $V_{I-J+K} \subset V_2(f_1)$. But $V_2(f_1)$ is a Peirce space relative to E by Theorem 4.9, whence $f_1 \in V_2(f_1) = V_{I-J+K}$, and the case 8 is done.

Finally we consider the cases 24-26 of the Local-Structure-Theorem 3.5. In case 24 we have $\{V_I V_J V_K\} = 0$ by assumption ii), whence Lemma 3.9.b) implies $\{V_J V_K V_I\} = 0 = \{V_K V_I V_J\}$. But this settles the cases 25 and 26, since (e_1,e_2,e_3) in those cases has the same relations as (e_2,e_3,e_1) resp. (e_3,e_1,e_2) in case 24. ⊞

As a supplement to Theorem 4.12 we note

<u>Corollary 4.15.</u> <u>Let</u> E <u>be a closed cog.</u>
a) <u>If every collinear pair in</u> $E^{(1)}$ <u>is rigid-imbedded in</u> V, <u>then</u> $C_V(E)$ <u>is a subsystem of</u> V.

b) *Conversely, if* $C_V(E)$ *is a subsystem, then every collinear pair in* $E^{(1)}$ *is rigid-imbedded at least in* $C_V(E)$.

§5 Atomic cogs with minimal tripotents

We introduce the notions of minimal tripotents and of atomic cogs both of which are fundamental for constructing Peirce-dense grids in special classes of Jordan triple systems.

5.1. We call a non-zero tripotent e of V minimal (in V), if every non-zero tripotent of $V_2(e)$ is associated to e. Sometimes we will also consider the more general situation where e belongs to a subsystem W of V and is minimal in W, but not necessarily in V. In this situation we call e W-minimal.

Examples of minimal tripotents are E_{ij} in Mat(p,q;k) for a domain k, also H_{ii} in $H_p(k)$, but not H_{ij} for $i \neq j$. Or, let V be the Jordan triple system of a quadratic form as defined in Example 2.4.a , then $V = V_2(e)$ for every non-zero tripotent e, hence any such e is minimal.

For a comparison with the sometimes also used notions of local and primitive tripotents the reader is refered to §7.

Looking at a cog E, some of its elements may be minimal and they necessarily all lie in $E^{(1)}$. If all tripotents in $E^{(1)}$ are minimal, i.e. if

$$E^{(1)} = \{f \in E;\ f \text{ is minimal in } V\}, \tag{5.1}$$

then we call E a cog with minimal tripotents. An equivalent condition for closed cogs will follow from

Lemma 5.1. Let E be a cog such that every connected component of E contains minimal tripotents. Then every closure of E is a cog with minimal tripotents.

Proof. Since every connected component of a closure E_c contains one of E, we may assume that E is closed. Then for every connected component $\tilde{E}$ of E the ortho-collinear system $\tilde{E}^{(1)}$ contains a minimal tripotent. Applying Corollary 4.8.b to $\tilde{E}$ shows that all tripotents in $\tilde{E}^{(1)}$ and thus in $E^{(1)}$ are minimal. ∎

Corollary 5.2. Let E be a closed cog. Then there are equivalent:

a) every connected component of E contains minimal tripotents,

b) E is a cog with minimal tripotents.

By Theorem 4.12 every closed cog E is a grid as soon as every collinear pair in $E^{(1)}$ is rigid-imbedded. As we will see in Corollary 5.4. this last condition is automatic for cogs with minimal tripotents in tripotential Jordan triple systems, where V is called <u>tripotential</u>, if for every pair (e,f) of Peirce-compatible tripotents every Peirce space $V_i(e) \cap V_j(f)$, $i,j \in \{0,1,2\}$, is either zero or contains a non-zero tripotent. Examples of tripotential Jordan triple systems will be given in the next section. Note that a non-zero tripotential Jordan triple system contains non-zero tripotents.

<u>Theorem 5.3.</u> <u>Let V be tripotential and let e $\in$ V be a minimal tripotent. Then every tripotent f which is Peirce-compatible with e satisfies the relation</u>

$$V_2(e) \subset V_j(f) \quad \text{for some } j \in \{0,1,2\}$$

<u>Proof.</u> Compatibility implies the decomposition

$$V_2(e) = V_2(e) \cap V_2(f) \oplus V_2(e) \cap V_1(f) \oplus V_2(e) \cap V_0(f) .$$

Let c be a non-zero tripotent in $V_2(e) \cap V_j(f)$ for $j = 2$ or $j = 0$. By minimality $c \approx e$, and since $V_2(c) \subset V_j(f)$ we also have $V_2(e) \subset V_j(f)$. If c does not exist, we have $V_2(e) \subset V_1(f)$. ∎

<u>Corollary 5.4.</u> <u>Let V be a tripotential Jordan triple system.</u>

a) <u>Every non-zero tripotent f $\in V_1(e)$, where e is a minimal tripotent, satisfies f $\top$ e or f $\vdash$ e.</u>

b) <u>Each two minimal collinear tripotents in V are rigid-imbedded.</u>

c) <u>Every closed cog with minimal tripotents is a grid.</u>

<u>5.2.</u> Every element e of a cog E belongs to a unique Peirce space $V_I(E)$ of E which is a subsystem of V such that $V_I(E) \subset V_2(e)$. In general, it can be possible to decompose $V_I(E)$ by adding new tripotents of $V_I(E)$ to the cog E. This leads us to the following notion which will also be useful to show the existence of Peirce-dense grids in Corollary 5.10.

A cog E is called <u>atomic</u> (<u>in</u> V) if every $e \in E$ is minimal in the subsystem $V_I(E)$ where $e \in V_I(E)$, i.e. every non-zero tripotent $f \in V_I(E)$ is associated to e. We illustrate this new notion by looking at some

<u>Examples 5.5.</u> a) Let k be a domain. Then the family of p × p hermitian matrix units $H = \{H_{ij};\ 1 \leq i \leq j \leq p\} \subset H_p(k)$ is an atomic grid with minimal tripotents. The ortho-collinear family $H^{(2)} = \{H_{ij};$

$1 \leq i < j \leq p\}$ is also atomic, but note that $H^{(2)}$ does not contain minimal tripotents.

b) In $[K,^{-},2m+1]$ for a field K the family $E = \{e_0\} \cup \{e_j^{\varepsilon};\quad \varepsilon=\pm,$ $1 \leq j \leq m\}$ is atomic, however it is not atomic in $[K,^{-},2n+1]$ for $n > m$, where $[K,^{-},2m+1] \subset [K,^{-},2n+1]$ in the natural way. All e_j^{ε} are minimal in $[K,^{-},2n+1]$, whereas e_0 is not. In particular, for all $n \geq m$ $E \subset [K,^{-},2n+1]$ is a grid with minimal tripotents. These examples also show that the two notions "cog with minimal tripotents" and "atomic cog" are independent.

The following lemma is a supplement to Theorem 4.11:

Lemma 5.6. *Let E be an atomic cog. Then there exists an atomic closure of E.*

Proof. We will show that the closure E_c constructed in the proof of Theorem 4.11 is atomic.

Recall, that we first constructed a cog $E_1 \supset E$. Since the Peirce spaces relative to E_1 are the same as the ones relative to E, every $e \in E$ is obviously minimal in its Peirce space $V_I(E_1)$. On the other hand every $c \in E_1 \setminus E$ belongs to one of the 4 cases of Lemma 4.10, in particular $c \top f$ for $f \in E$ or $c \dashv e \vdash e_1 \perp c$ for $e, e_1 \in E$. Thus by Lemma 3.12 and 3.13 there always exists an isomorphism of $V_I(E_1) \ni c$ onto a Peirce space $V_J(E_1)$ mapping c onto a $V_J(E_1)$-minimal tripotent. Altogether, E_1 is atomic. In the next step we put $E_{n+1} = (E_n)_1$ which again is atomic, thus $E_c = \bigcup_{n \geq 0} E_n$ is atomic too. ∎

Later on, the following theorem will be used to prove the existence of Peirce-dense or even covering grids in special types of Jordan triple systems (see Corollary 5.11). Note that the hypothesis (5.2) holds for example if V is tripotential.

Extension Theorem 5.7. *Assume V has the property:*

(5.2) *If $e \in V$ is a minimal tripotent and $0 \neq h \in V_1(e)$ is another tripotent, then $e \top h$ or $e \dashv h$.*

Let $E \subset V$ be an atomic closed cog with minimal tripotents and assume the Peirce space $V_I(E)$, where $I \neq (0)$, contains non-zero tripotents. Then

a) *$V_I(E)$ contains $V_I(E)$-minimal tripotents.*

b) *If $f \in V_I(E)$ is a $V_I(E)$-minimal tripotent, every $e \in E$ lies in some Peirce space of f (so $f \approx E$ or $E \cup \{f\}$ is an atomic cog).*

Proof. a) Since $I \neq (0)$ we have $V_I(E) \subset V_j(e)$ for $j = 1$ or 2 and some $e \in E$. If $e \in E^{(2)}$ then $V_I(E) \subset V_k(c)$ for $k = 1$ or 2 and a $c \in E_e$, so in any case $V_I(E) \subset V_\ell(e)$, $\ell = 1$ or 2, for some $e \in E^{(1)}$. If $\ell = 2$ every non- zero tripotent $f \in V_I(E) \subset V_2(e)$ is associated to the minimal e forcing $e \in V_I(E)$ by Corollary 3.3, and we are done.

Assume $\ell = 1$. By hypothesis every non-zero tripotent $f \in V_I(E) \subset V_1(e)$ satisfies $f \top e$ or $f \vdash e$. If there is a f with $f \top e$, we choose such an f, which is then minimal in V because $V_2(f) \cong V_2(e)$. Otherwise, all f have $e \dashv f$. We choose one of them and claim that f is $V_I(E)$-minimal. Indeed, any other non-zero tripotent $g \in V_I(E) \cap V_2(f)$ is compatible with f so $f = f_2 + f_1 + f_0$ for $f_i \in V_2(f) \cap V_i(g)$. Clearly $f_2 = P(g)^2 f \in V_I(E)$ and $f_1 = \{ggf\} - 2f_2 \in V_I(E)$, so also $f_0 \in V_I(E) \subset V_1(e)$. But g must satisfy $e \dashv g$, thus $V_0(g) \subset V_0(e)$ whence $f_0 = 0$ and $f = f_2 + f_1$ is a sum of orthogonal tripotents, possibly zero, by the Compatibility Criterion 1.8. If both $f_i \neq 0$ then $f_2 \vdash e \dashv f_1$ shows $e \in V_2(f_1) \cap V_2(f_2)$, contradiction. If $f = f_1$ then $e \dashv g \dashv f \vdash e$ contradicting the No-Tower-Lemma 3.4. Thus $f = f_2$, $f \approx g$, proving $V_I(E)$ - minimality of f.

b) We will use the following consequence of Corollary 3.3 several times:

(*) If $f \approx E$, then b) holds.

The case $e \in E^{(1)}$ is easy: $f \in V_2(e)$ implies $f \approx e$ and we are done, so either $f \in V_1(e)$, in which case $e \top f$ or $e \dashv f$ by our hypothesis, or $f \in V_0(e)$ in which case $f \perp e$.

Thus let $e \in E^{(2)}$ and assume $f \in V_2(e)$. If $f \perp d$ for a $d \in E_e \subset E^{(1)}$ we are done by (*) since, by Theorem 4.9.a, there exists a unique $\bar{d} \in E_e$ with $d \perp \bar{d}$, and $f \in V_0(d) \cap V_2(e) = V_2(\bar{d})$ gives $f \approx \bar{d}$. Therefore we may assume $f \in V_1(d)$ for all $d \in E_e$, whence $d \top f$ or $d \dashv f$ by (5.2). If $d \dashv f$ for some $d \in E_e$, then $P(f)d \in V_2(e) \cap V_0(d) = V_2(\bar{d})$ for the unique $\bar{d} \in E_e$ satisfying $d \perp \bar{d}$, hence $P(f)d \approx \bar{d}$ by minimality of $\bar{d}$ and $f \approx d + P(f)d \approx d + \bar{d} \approx e$, so (*) applies again. Finally, if $d \top f$ for all $d \in E_e$, then $e, f \in V_2(e) \cap (\cap\{V_1(d);\ d \in E_e\})$ which is a Peirce space relative to E, as follows from Theorem 4.9.b,c. By atomicity $e \approx f$, contradicting $d \top f$.

The case $f \in V_0(e)$ for $e \in E^{(2)}$ being clear, we are left with $f \in V_1(e)$ for $e \in E^{(2)}$. Let $(c_1, c_2) \subset E_e$ be an orthogonal pair. Then f lies in a Peirce space V_{ij} relative to (c_1, c_2), namely in V_{10} or in V_{20}, say $f \in V_{10}$. Consequently $P(f)e \in V_{-1}(c_2) = 0$ showing $e \in V_1(f) \oplus V_0(f)$, hence by the special case of the Compatibility Criterion 1.8 we have $e = e_1 + e_0$ with $e_i \in V_2(e) \cap V_i(f)$ and (e_1, e_0) are orthogonal tripotents. For $e \in V_J(E)$ we get $e_1 = \{ffe\} \in V_J(E)$ and so also $e_0 \in$

$V_J(E)$. Now minimality of e in $V_J(E)$ implies $e_0 = 0$ (otherwise $e = e_0$ contradicting $f \in V_{10} \subset V_1(e)$), thus $e \top f$, and b) is proven. ∎

5.3. The set of cogs in V is ordered by inclusion. Under a maximal element of this set or a subset thereof we understand a maximal element relative to inclusion. The existence of maximal elements is proven as usually, using the Lemma of Zorn:

Lemma 5.8. Let $\mathbb{E}_i$, $i = 1,\ldots,7$ be the set of all

i = 1) cogs in V,

i = 2) closed cogs,

i = 3) grids,

i = 4) cogs with minimal tripotents,

i = 5) atomic cogs,

i = 6) atomic closed cogs with minimal tripotents,

i = 7) elements of $\mathbb{E}_6$ containing a given $E \in \mathbb{E}_6$.

If $\mathbb{E}_i \neq \emptyset$ there exist maximal elements in $\mathbb{E}_i$.

Proof. Since in all seven cases the proof follows the same pattern, we will only do the last case. By Zorn, it is enough to prove that $\mathbb{E}_7$ is inductively ordered: Given a totally ordered chain $\mathbb{F} \subset \mathbb{E}_7$ we have to show $F^* = \bigcup_{F \in \mathbb{F}} F \in \mathbb{E}_7$. Because every finite subset of F^* lies in some $F \in \mathbb{F}$ it immediately follows that F^* is a closed cog. Also, the Peirce spaces are getting smaller when passing from F to F^*, so F^* is atomic. Finally, every connected component of F^* contains a connected component of some $F \in \mathbb{F}$ and thus, by Corollary 5.2, F^* is a cog with minimal tripotents, showing $F^* \in \mathbb{E}_7$. ∎

Theorem 5.9. Assume V has the property

(5.2) If $e \in V$ is a minimal tripotent and $0 \neq h \in V_1(e)$ is another tripotent, then $e \top h$ or $e \dashv h$.

Then every maximal atomic closed cog E with minimal tripotents (whose existence follows from Lemma 5.8) satisfies:

a) $V_{(0)}(E)$ contains no minimal tripotent,

b) $V_I(E)$ for $I \neq (0)$ contains no non-zero tripotent or contains a tripotent of E.

Proof. a) If there would exist a minimal $f \in V_{(0)}(E)$, we could enlarge the maximal E: $E \cup \{f\}$ would be an atomic (since f is minimal) closed cog with minimal tripotents.

b) We assume that $V_I(E)$, $I \neq (0)$, contains non-zero tripotents. Then, by Theorem 5.7, we can choose a $V_I(E)$-minimal tripotent f and we have $f \approx E$ or $F = E \cup \{f\}$ is a cog. In the first case, $E \cap V_I(E) \neq \emptyset$ follows from Corollary 3.3. In the second, we note that F is an atomic cog since f is minimal in $V_I(E)$. By Lemma 5.6. we can imbed F into an atomic closed cog F^* such that each component of F^* meets one of F hence one of E (because $I \neq (0)$). Therefore, by Corollary 5.2, F^* is a cog with minimal tripotents. But then $E \subsetneq F^*$ contradicts the maximality of E proving $E \cap V_I(E) \neq \emptyset$ in all cases. ∎

Remark 5.10. By Corollary 5.4, Theorem 5.9 in particular applies to tripotential Jordan triple systems. In this case E is even a grid.

As another application of Theorem 5.9 we obtain the existence of a Peirce-dense (defined in 1.4.) grid in a special class of Jordan triple systems:

Corollary 5.11. Let V be a Jordan triple system having the following 4 properties:

(5.2) If $e \in V$ is a minimal tripotent and $0 \neq h \in V_1(e)$ is another tripotent, then $e \top h$ or $e \dashv h$.

(5.3) Each two minimal collinear tripotents of V are rigid-imbedded.

(5.4) V contains an orthogonal system $\mathcal{O}$ of minimal tripotents such that

a) $V_{(0)}(\mathcal{O}) = \bigcap_{e \in \mathcal{O}} V_0(e) = 0$.

b) If $E \supset \mathcal{O}$ is an atomic closed cog with minimal tripotents and if $V_I(E) \neq 0$ for $I \neq (0)$, then $V_I(E)$ contains a non-zero tripotent.

Then V contains a Peirce-dense atomic grid E^* with minimal tripotents and $\mathcal{O} \subset E^*$.

Proof. $\mathcal{O}$ is an atomic closed cog with minimal tripotents, so, by Lemma 5.8, we may choose a maximal atomic closed cog $E^* \supset \mathcal{O}$ with minimal tripotents. By Theorem 5.9. and hypotheses (5.2), (5.4) this E^* is Peirce-dense. It is also a grid by Theorem 4.12 and (5.3). ∎

Remark 5.12. The conditions (5.2) and (5.3) hold in tripotential Jordan triple systems by Corollary 5.4, or in triple systems where every minimal tripotent is a domain tripotent (by Criterion 1.12). In II §4 we will reconsider Jordan triple systems satisfying (5.2)-(5.4) and will prove the existence of a grid with better density properties.

§6 Examples

In this section we consider some important classes of Jordan triple systems and give necessary conditions under which they are tripotential and/or covered by a grid.

The necessary condition, we will use to insure the existence of tripotents, is semisimplicity. In this context the question arises: Does the Peirce space $V_I(E)$ inherit semisimplicity from V? To decide this we use the concept of an <u>involutive grading</u> of a Jordan triple system V as defined in [46]. This is a decomposition $V = V_+ \oplus V_-$ into submodules V_+ and V_- such that

$$P(V_\varepsilon)V_\mu \subset V_\mu \text{ and } \{V_\varepsilon V_\varepsilon V_\mu\} \subset V_\mu$$

for $\varepsilon, \mu = \pm$. A special example of an involutive grading is a <u>Peirce grading</u> $V = V_0 \oplus V_1 \oplus V_2$ where V_i are submodules satisfying for $i,j,k \in \{0,1,2\}$:

$$P(V_i)V_j \subset V_{2i-j}, \ \{V_i\ V_j\ V_k\} \subset V_{i-j+k}$$
$$\text{and } \{V_2\ V_0\ V\} = 0 = \{V_0\ V_2\ V\}.$$

Every Peirce grading induces an involutive grading by putting $V_+ = V_0 \oplus V_2$ and $V_- = V_1$. Obviously, the Peirce decomposition relative to a single tripotent gives an example of a Peirce grading - see the formulas (1.20) - (1.22). We remark that in general not every Peirce grading arises in this way.

<u>Lemma 6.1.</u> a) <u>If</u> $V = V_+ \oplus V_-$ <u>is an involutive grading, then</u> Rad $V_\varepsilon = V_\varepsilon \cap$ Rad V <u>for</u> $\varepsilon = \pm$.

b) <u>If</u> $V = V_0 \oplus V_1 \oplus V_2$ <u>is a Peirce grading, then</u> Rad $V_j = V_j \cap$ Rad V <u>for</u> $j = 0,1,2$.

c) <u>If</u> E <u>is a finite compatible family, then</u> Rad $V_I(E) = V_I(E) \cap$ Rad V.

<u>Proof.</u> a) was proven in [43] Lemma 4.3 for characteristic $\neq 2$, but the argument given there holds in general; b) is a consequence of a), since $V_+ = V_0 \oplus V_2$ is a direct sum of ideals, so Rad $V_+ =$ Rad $V_0 \oplus$ Rad V_2, and finally, c) follows by repeated application of b). ∎

We note that Rad $V_I(E) = V_I(E) \cap$ Rad V does not hold for an infinite compatible family E - see Example 6.5(β).

Besides the question of existence of a tripotent in general we also need to look at the question of the existence of minimal tripotents. A necessary condition is given by

Lemma 6.2. Assume V contains a tripotent $\neq 0$ and satisfies the dcc for principal inner ideals. Then V also contains minimal tripotents and a maximal atomic closed cog with minimal tripotents.

Proof. The set of all modules $P(e)V$ where e ranges over all non-zero tripotents in V consists of inner ideals, and thus contains a minimal element $P(f)V = V_2(f)$. This f is a minimal tripotent. The second assertion follows from Lemma 5.8. ∎

The first class of examples we want to consider are Jordan pairs which we view as polarized Jordan triple systems (see Example 1.2).

Theorem 6.3. Let $V \neq 0$ be a semisimple polarized Jordan triple system satisfying the dcc on principal inner ideals. Then V is tripotential and contains minimal tripotents. Hence (by Remark 5.10) V contains an atomic grid $\mathcal{E}$ with minimal tripotents such that

a) $V_{(0)}(\mathcal{E})$ contains no minimal tripotent,

b) $V_I(\mathcal{E})$ for $I \neq (0)$ contains no non-zero tripotent or contains a tripotent of $\mathcal{E}$.

Proof. In general one knows for a polarized Jordan triple system $V = V^+ \oplus V^-$:

1) An element $e = e^+ \oplus e^-$ is a tripotent of V iff (e^+,e^-) is an idempotent of the Jordan pair $\mathcal{V} = (V^+,V^-)$.

2) V is semisimple iff $\mathcal{V}$ is semisimple.

3) ([31]10.17) Assume $\mathcal{V}$ has dcc on principal inner ideals. Then $\mathcal{V}$ is (von Neumann) regular iff it is semisimple.

4) ([31]5.2) If $x \in V^+$ is regular, $x^+ = Q(x^+)y^-$, then $(x^+,Q(y^-)x^+)$ is an idempotent of $\mathcal{V}$.

Clearly, 1)-4) show that V contains tripotents, hence minimal ones by Lemma 6.2.

To prove that V is tripotential we use

5) ([46]1.11) If $\mathcal{V}$ has dcc on principal inner ideals and is von Neumann regular, the same holds for $\mathcal{V}_\varepsilon$, $\varepsilon = \pm$, where $\mathcal{V} = \mathcal{V}_+ \oplus \mathcal{V}_-$ is an involutive grading of the Jordan pair $\mathcal{V}$.

Now let $e,f \in V$ be Peirce-compatible tripotents. Then all Peirce spaces $V_i(e) \cap V_j(f)$ for $i,j \in \{0,1,2\}$ are polarized. Applying 5) twice shows that $V_i(e) \cap V_j(f)$ is regular, hence vanishes or contains a non-zero tripotent. ∎

Corollary 6.4. Let V be a semisimple polarized Jordan triple

<u>system satisfying the dcc on principal inner ideals and</u> E <u>a maximal atomic grid with minimal tripotents. If</u> E <u>is finite, then</u> E <u>covers</u> V.

<u>Proof.</u> Since E is finite we have $V = \oplus V_I(E)$. It also follows from Lemma 6.1 and 5) in the proof above that all Peirce spaces $V_I(E)$ are regular and have dcc on principal inner ideals, hence they contain minimal tripotents, and $E \cap V_I(E) \neq \emptyset$ by Theorem 5.9. ∎

We note that E will be finite as soon as V has dcc on all inner ideals (see Theorem 6.6 below). If E is infinite it does not always cover V, as the following example shows.

<u>Example 6.5.</u> Let k be a field, I an infinite index set, and put

$$V^{\varepsilon} = (\oplus_{i \in I} ke_i^{\varepsilon}) \oplus kd^{\varepsilon} \oplus kc^{\varepsilon}, \quad \varepsilon = \pm$$

We define a k-bilinear form

$$\phi: \quad V^+ \times V^- \to k$$

by

ϕ	e_j^-	d^-	c^-
e_i^+	δ_{ij}	0	1
d^+	0	0	1
c^+	1	1	0

It is easy to check ([31]6.4 and 7.1), that $V = V^+ \oplus V^-$ together with

$$P(x^+ \oplus x^-)(y^+ \oplus y^-) = \phi(x^+,y^-)x^+ \oplus \phi(y^+,x^-)x^-$$

is a polarized Jordan triple system. Since ϕ is non-degenerate, V is semisimple and has dcc and acc on principal inner ideals ([31]12.9). The elements $e_i = e_i^+ \oplus e_i^-$ are minimal tripotents of V which are pairwise collinear. The Peirce spaces relative to the collinear system $E = \{e_i;\ i \in I\}$ are

$$V_2(e_i) = ke_i^+ \oplus ke_i^- \subset V_1(e_j) \text{ for } j \neq i$$

$$V_{(1)}(E) = \bigcap_{i \in I} V_1(e_i) = kd^+ \oplus kd^-,$$

and other Peirce spaces do not occur, thus

(α) $\quad \oplus_I V_I(E) \subsetneq V$, in particular E does not cover V.

Since $\phi | kd^+ \oplus kd^- = 0$ we see that

(β) $\quad 0 \neq V_{(1)}(E)$ is a trivial system, although V is semisimple.

Because of (β) we also have

(γ) $\quad E$ is a maximal grid with minimal tripotents such that $E \cap V_{(1)}(E) = \emptyset$.

Theorem 6.6. In a tripotential Jordan triple system with dcc on all inner ideals every closed cog with minimal tripotents is finite.

Proof. Let $E \subset V$ be such a cog. First, we prove some auxiliary claims:

(1) E does not contain an infinite orthogonal system. Indeed, if $(e_i)_{i\in\mathbb{N}}$ is an orthogonal system, then $V_0(e_1) \supsetneq V_0(e_1+e_2) \supsetneq V_0(e_1+e_2+e_3) \supsetneq \dots$ is an infinite descending chain of inner ideals.

(2) $E^{(1)}$ does not contain an infinite collinear system.

Suppose $(f_i)_{i\in\mathbb{N}}$ is an infinite collinear system. Then $\sum_{i\geq 1} V_2(f_i) \supsetneq \sum_{i\geq 2} V_2(f_i) \supsetneq \dots$ is a descending chain of inner ideals since every Peirce space $V_2(f_i)$ is an inner ideal and for $i \neq j$ (*) $P(V_2(f_i), V_2(f_j))V \subset V_2(f_i) \oplus V_2(f_j)$. For f_i, f_j are minimal tripotents and therefore by Corollary 5.4 rigid-imbedded, i.e. $V = V_{(21)} \oplus V_{(12)} \oplus V_{(11)} \oplus V_{(10)} \oplus V_{(01)} \oplus V_{(00)}$ is the Peirce decomposition of V relative to (f_i, f_j). Now (*) easily follows.

(3) A subcog $\tilde{E} \subset E^{(1)}$ is finite, if it has the property that to every $c \in \tilde{E}$ there exists a unique $c^{\perp} \in \tilde{E}$ with $c \perp c^{\perp}$.

Indeed, if $\tilde{E}$ is not finite we can construct an infinite collinear system in $\tilde{E}$ using induction: If $(f_1, \dots f_n) \subset \tilde{E}$ is a collinear system, every element of $\tilde{E}\setminus(f_1, \dots, f_n, f_1^{\perp}, \dots, f_n^{\perp})$ is collinear to all f_i.

By (1) there exists a finite maximal orthogonal system $(e_1, \dots, e_r)$. Let V_{ij} be its Peirce spaces as described in Theorem 1.10 and put $E_{ij} = E \cap V_{ij}$. We will show that all E_{ij} are finite. By maximality, $E_{00} = \emptyset$ and $\#E_{ii} = 1$ for $i \geq 1$. We now consider E_{12}. There can only be one $e_{12} \in E_{12}$ with $e_1 \dashv e_{12} \vdash e_2$, because if also $e_1 \dashv c \vdash e_2$ then $e_{12} \approx e_1 + e_2 \approx c$ by the Triangle Criterion 2.5.b. Therefore $\{e_1, e_2\} \cup (E_{12}\setminus\{e_{12}\})$ is an ortho-collinear system. (We remark that the example of an odd-dimensional quadratic form grid shows that $E_{12}\setminus\{e_{12}\}$ may be non-empty.) For every $c \in E_{12}\setminus\{e_{12}\}$ there exists a unique $c^{\perp} \in E\setminus\{e_{12}\}$ with $c^{\perp} \perp c$, namely the element of E which is associated to $\{e_1 c e_2\}$. Hence $E_{12}\setminus\{e_{12}\}$ is finite by (3). The same argument shows finiteness of E_{12} in case no $e_{12} \in E_{12}$ satisfies $e_1 \dashv e_{12} \vdash e_2$. It remains to consider E_{i0}. By maximality of $(e_1, \dots, e_r)$ E_{i0} does not contain orthogonal tripotents, hence is a collinear system. It is contained in $E^{(1)}$, since otherwise $e_1 \dashv e_{10}$ and $P(e_{10})e_1$ would be associated to a non-zero tripotent in E_{00}. Therefore, by (2) E_{i0} is finite. ∎

Corollary 6.7. Let V be a semisimple polarized Jordan triple system with dcc on all inner ideals. Then V is covered by a finite atomic grid with minimal tripotents.

Proof. Corollary 6.4 and Theorem 6.6. ∎

The next class of examples we consider are finite-dimensional semisimple Jordan triple systems over an algebraically-closed field of characteristic $\neq 2$, which were classified in [30].

Theorem 6.8. *Let $V \neq 0$ be a finite-dimensional semisimple Jordan triple system over an algebraically-closed field of characteristic $\neq 2$. Then*

a) *V is tripotential,*

b) *V is covered by a finite atomic grid E with minimal tripotents.*

Proof. It follows from [30] 3.8 and 8.4 that any such triple system contains a non-zero tripotent, whence also a minimal tripotent by Lemma 6.2. Since every cog E in V is finite, every Peirce space $V_I(E)$ is zero or contains a minimal tripotent. In particular, V is tripotential. The existence of a covering grid as claimed in b) follows now from Remark 5.10. ∎

We remark that the characteristic $\neq 2$ assumption is needed to ensure the existence of non-zero tripotents, since there are quadratic form examples $[V,q,S]$ in characteristic 2 which do not contain non-zero tripotents: By [54] §5 Satz 2 this is the case iff V is the orthogonal sum of an even number of hyperbolic planes which are exchanged by S.

The last class of examples we consider are compact Jordan triple systems, i.e. finite-dimensional Jordan triple systems over $\mathbb{R}$ where the trace form σ, defined by $\sigma(x,y) = \frac{1}{2}$ trace $(L(x,y) + L(y,x))$, is positive-definite.

Theorem 6.9. *The assertions of Theorem 6.8 also hold for compact Jordan triple systems.*

Proof. Similar to the previous case the result follows from the following two facts shown in [32] §11:

1) Every non-zero compact Jordan triple system contains a non-zero tripotent.

2) Every subsystem of a compact Jordan triple system is again compact. ∎

§7 Local, minimal and primitive tripotents

In this section we clarify the relations between local, minimal and primitive tripotents.

A primitive tripotent is a non-zero tripotent which cannot be written as the sum of two non-zero orthogonal tripotents. One calls $0 \neq e \in V$ a local tripotent if $V_2(e)$ is a local Jordan triple system, i.e. the set N of non-invertible elements of $V_2(e)$ form a proper ideal of $V_2(e)$. Note that in this case $V_2(e)/N$ is a division system.

Lemma 7.1. Let e be a non-zero tripotent. Then

$$e \text{ local} \rightarrow e \text{ minimal} \rightarrow e \text{ primitive}.$$

Proof. Let e be local and $0 \neq f \in V_2(e)$ a tripotent. If f is not associated to e, then, by Theorem 2.3., it is not invertible in $V_2(e)$, and hence by [31]4.4 belongs to the radical of $V_2(e)$, in particular $B(f,f)|V_2(e)$ is invertible. But $B(f,f)|V_2(e)$ is the projection onto $V_2(e) \cap V_0(f) \subsetneq V_2(e)$, contradiction. The second implication is obvious. ∎

Examples 7.2. a) (primitive $\not\rightarrow$ minimal) Let k be a field of characteristic 2 and consider $V = H_2(k)$. As it was pointed out in [31]5.12 the tripotent $e = E_{12} + E_{21}$ is primitive in V, but it is not minimal since $V = V_2(e)$ and E_{11} is not associated to e.

b) (minimal $\not\rightarrow$ local) Let V be the Jordan triple system of Example 2.4.a) and suppose that V contains a non-zero tripotent e. By (2.7) e is minimal. However, V is not local in case q non-degenerate and isotropic.

Although a primitive tripotent is not a minimal tripotent in general, the two notions coincide for idempotents in a Jordan algebra:

Lemma 7.3. For an idempotent $c \neq 0$ in a Jordan algebra J there are equivalent:

(1) c is a primitive idempotent, i.e. it is impossible to decompose $c = c_1 + c_2$ as a sum of two orthogonal idempotents $c_i \neq 0$,

(2) c is a minimal idempotent, i.e. every idempotent $0 \neq d \in J_2(c)$ is associated to c (thus $d = c$ by Example 2.4.b),

(3) c is a minimal tripotent of V(J),

(4) c is a primitive tripotent of V(J).

Proof. (1) → (2): c - d is an idempotent orthogonal to d, and $c = d + (c-d)$. (2) → (3): For every tripotent $0 \neq d \in J_2(c)$ d^4 is a nonzero idempotent in $J_2(c)$ which is associated to d (Example 2.4.d), whence $d \approx d^4 \approx c$. (3) → (4) → (1) is obvious. ∎

We note that Example 7.2.a) shows that Lemma 7.3 is not true for a tripotent e in a Jordan algebra: $H_2(k)$ is a Jordan algebra, $e = E_{12} + E_{21}$ is primitive but not minimal.

An idempotent of a Jordan algebra J is called local, if it is a local tripotent of the Jordan triple system V(J). It is well-known that in characteristic ≠ 2 and under suitable finiteness conditions the implication "e local idempotent → e primitive idempotent" actually is an equivalence (see e.g. [4]IV5.9 for the finite-dimensional case). However, this is not so in characteristic 2: The Jordan algebra of a traceless quadratic form Q over a field of characteristic 2 satisfies $x^2 = Q(x)1$ for all x, so 1 is a primitive idempotent which is not local in general. But this is essentially the only exception as the following theorem says.

We recall that a Jordan algebra is called an I-algebra if every non-nil inner ideal contains a non-zero idempotent. We note that any algebra with dcc on principal inner ideals and any algebraic algebra is an I-algebra ([21]6.5.2).

Theorem 7.4.([36]) *In an I-algebra J an idempotent e is primitive iff $J_2(e)/\mathrm{Nil}\, J_2(e)$ is a division algebra (so e is local) or is a Jordan algebra determined by a traceless nondegenerate quadratic form Q over a field of characteristic 2.*

Proof. Using primitivity of e and the fact that $J_2(e)$ is also an I-algebra we see that every element of $J_2(e)$ is either invertible or nilpotent. Thus [36] can be applied and gives the structure of $J_2(e)$ as described in the theorem. Conversely, if $J_2(e)/\mathrm{Nil}\, J_2(e) = J(Q,1)$ then any nonzero idempotent c of $J_2(e)$ becomes 1 in J(Q,1), whence c = 1 in $J_2(e)$. ∎

We will use the results for Jordan algebras to characterize primitivity of a tripotent e in a Jordan triple system V. To this end we recall the definition of the subsystem

$$V_2^+(e) = \{v \in V_2(e);\ P(e)v = v\}$$

and the notion of Jordan algebras associated to Jordan triple systems: Let W be a Jordan triple system and $u \in W$. Putting for $w \in W$

$$w^2 = w^{(2,u)} = P(w)u, \quad U(w) = U(w)^{(u)} = P(w)P(u)$$

the underlying module of W becomes a Jordan algebra denoted by $W^{(u)}$ ([31]1.9).

Theorem 7.5. Let $0 \neq e$ be a tripotent of V.

a) There are equivalent:

(1) e is a primitive tripotent of V,

(2) e is a primitive idempotent of the Jordan algebra $V_2^+(e)^{(e)}$,

(3) e is $(V_2^+(e))$-minimal.

b) Assume V is semisimple and has dcc on principal inner ideals. Then e is primitive iff $V_2^+(e)$ is a division system or $V_2^+(e)^{(e)}$ is the Jordan algebra determined by a traceless nondegenerate quadratic form Q over a field of characteristic 2.

c) Assume $\frac{1}{2} \in k$. Then e is primitive iff 0 and $\pm e$ are the only tripotents of $V_2^+(e)$.

Proof. a) (1) $\leftrightarrow$ (2): In any decomposition $e = e_1 + e_2$ with orthogonal tripotents e_i the e_i are orthogonal idempotents of $V_2^+(e)^{(e)}$. (2) $\leftrightarrow$ (3) follows from Lemma 7.3.

b) Under the descending chain condition semisimplicity of V is equivalent to von Neumann regularity. Applying therefore [46]1.11 first to the Peirce grading $V = V_2(e) \oplus V_1(e) \oplus V_0(e)$ and then to the involutive grading $V_2(e) = V_2^+(e) \oplus V_2^-(e)$ shows that $V_2^+(e)$ is von Neumann regular (and hence nondegenerate) and has dcc on principal inner ideals. Therefore $J = V_2^+(e)^{(e)}$ is a semisimple I-algebra, so the result follows from Theorem 7.4.

c) Let e be primitive and $0 \neq d$ a tripotent of $V_2^+(e)$. Since $V_2^+(e)$ is the Jordan triple system associated to the Jordan algebra $V_2^+(e)^{(e)}$ we know $d = d_1 - d_2$ where d_1, d_2 are orthogonal idempotents of $V_2^+(e)^{(e)}$. But e is primitive in $V_2^+(e)^{(e)}$, and thus 0 and e are the only idempotents in $V_2^+(e)^{(e)}$. If $d_2 = 0$, then $d = d_1 = e$, otherwise $d_1 = 0$ and $-d = d_2 = e$. ∎

We have seen that the notions of minimal and primitive tripotents are not the same in general, but that they coincide for idempotents in Jordan algebras. Another instance where "minimal" = "primitive" is described in

Theorem 7.6. Let V be a Jordan triple system over a field k of characteristic $\neq 2$. Then every tripotent $e \neq 0$ satisfying

(7.1) $$V_2^+(e) = ke$$

is minimal as well as primitive. If V is semisimple and finite-dimensional and k is algebraically closed, (7.1) holds for every primitive tripotent, hence in this case

$$e \text{ primitive} \leftrightarrow \text{minimal}.$$

Proof. For the first assertion it is enough to show (7.1) $\rightarrow$ minimality. So let $c \in V_2(e)$ be a non-zero tripotent. Then $P(e)\{cce\} = -\{cce\} + \{\{cce\}ee\}$ (by (1.12)) $= \{cce\}$ shows $\{cce\} = \mu e$ for some $\mu \in k$, whence $\mu = 0,1$ or 2. The case $\mu = 0$ is impossible since then $c \perp e$, contradiction. If $\mu = 1$ we have $c \dashv e$, but then $c + P(e)c$ is an element of $V_2^+(e)$ which is linear independent of e. Thus $\mu = 2$, and $e \approx c$ follows.

For the second assertion we know from Theorem 7.5.b that the Jordan algebra $V_2^+(e)^{(e)}$ is a division algebra whenever e is primitive, hence $V_2^+(e) = ke$. ∎

We note that the condition (7.1) is equivalent to $V_2(e)$ being the Jordan triple of a quadratic form as in Example 2.4.a. This has been proven in [50]2.1.

Remark 7.7. Let e be a primitive tripotent in a compact Jordan triple system. Then $V_2^+(e) = \mathbb{R}e$ ([32]11.6), whence the same argument as in the proof above yields that $V_2(e)$ is isomorphic to the Jordan triple system of a positive-definite quadratic form, and therefore also in this case: e primitive $\leftrightarrow$ e minimal.

So far we looked at the connection between minimal and primitive tripotents. Another question of interest is: When are the "standard" tripotents in our "standard" example minimal? E.g., when is $E_{ij} \in \mathrm{Mat}(I,J;D,{}^-)$ a minimal tripotent? Obviously, the answer to this question only depends on the Peirce-2-space of E_{ij}, i.e. on the Jordan triple D with product $P(a)b = a\bar{b}a$. We note that this setting also covers the standard tripotents in symplectic matrix systems and in quadratic form triples (and also in Bi-Cayley and Albert triples - see III §3).

Lemma 7.8. Let D be an alternative algebra with involution $^-$ and unit 1. Then 1 is a minimal tripotent in $(D,{}^-)$ iff 0 and 1 are the only idempotents of D which are invariant under the involution $^-$. This will *be the case if* $H(D,\bar{\ }) = \{d \in D;\ \bar{d} = d\}$ *is a domain.*

Proof. Suppose 1 is minimal and let $c = \bar{c} = c^2$. Then $c = c\bar{c}c$ is a tripotent in $(D,\bar{\ })$. If $0 \neq c$ we get $D = cDc$, in particular $1 = cac$ for some a, hence c is invertible. But then $c = 1$. Conversely, for $d = d\bar{d}d$ we obtain $(d\bar{d})(d\bar{d})$ = (by Artin's theorem) $(d\bar{d}d)\bar{d} = d\bar{d}$ whence $d\bar{d} = 0$ or 1. Similarly, $\bar{d}d = 0$ or 1. If $d\bar{d} = 0$, then $d = d\bar{d}d = 0$, and if $d\bar{d} = 1$, then also $\bar{d}d = 1$, so d is invertible and $D = dDd$ showing d is associated to 1. ∎

If we look at polarized standard examples the coordinate algebra D is of the form $C \oplus C^{op}$:

Corollary 7.9. *Let C be an alternative algebra with unit 1. Then* $1 \oplus 1$ *is a minimal tripotent in* $(C \oplus C^{op}$, *exchange) iff 0 and 1 are the only idempotents of* C.

In particular for a composition algebra C $1 \oplus 1$ *is minimal iff* C *is a division algebra.*

Proof. The first statement is obvious whereas the second one uses the well-known fact that C is a division algebra iff 0 and 1 are the only idempotents of C. ∎

Remark 7.10. (W. Burgess) Whenever an alternative C is a domain, it satisfies the condition of Corollary 7.9. However, even in nice situations C need not be a division algebra: There exist simple associative domains which are not division algebras.

CHAPTER II. CLASSIFICATION OF GRIDS

The development of Chapter I led us from cogs over closed cogs to grids, the most structured family of tripotents we considered. We have also seen (I§6) that there are important classes of Jordan triple systems which are covered by grids. It is therefore a natural aim to classify grids, which will be done in the first 3 sections of this chapter.

We have to make precise what we understand by classifying grids: We call two grids E_1, E_2 in a Jordan triple system V associated (symbolically $E_1 \approx E_2$) if there is a bijection ϕ: $E_1 \to E_2$ such that $\phi e \approx e$ for every $e \in E_1$. This clearly defines an equivalence relation on the set of all grids in V, and under classification of grids we will understand classification up to association.

That association is the right equivalence relation for dealing with grids or more generally with cogs will follow from our final classification theorem which is stated as résumé at the end of §3. But even without the classification it is obvious that association is the natural equivalence relation: Associated cogs induce the same Peirce decomposition. Association preserves connectedness, or more generally: If $E_j = \dot{\cup} \{E_j^{(i)}; i \in I_j\}$ $j = 1,2$, is the decomposition of E_j into the connected components of E_j, then $E_1 \approx E_2$ iff every $E_1^{(i)}$ is associated to exactly one $E_2^{(k)}$ and conversely. In particular, this observation shows that it is enough to classify connected grids.

Changing a tripotent by an appropriate associated one (see Corollary I.3.3) produces a new associated grid. This indicates that the equivalence classes of associated grids are rather big and raises the question according to which principle we want to choose the representatives in our classification list. To answer this we recall the essential property of a grid E: with one exception all products $P(e_1)e_2$ and $\{e_1e_2e_3\}$ for $e_i \in E$ are zero or associated to some tripotent in E. Our aim now is to associate to a connected grid E a grid $\tilde{E}$ where the products $P(e_1)e_1$ and $\{e_1e_2e_3\}$ for $e_i \in \tilde{E}$ are zero or lie up to sign in $\tilde{E}$. This approach is justified by the examples of grids we already know: A rectangular, symplectic, hermitian or even- or odd-dimensional quadratic form grid (defined in the standard examples in I § 1 and in an abstract setting in the following §1 and §2) has this property. In fact, the final classification theorem states that besides these 5 examples, which are all non-associated, there are only 2 more types:

the Bi-Cayley grid and the Albert grid. In the following we will use the term <u>standard grid</u> to denote one of these 7 types.

An essential part in the classification consists in showing that every grid contains a subfamily which generates a standard grid. This result is used in §4 to prove the existence of certain Peirce-dense grids in Jordan triple systems satisfying (I.5.2)-(I.5.4). This serves as preparation for the classification of Hilbert triples and atomic JBW*-triples in IV §§ 2,3.

The chapter concludes with an exposition of the connection between the 27 tripotents of an Albert grid and the 27 lines upon a cubic surface.

Throughout this chapter we consider Jordan triple systems over an arbitrary ring of scalars as defined in I§1.

§1 Non-ortho-collinear grids

We prove that every connected grid E with $E^{(2)} \neq \emptyset$ is associated to a hermitian grid or an odd-dimensional quadratic form grid.

1.1. An odd-dimensional quadratic form grid is a family

$$\mathcal{Q}_0 = \mathcal{Q}_0(I) = \{e_0\} \dot{\cup} \{e_i^{\varepsilon};\ i \in I,\ \varepsilon = \pm\}$$

of non-zero tripotents where I is an arbitrary index set and $\mathcal{Q}_0$ has the following properties ($j,k \in I$ with $j \neq k$ and $\varepsilon, \mu = \pm$):

(1.1) (e_0, e_j^+, e_j^-) is a triangle, i.e.

$e_j^+ \dashv e_0 \vdash e_j^- \perp e_j^-$ and $P(e_0)e_j^{\varepsilon} = e_j^{-\varepsilon}$, $\{e_j^+ e_0 e_j^-\} = e_0$,

(1.2) $(e_j^{\varepsilon}, e_k^{\mu}, e_j^{-\varepsilon}, -e_k^{-\mu})$ is a quadrangle, i.e.

$e_j^+ \perp e_j^-$, $e_j^{\varepsilon} \top e_k^{\mu}$ and $\{e_j^{\varepsilon}\, e_k^{\mu}\, e_j^{-\varepsilon}\} = -e_k^{-\mu}$,

(1.3) all other products involving three different elements of $\mathcal{Q}_0$ are zero.

The reason for the name "odd-dimensional quadratic form grid" is Example I.1.6.b: The standard basis $\{e_0\} \cup \{e_i^{\varepsilon};\ i \in I,\ \varepsilon = \pm\}$ of the Jordan triple system $[K, ^-, 1+2I]$ defined there gives an example of an odd-dimensional quadratic form grid. In the case of a finite index set we define

$$\mathcal{Q}(2m+1) = \mathcal{Q}_0(I) \text{ for } \#I = m < \infty$$

and call it a (2m+1) - dimensional quadratic form grid.

It is easily seen that an odd-dimensional quadratic form grid is a connected grid.

By (1.2) and (1.3) the subfamilies $\{e_j^{\varepsilon};\ j \in I\}$, $\varepsilon = \pm$, are pure collinear systems in the following sense: A collinear system, or more generally an ortho-collinear system E is called pure if $\{efg\} = 0$ for all triplets (e,f,g) with $e \top f \top g \top e$. That not every collinear system is pure follows by looking at diamonds: In a diamond (e_0, e_1, e_2, e_3) the subfamily (e_1, e_2, e_3) is a collinear system, but $\{e_1 e_2 e_3\} = 2e_0$ is in general non-zero.

The following result is a construction theorem for an odd-dimensional quadratic form grid.

Theorem 1.1. Let $\{e_i;\ i \in I\}$ be a pure collinear system and assume there exists e_0 such that for all $i, j \in I$, $i \neq j$

$$e_0 \vdash e_i \text{ and } \{e_0 e_i e_j\} = 0$$

<u>Put</u> $e_i^+ = e_i$ <u>and</u> $e_i^- = P(e_0)e_i$. <u>Then</u> $\mathcal{Q}_0 = \{e_0\} \cup \{e_i^\varepsilon;\ i \in I,\ \varepsilon = \pm\}$ <u>is an odd-dimensional quadratic form grid</u>.

<u>Proof</u>. (1.1) is obvious from the Triangle Criterion I.2.5.

Regarding (1.2) we use $e_k^+ \top e_j^+$ and (I.1.20) to deduce e_k^+, $e_k^- = P(e_0)e_k^+ \in V_2(e_0) \cap V_1(e_j^+) = V_1(e_j^+) \cap V_1(e_j^-)$, where the last equality follows from the Triangle Decomposition Theorem I.2.6. By symmetry we get $e_j^\varepsilon \top e_k^\mu$. Now (I.1.12) implies $\{e_j^+\ e_k^+\ e_j^-\} = \{e_j^+\ e_k^+\ P(e_0)e_j^+\} = -P(e_0)\{e_k^+\ e_j^+\ e_j^+\} + \{e_0\ e_j^+\{e_j^+\ e_k^+\ e_0\}\} = -P(e_0)e_k^+ = -\ e_k^-$, whence $(e_j^+, e_k^+, e_j^-, -e_k^-)$ is a quadrangle by the Quadrangle Criterion I.2.1. By symmetry, (1.2) follows.

To show (1.3) we first consider products with one factor e_0 and note that Lemma I.3.9 implies $\{e_i\ e_0\ e_j\} = 0$: By (I.1.25) we have $\{e_0\ e_j^-\ e_k^+\} = \{P(e_0)e_j^-\ e_0\ e_k^+\} = \{e_j^+\ e_0\ e_k^+\} = 0$, and applying the automorphism $P(e_0)|V_2(e_0)$ gives $\{e_0\ e_j^+\ e_k^-\} = P(e_0)\{e_0\ e_j^-\ e_k^+\} = 0$, $\{e_0\ e_j^- e_k^-\} = P(e_0)\{e_0\ e_j^+\ e_k^+\} = 0$, whence $\{e_0\ e_j^\varepsilon\ e_k^\mu\} = 0$. Using again (I.1.25) shows $\{e_j^\varepsilon\ e_0\ e_k^\mu\} = \{e_0\ e_j^{-\varepsilon}\ e_k^\mu\} = 0$. Therefore the only non-zero product with one factor e_0 is $\{e_j^+\ e_0\ e_j^-\}$ which stands in (1.1).

By applying $P(e_0)$, products of type $\{e_i^\delta\ e_j^\varepsilon\ e_k^\mu\}$ for three different elements $e_i^\delta, e_j^\varepsilon, e_k^\mu$ can be reduced to the following three types: $\{e_i^+\ e_j^+\ e_k^+\}$, $\{e_i^+\ e_j^+\ e_k^-\}$ and $\{e_i^+\ e_j^-\ e_k^+\}$. The first one vanishes by purity. For the second we may assume $i,j,k \neq$, because $\{e_i^+\ e_j^+\ e_i^-\} = -e_j^-$ stands in (1.2). Then $\{e_i^+ e_j^+\ e_k^-\} = \{e_i^+\ e_j^+\ P(e_0)e_k^+\}$ = (by (I.12)) $-P(e_0)\{e_j^+ e_i^+ e_k^+\} + \{e_0 e_\kappa^+\{e_i^+ e_j^+ e_k^+\}\} = 0$. Also for the third type it is enough to consider $i,j,k \neq$. Applying Lemma I.3.7.a shows $\{e_i^+\ e_j^-\ e_k^+\} = 0 \leftrightarrow \{e_i^+ e_k^+ e_j^-\} = 0$, and (1.3) follows. ∎

Our first classification result is

<u>Theorem 1.2.</u> <u>Let</u> E <u>be a connected grid which contains a tripotent</u> e <u>such that</u>

$$\# E_e = \#\{f \in E;\ f \dashv e\} > 3.$$

<u>Then</u> E <u>is associated to an odd-dimensional quadratic form grid</u> $\mathcal{Q}_0(I)$ <u>where</u> $\#I = \frac{1}{2}\ \#\ E_e > 1$.

<u>Proof</u>. By Theorem I.4.9 we know $E = \{e\}\ \dot\cup\ E_e$. Let $\{e_i; i \in I\} \subset E_e$ be a maximal collinear system. It is pure because of (I.4.8) and $e \vdash E_e$. Moreover, by (I.4.9) we have $\{e_i\ e\ e_j\} = 0$ for $i,j \in I$, $i \neq j$. We

therefore can apply Theorem 1.1 and see that $\mathcal{Q}_0(I) = \{e_0\} \cup \{e_i^{\varepsilon}; i \in I, \varepsilon = \pm\}$, where $e_0 = e$, $e_i^+ = e_i$, $e_i^- = P(e_0)e_i$, is an odd-dimensional quadratic form grid. By closedness there exists exactly one $\bar{e}_i \in E$ with $e_i^- \approx \bar{e}_i$. Thus the map

$$\phi: \quad \mathcal{Q}_0(I) \to E : e_0 \to e,\ e_i^+ \to e_i,\ e_i^- \to \bar{e}_i$$

is injective. For $c \in E_e \setminus \{e_i;\ i \in I\}$ there exists e_i with $c \perp e_i$ because otherwise $\{e_i;\ i \in I\}$ would not be maximal. But then Theorem I.4.9.a shows $c = \bar{e}_i$, and ϕ is surjective. ∎

1.2. In I §2.4 we introduced the notion of a hermitian grid of size 3. This will now be generalized in the following way: Let I be an arbitrary index set. A hermitian grid (of size I) is a family

$$H = H(I) = \{h_{ij} = h_{ji};\ i,j \in I\}$$

of non-zero tripotents, which satisfies for distinct $i,j,k \in I$:

(1.4) $\quad (h_{ii}, h_{ij}, h_{jk}, h_{ki})$ is a diamond,

(1.5) $\quad P(h_{ij})h_{ii} = h_{jj}$.

An example of an hermitian grid is provided by the usual hermitian matrix units $H_{ij} = 1[ij]$ in the hermitian Jordan triple system $H_I(k)$.

We mention that our notion of a family of hermitian matrix units is a straightforward generalization of the corresponding notion used in [42] §5.

The following result, which generalizes Lemma I.2.10, in particular shows that every hermitian grid is a connected grid.

Lemma 1.3. *A family $(h_{ij} = h_{ji};\ i,j \in I)$ of non-zero tripotents is a hermitian grid iff the following rules hold for $i,j,k,\ell \neq$:*

(1.6) $\quad h_{jj} \perp h_{ii} \perp h_{jk},\ h_{ii} \dashv h_{ij} \top h_{ik}$ *and* $h_{ij} \perp h_{k\ell}$,

(1.7) $\quad P(h_{ij})h_{ii} = h_{jj}$

(1.8) $\quad \{h_{im}\ h_{mn}\ h_{nj}\} = h_{ij} \qquad$ *(m arbitrary)*

(1.9) $\quad \{h_{im}\ h_{mn}\ h_{ni}\} = 2h_{ii} \qquad$ *(m,n arbitrary)*

(1.10) $\quad \{h_{mn}\ h_{pq}\ h_{rs}\} = 0, \qquad$ *if the indices cannot be linked.*

Proof. Since every family with (1.6)-(1.8) obviously is a hermitian grid, it is enough to prove the converse. By Lemma I.2.10 all rules, which involve at most three different indices, are clear. So it remains to show $h_{ij} \perp h_{k\ell}$, $\{h_{ij}\ h_{jk}\ h_{k\ell}\} = h_{i\ell}$ (a subcase of (1.8)) and $\{h_{ij}\ h_{ik} h_{i\ell}\} = 0$ (a subcase of (1.10)). For $h_{ij} \perp h_{k\ell}$ one only has to realize $h_{ij} \in V_2(h_{ii} + h_{jj})$, $h_{k\ell} \in V_2(h_{kk} + h_{\ell\ell})$ and $(h_{ii} + h_{jj}) \perp$

$(h_{kk} + h_{\ell\ell})$. Further, $h_{k\ell} = \{h_{ki}\ h_{ii}\ h_{i\ell}\}$, which together with (I.1.14) shows $\{h_{ij}h_{jk}\ h_{k\ell}\} = \{\{h_{ij}\ h_{jk}\ h_{ki}\}h_{ii}\ h_{i\ell}\} - \{h_{ki}\{h_{jk}\ h_{ij}\ h_{ii}\}h_{i\ell}\} + \{h_{ki}\ h_{ii}\{h_{ij}\ h_{jk}\ h_{i\ell}\}\} = 2\{h_{ii}\ h_{ii}\ h_{i\ell}\} - \{h_{ki}\ h_{ki}h_{i\ell}\}$ (since $h_{jk} \perp h_{i\ell}$) $= h_{i\ell}$. Finally, $\{h_{ij}\ h_{ik}h_{i\ell}\} \in V_{0-1+0}(h_{kk}) = 0$. ∎

A construction theorem for a hermitian grid is given in the following

<u>Theorem 1.4.</u> <u>Let</u> $\{h_{1i} = h_{i1};\ i \in I\}$ <u>be a collinear system where</u> I <u>is an index set with</u> $1 \notin I$. <u>Assume there exists a tripotent</u> h_{11} <u>such that</u> $h_{11} \dashv h_{1i}$ <u>for all</u> $i \in I$. <u>Put</u> for $i,j \in I$, $i \neq j$

$$h_{ii} = P(h_{1i})h_{11},\ h_{ij} = h_{ji} = \{h_{1i}\ h_{11}\ h_{1j}\}.$$

<u>Then</u> $H = \{h_{ij};\ i,j \in I \cup \{1\}\}$ <u>is a hermitian grid</u>.

<u>Proof.</u> From the Diamond Criterion I.2.7.b we know that $(h_{11}, h_{1i}, h_{ij}, h_{1j})$ is a diamond, hence Theorem I.2.11 in particular shows that (h_{ij}, h_{ii}, h_{jj}) is a triangle, which proves (1.5), and that $(h_{ii}, h_{1i}, h_{1j}, h_{ij})$ is also a diamond. It follows $h_{ij} \vdash h_{ii} \dashv h_{ik}$ for $i,j,k \neq$. Since $h_{ik} = \{h_{i1}\ h_{11}\ h_{1k}\} \in V_0(h_{jj}) \cap V_1(h_{ii}) = V_1(h_{jj} + h_{ii}) = V_1(h_{ij})$ we get by symmetry $h_{ik} \top h_{ij}$. Therefore $(h_{ii}, h_{ij}, \{h_{ij}\ h_{ii}\ h_{ik}\}, h_{ik})$ is a diamond, but $\{h_{ij}\ h_{ii}\ h_{ik}\} = \{h_{ij}h_{ii}\{h_{i1}\ h_{11}\ h_{1k}\}\} = \{\{h_{ij}\ h_{ii}\ h_{i1}\}h_{11}\ h_{1k}\} - \{h_{i1}\{h_{ii}\ h_{ij}\ h_{11}\}h_{1k}\} + \{h_{i1}\ h_{11}\{h_{ij}\ h_{ii}\ h_{1k}\}\} = \{h_{1j}h_{11}\ h_{1k}\}$ (since $\{h_{ij}\ h_{ii}\ h_{i1}\} = h_{1j}$ and $h_{ij} \perp h_{11}$, $h_{ii} \perp h_{1k}$) $= h_{jk}$. Thus (1.4) holds in general. ∎

In analogy to Theorem 1.2 we can now prove

<u>Theorem 1.5.</u> <u>Let</u> E <u>be a connected grid such that</u> $\# E_e = 2$ <u>for all</u> $e \in E^{(2)}$. <u>Then</u> E <u>is associated to a hermitian grid</u>.

<u>Proof.</u> We choose a tripotent in $E^{(1)}$, call it h_{11}, and write $\{e \in E; e \vdash h_{11}\} = \{h_{1i}, i \in I\}$ for an index set I with $1 \notin I$. Then $\{h_{1i}; i \in I\}$ is a collinear system, because $h_{1i} \vdash h_{11} \dashv h_{ij}$ for $i \neq j$ implies $h_{1i} \top h_{1j}$. So we can apply Theorem 1.4. and get a hermitian grid $H = \{h_{ij}; i,j \in I \cup \{0\}\}$. Since $h_{11}, h_{1i} \in E$, closedness of E implies $h_{ij} \approx E$ for all $i,j \in E$. So $H \approx E$, as soon as every $e \in E$ is associated to some $h_{mn} \in H$. To prove this let $e \in E$ and $e \neq h_{1i}$ for all $i \in I$. First we consider the special case $e \top h_{ij}$ for some $j \in I$. Then $e \in E^{(2)}$ and $e \perp h_{11}$. By assumption $\{f \in E; f \dashv h_{1j}\} = \{h_{11}, c\}$ where $c \approx h_{jj}$. Applying Theorem I.4.10.b shows $c \in V_2(e)$ since otherwise $c \in V_0(e)$ and $h_{1j} \in V_2(c + h_{11}) \subset V_0(e)$. Therefore $c \dashv e$. Moreover, $(h_{11}, h_{1j}, e, \{h_{11}\ h_{1j}\ e\})$ is a diamond, in particular $\{h_{11}\ h_{1j}\ e\} \vdash h_{11}$. By

closedness $\{h_{11}\ h_{1j}\ e\} \approx E$ and by Corollary I.3.3 we have $\{h_{11}\ h_{1j}\ e\} \approx h_{1k}$ for some $k \in I$, $k \neq j$. It follows $e \top h_{1k}$, thus $e \vdash h_{kk}$ by what we have already shown. But now the Triangle Criterion I.2.5.b implies $e \approx h_{jj} + h_{kk} \approx h_{jk}$.

In general, let $e \in E^{(2)}$. Since $E^{(2)}$ is connected, there is a connecting sequence $(h_{1i}, g, e) \subset E^{(2)}$ as follows from Lemma I.4.6. By what we already proved we can assume $g \approx h_{ik}$ and $e \perp h_{1i}$. Because $e \top g$ and $e \perp h_{ii}$ we must have $e \in V_1(h_{kk})$ whence $h_{kk} \dashv e$ and $e \top h_{1k}$, thus $e \approx h_{k\ell}$ for some ℓ.

Finally, let $e \in E^{(1)}$, $e \neq h_{11}$. Then $e \perp h_{11}$ by Theorem I.4.10.b, and we can apply Theorem I.4.7.b to conclude that there exists $h_{1i} \in E$ such that $h_{1i} \vdash e$. But then $e \approx h_{ii}$, and we are done. ⊞

We note in passing that every closed cog E with the property $\# E_e = 2$ for all $e \in E^{(2)}$ is already a grid.

Combining Theorems 1.2 and 1.5 with Theorem I.4.9 shows

Corollary 1.6. *A connected grid E with $E^{(2)} \neq \emptyset$ is associated to an odd-dimensional quadratic form grid or to a hermitian grid.*

Because of Corollary 1.6. we consider in the next sections grids E with $E^{(2)} = \emptyset$. By Lemma I.3.10 these are ortho-collinear systems.

§2 Ortho-collinear grids I (the special cases)

We describe connected ortho-collinear grids which are associated to rectangular grids, symplectic grids or even-dimensional quadratic form grids.

2.1. Our first aim is to characterize collinear systems using the following notion. In general, the rank of a family of tripotents is the supremum over all cardinalities of orthogonal systems in that family. Clearly, the cardinality of a maximal orthogonal subsystem need not be equal to the rank of the family. We also point out that the rank may be infinite.

It will be very useful to have the following abbreviations at our disposal: For a cog E in the Jordan triple system V and a tripotent $e \in E$ we put

$$E_j(e) = E \cap V_j(e), \qquad j = 0,1,2.$$

By the rules of a cog we have $E_2(e) = \{e\}$, $E_1(e) = \{f \in E \; ; f \top e$ or $f \vdash e\}$ and $E_0(e) = \{f \in E \; ; f \perp e\}$. More generally, if $(e_1,\ldots,e_n) \subset E$ is an orthogonal system we put

$$E_{ij}(e_1,\ldots,e_n) = E_{ij} = E \cap V_{ij},$$

where the V_{ij} are the usual Peirce space of V relative to $(e_1,\ldots,e_n)$.

Lemma 2.1. Let E be a closed ortho-collinear system.

a) For every $e \in E$ we have rank $E_1(e) \leq 2$, and rank $E_1(e) = 2$ implies $E_0(e) \neq \emptyset$.

b) If E is connected, then

$$\text{rank } E_1(e) = 2 \leftrightarrow E_0(e) \neq \emptyset.$$

Proof. a) Let $(c_i ; i = 1,2,3) \subset E_1(e)$ be an orthogonal system. Then $e \in V_1(c_1) \cap V_1(c_2) \cap V_1(c_3)$, but $V_1(c_1) \cap V_1(c_2) \subset V_0(c_3)$ by Theorem I.1.10. Thus rank $E_1(e) \leq 2$. If $(c_1,c_2) \subset E_1(e)$ is an orthogonal system, then $(c_1,e,c_2,\{c_1ec_2\})$ is a quadrangle and there exists $e_2 \in E$ with $e_2 \approx \{c_1ec_2\}$, in particular $e \in V_0(\{c_1ec_2\}) = V_0(e_2)$, so $e \perp e_2$.
b) Assume $e_2 \in E$ such that $e_2 \perp e$. By Lemma I.4.6 there exists a connecting chain $e \top f \top e_2$. Let $f_2 \in E$ such that $\{e f e_2\} \approx f_2$. Then $(f,f_2) \subset E_1(e)$ is an orthogonal system. ∎

Corollary 2.2. Let E be a connected closed ortho-collinear system and assume $e_1 \perp e_2$ for $e_i \in E$. Then rank $E_{12}(e_1,e_2) = 2$.

Proof. This follows from the proof of Lemma 2.1.b. ∎

Corollary 2.3. Let E be a connected closed ortho-collinear system with $\# E > 2$. Then there are equivalent:

1) There exists $e \in E$ such that rank $E_1(e) = 1$.
2) E is a collinear system.

Proof. If rank $E_1(e) = 1$, then $E_1(e)$ is a collinear system and $E = \{e\} \cup E_1(e)$ by Lemma 2.1.b. Hence E is a collinear system too. The converse is trivial. ∎

We note that the last corollary in particular applies to ortho-collinear grids.

2.2. We will need a result on the symmetries of an ortho-collinear grid. In general, a symmetry of a cog E is a bijection $\tau: E \to E$ which preserves all relations within E. So, a symmetry of an ortho-collinear system E is a bijection $\tau: E \to E$ such that $e \perp f \leftrightarrow \tau e \perp \tau f$.

Theorem 2.4. Let $G = G(E)$ be the subgroup of the automorphism group of V generated by

$$\{ T_{e,f} = B(e + f, e + f);\ e, f \in E,\ e \top f\}$$

where E is an ortho-collinear grid.

a) $G(E)$ preserves E up to association, i.e. for every $\gamma \in G$ there exists a unique symmetry $\hat\gamma$ of E such that $\gamma e \approx \hat\gamma e$ for all $e \in E$.

b) $G(E)$ permutes the Peirce spaces relative to E.

c) Let E be connected. Then for each pair $(e,f) \subset E$ there exists $\gamma \in G(E)$ such that

$$\gamma e = f \text{ and } \hat\gamma E_j(e) = E_j(f),\ j = 0,1,2.$$

For two orthogonal systems $(e_i;\ 1 \leq i \leq m)$ and $(\tilde e_i;\ 1 \leq i \leq n)$ in E with $m<n$ there exists $\gamma \in G(E)$ such that

1) $\hat\gamma e_i = \tilde e_i$ for $1 \leq i \leq m$ and
2) $\hat\gamma E_{ij}(e_1,\dots,e_m) = E_{ij}(\tilde e_1,\dots,\tilde e_n),\ 1 \leq i \leq j \leq m.$

Proof. a) We start by looking at $\gamma = B(e + f, e + f)$ for $e \top f$. We know from Theorem I.1.13 that γ exchanges e and f. For $c \in E\setminus\{e,f\}$ there are the following possibilities:

1) $c \in V_0(e) \cap V_0(f)$: Then $\gamma c = c$.
2) $c \in V_0(e) \cap V_1(f)$: Then $\gamma c = \{e\ f\ c\}$ which is associated to some (hence to a unique) $d \in E$.
3) $c \in V_1(e) \cap V_0(f)$: This case is analogous to 2).
4) $c \in V_1(e) \cap V_1(f)$: Then $\gamma c = -c + \{e\ f\{f\ e\ c\}\} = -c$ since $\{f\ e\ c\} = 0$ by (I.4.8).

So far we proved the assertion for $\gamma = B(e+f,e+f)$. Note that $\hat{\gamma}$ is a symmetry since clearly $e \perp f \leftrightarrow \gamma e \perp \gamma f \leftrightarrow \hat{\gamma}e \perp \hat{\gamma}f$. In general, an element of $G(E)$ is a finite product of elements of type $B(e+f,e+f)$. Thus it is enough to show: If γ_i, $i = 1,2$, preserve E up to association, so does $\gamma_1\gamma_2$. But this is obvious: Since γ_1 is an automorphism, $\gamma_2 e \approx \hat{\gamma}_2 e$ implies $\gamma_1\gamma_2 e \approx \gamma_1\hat{\gamma}_2 e \approx \hat{\gamma}_1\hat{\gamma}_2 e$.
b) and the assertion $\gamma e = f$ in c) follow from Theorem I.3.11 and its proof. For this γ and for $c \in E_0(e)$ we have $\hat{\gamma}c \approx \gamma c \in V_0(f)$, whence $\hat{\gamma}E_0(e) \subset E_0(f)$. Applying $\hat{\gamma}^{-1}$ shows $\hat{\gamma}E_0(e) = E_0(f)$. Since $\hat{\gamma}E_2(e) = E_2(f)$ is trivial we proved $\hat{\gamma}E_j(e) = E_j(f)$ for $j = 0,1,2$.

To prove the assertion about the two orthogonal systems we first choose $\gamma_1 \in G$ with $\gamma_1 e_1 = \tilde{e}_1$. Then $\gamma_1 e_i \in V_0(\tilde{e}_1)$ for $i \geq 2$, which are associated to $c_2, \ldots, c_m \in E_0(\tilde{e}_1)$. Now there exists $\gamma_2 \in G(E)$ with $\gamma_2 c_2 = \tilde{e}_2$. Since γ_2 is constructed by using elements from $E_0(\tilde{e}_1)$, we have $\gamma_2\tilde{e}_1 = \tilde{e}_1$. Proceeding in this way we obtain an element γ of G with $\gamma e_i \approx \tilde{e}_i$ for $1 \leq i \leq m$. Then $\hat{\gamma}E_j(e_i) = E_j(\tilde{e}_i)$, $j = 0,1,2$, $1 \leq i \leq m$, which implies $\hat{\gamma}E_{ij}(e_1,\ldots,e_m) = E_{ij}(\tilde{e}_1,\ldots,\tilde{e}_n)$ for $1 \leq i \leq j \leq m$. ∎

Remark 2.5. a) By Theorem 2.4.a there is a map $G(E) \to \mathrm{Sym}(E)$: $\gamma \to \hat{\gamma}$. We want to point out that this map is neither injective nor surjective. To see non-injectivity we look at the example Mat(2,2;k) of 2×2-matrices with the grid $E = \{E_{11}, E_{12}, E_{22}, -E_{21}\}$. Here $\gamma = B(E_{22}-E_{21}, E_{22}-E_{21})B(E_{11}+E_{12}, E_{11}+E_{12})$ is not the identity, since e.g. $\gamma E_{11} = -E_{11}$, but $\hat{\gamma} = \mathrm{Id}$. To see non-surjectivity we look at a pure collinear system $E = (e_i; i \in I)$: the symmetry group of E coincides with the group of bijections of I, but any element of $G(E)$ only moves a finite number of tripotents.

b) An immediate consequence of Theorem 2.4.c) is that two maximal orthogonal systems in E have the same size (finite or infinite).

2.3. For arbitrary index sets I and J a family

$$R = R(I,J) = \{e_{ij}; i \in I, j \in J\}$$

of non-zero tripotents is called a rectangular grid (of size $I \times J$) if it satisfies:

(2.1) for $i \neq k$ and $j \neq \ell$ is $(e_{ij}, e_{i\ell}, e_{k\ell}, e_{kj})$ a quadrangle.

(2.2) R is pure, i.e. $\{efg\} = 0$ for all collinear systems $(e,f,g) \subset R$.

We state some immediate consequences:

(2.3) For $i \neq k$ and $j \neq \ell$ we have $e_{ij} \top e_{i\ell} \top e_{k\ell} \perp e_{ij}$

In particular, R is a connected ortho-collinear cog which in case $\#I = 1$

or $\#J = 1$ is nothing else but a pure collinear system. It also follows that all products between three different tripotents vanish, unless they can be imbedded into a quadrangle (2.1). Hence every rectangular grid is a connected ortho-collinear grid.

It is a straightforward consequence of the formulas of Example I.1.3 that the rectangular matrix units E_{ij}, $i \in I$, $j \in J$ form a rectangular grid in Mat(I,J;D).

We finally mention that our notion of a rectangular grid differs from that used in [42] §3 as far as we consider families of arbitrary size and do not make any assumptions about rigidity and covering.

In analogy to the procedure of the previous section we now prove a construction theorem for a rectangular grid.

Theorem 2.6. Let I and J be index sets with $1 \in I$ and $1 \in J$ and assume

i) $\{e_{i1}, i \in I\}$ and $\{e_{1j}; j \in J\}$ are pure collinear systems,

ii) $e_{i1} \perp e_{1j}$ for all $(i,j) \in I \times J$, $i \neq 1$ and $j \neq 1$.

For $(i,j) \in I \times J$ with $i \neq 1$, $j \neq 1$ put

$$e_{ij} = \{e_{i1}\, e_{11}\, e_{1j}\}$$

Then $R = R(I,J) = \{e_{ij}; i \in I, j \in J\}$ is a rectangular grid.

Proof. In the following let $1,i,k \neq$ and $1,j,\ell \neq$. We first note that by the Quadrangle Criterion I.2.1 $(e_{i1}, e_{11}, e_{1j}, e_{ij})$ is a quadrangle. This in particular implies $e_{1j} \perp e_{k\ell} \perp e_{i1}$ and $e_{i1} \top e_{i\ell} \top e_{1\ell}$. Applying the Quadrangle Decomposition Theorem I.2.2 relative to $(e_{i1}, e_{11}, e_{1j}, e_{ij})$ shows $e_{k\ell} \in V_0(e_{i1}) \cap V_0(e_{11}) \cap V_0(e_{1j}) \subset V_0(e_{ij})$, whence $e_{ij} \perp e_{k\ell}$, and $e_{i\ell} \in V_1(e_{i1}) \cap V_0(e_{11}) \cap V_0(e_{1j}) \subset V_1(e_{ij})$, whence, by symmetry, $e_{ij} \top e_{i\ell}$. Analogously $e_{i\ell} \top e_{k\ell}$, so (2.3) as a first step towards (2.1) is proven. By definition of e_{ij}, (2.1) is known if one of the indices in I and in J equals 1. Because $(e_{k\ell}, e_{1\ell}, e_{11}, e_{k1})$ is a quadrangle we have $L(e_{k\ell}, e_{1\ell}) = L(e_{k1}, e_{11})$ by (I.2.3) and thus $\{e_{k\ell}\, e_{1\ell}\, e_{1j}\} = \{e_{k1}\, e_{11}\, e_{1j}\} = e_{kj}$ and $(e_{1j}, e_{1\ell}, e_{k\ell}, e_{kj})$ is a quadrangle. Analogously, $(e_{i1}, e_{i\ell}, e_{k\ell}, e_{k1})$ is a quadrangle. It remains to show that $(e_{ij}, e_{i\ell}, e_{k\ell}, e_{kj})$ with the restrictions on the indices valid for this proof is a quadrangle. This follows from $\{e_{ij}\, e_{i\ell}\, e_{k\ell}\} = \{e_{1j}\, e_{1\ell}\, e_{k\ell}\} = e_{kj}$. So we have proven (2.1) in general.

There are only 2 types of collinear systems $(e,f,g) \subset R$: For all three tripotents the first index coincides or the second. By symmetry we may assume the first index coincides. If it equals 1, (2.2) follows from assumption i). So it is enough to show $\{e_{im}\ e_{in}\ e_{ip}\} = 0$ for m,n,p $\neq$ and $i \neq 1$. But this follows from $\{e_{im}\ e_{in}\ e_{ip}\} = \{e_{1m}\ e_{1n}\ e_{ip}\} = 0$ by (2.3). ∎

Let E be a connected ortho-collinear grid. If rank $E = 1$, then E is a pure collinear system which is a special case of a rectangular grid. So in our further investigations we may assume rank $E \geq 2$, i.e. there exists an orthogonal system $(e_1,e_2) \subset E$. Then, by Corollary 2.2, we know rank $E_{12}(e_1,e_2) = 2$. The next result handles the extreme case $\# E_{12}(e_1,e_2) = 2$:

<u>Theorem 2.7.</u> <u>Let</u> E <u>be a connected ortho-collinear grid which contains an orthogonal system</u> (e_1,e_2) <u>such that</u> $\#E_{12}(e_1,e_2) = 2$. <u>Then</u> E <u>is associated to a rectangular grid</u>.

<u>Proof</u>. We put $e_1 = e_{11}$ and $E_{12}(e_1,e_2) = \{e_{21},e_{12}\}$. Then necessarily $e_{21} \top e_{11} \top e_{12} \perp e_{21}$.

Let $f \in E_1(e_1) \cap V_1(e_{12})$ and assume $f \top e_{21}$. Then there exists $\tilde{f} \in E$ with $\tilde{f} \approx \{e_{12}\ f\ e_{21}\}$, in particular $\tilde{f} \in V_1(e_{12}) \cap V_1(e_{21}) \cap V_1(e_1) \cap V_1(e_2)$ which shows $\{e_1,e_2,f,\tilde{f}\} \subset E_{12}(e_{12},e_{21})$. But $\# E_{12}(e_{12},e_{21}) = 2$ by Theorem 2.4.c.2). Therefore $f \perp e_{21}$ by ortho-collinearity and

$$\{e_{12}\} \cup (\ E_1(e_1) \cap V_1(e_{12})) \subset V_0(e_{21}).$$

Since rank $E_1(e_1) = 2$ this implies that $E_1(e_1) \cap V_1(e_{12})$ is a collinear system, and the same holds for

$$\{e_{11},e_{12}\} \cup (\ E_1(e_{11}) \cap V_1(e_{12})) =: \{e_{1j};j \in J\}.$$

It is pure because of (I.4.8). In the same way we get that

$$\{e_{11},e_{21}\} \cup (\ E_1(e_{11}) \cap V_1(e_{21})) =: \{e_{i1};i \in I\} \subset V_0(e_{12})$$

is a pure collinear family. Let $\{i,j\} \neq \{1,2\}$ and assume $e_{i1} \top e_{1j}$. Then $\{e_{21}\ e_{i1}\ e_{1j}\} \approx c \in E_1(e_{11}) \cap E_{12}(e_{21},e_{12})$ which gives a contradiction. So $e_{i1} \perp e_{1j}$ and the assumptions of Theorem 2.6 are fulfilled. Therefore $R = \{e_{ij};i \in I,j \in J\}$ is a rectangular grid where $e_{ij} = \{e_{i1}\ e_{11}\ e_{1j}\}$ for $i,j \neq 1$.

By closedness every $e_{ij} \in R$ is associated to some tripotent of E. Conversely, let $f \in E_1(e_{11})$, $f \neq e_{21}$, e_{12}. Since rank $E_1(e_{11}) = 2$ we have $f \top e_{21}$ or $f \top e_{12}$, i.e. $f = e_{i1}$ or $f = e_{1j}$ for suitable i or j. For $f \in E_0(e_{11})$ there exists $g \in E$ such that $e_{11} \top g \top f$. We may assume $g = e_{1j}$. Then $\{e_{11}\ g\ f\} \approx e_{i1}$ for some $i \in I$ and $f \approx e_{ij}$. Altogether we proved $E \approx R$. ∎

<u>2.4.</u> Let I be totally ordered by $<$ and assume $\#I \geq 4$. (Note that any set can be totally ordered.) A family

$$S = S(I) = \{f_{ij}; i,j \in I \text{ with } i < j\}$$

of non-zero tripotents is called a <u>symplectic grid (of size #I)</u> if it satisfies:

(2.4) Put $f_{ji} = -f_{ij}$ for $i < j$, then for $i,j,k,\ell \neq$ $(f_{ij}, f_{kj}, f_{k\ell}, f_{i\ell})$ is a quadrangle;

(2.5) S is pure.

We remark that $\hat{S} = \{f_{ij}; i,j \in I,\ i \neq j\}$ with $f_{ij} = -f_{ji}$ for $i < j$ is not a cog since $f_{ij} \approx f_{ji}$. This is the reason, why we need a total order on I. On the other hand, (2.4) and some of the results below are more naturally described within $\hat{S}$. We will therefore mostly work with $\hat{S}$ without distinguishing it from S.

Further multiplication rules for symplectic grids are given in

<u>Lemma 2.8.</u> <u>Every symplectic grid satisfies the following rules:</u>

(2.6) $f_{ij} \top f_{ik} \perp f_{j\ell}$ <u>for</u> $i,j,k,\ell \neq$, i.e. $f_{mn} \top f_{pq}$ <u>if they share exactly one index and</u> $f_{mn} \perp f_{pq}$ <u>if they have no index in common.</u>

(2.7) $\{f_{ij}\ f_{k\ell}\ f_{mn}\} = 0$ <u>unless</u> (k,ℓ) <u>or</u> $(\ell,k) \in \{i,j\} \times \{m,n\}$.

<u>Proof.</u> (2.6) is an immediate consequence of (2.4). For (2.7) we first note that the assumption can be written in the form

(*) $[(k = i \text{ or } k = j) \to \ell \neq m,n]$ and $[(\ell = i \text{ or } \ell = j) \to k \neq m,n]$.

Clearly, (*) is fulfilled if $\{i,j\} \cap \{k,\ell\} = \emptyset$. But in this case $f_{ij} \perp f_{k\ell}$ and the product vanishes. So we may assume $\#(\{i,j\} \cap \{k,\ell\}) \geq 1$. Without loss of generality it is enough to consider the case $k = i$, thus $\ell \neq m,n$. If $k \notin \{m,n\}$ then $f_{k\ell} \perp f_{mn}$ and the product is zero. Otherwise $i = k = m$ or $i = k = n$. Then (*) implies $\ell \neq i,j$ and we have to consider the products $\{f_{ij}\ f_{i\ell}\ f_{in}\}$ with $j \neq \ell \neq n$ and $\{f_{ij}\ f_{i\ell}\ f_{mi}\}$ with $j \neq \ell \neq m$. It is obviously enough to look at the first case, for which there are two possibilities: $j = n$, in which case the product equals $2P(f_{ij})f_{i\ell} = 0$ because $f_{ij} \top f_{i\ell}$, or $j \neq n$, in which case the product vanishes by (2.5). ∎

We remark that it is straightforward to check that a symplectic grid is a connected ortho-collinear grid. Also, Lemma 2.8. shows that our notion of a symplectic grid generalizes the one used in [42] §4: All conditions of [42] (4.1) are obviously fulfilled except

$$\{f_{ij}\ f_{kj}\ f_{k\ell}\} = f_{i\ell} \quad \text{for } k \neq i,\ \ell \neq j.$$

This equation follows from (2.4) in case $i \neq \ell$ and $j \neq \ell$. But for $i = \ell$ resp. $j = \ell$ the equation becomes $\{f_{ij}\ f_{kj}\ f_{ki}\} = 0$ resp. $\{f_{ij}\ f_{kj}\ f_{kj}\}$

$= f_{ij}$ which follows from (2.5) resp. (2.6).

The connection with [42] provides us also with examples: The multiplication rules stated in Example I.1.4 easily show that the symplectic matrix units F_{ij} form a symplectic grid in the symplectic triple system $A(I;K)$.

As in §2.3 we now prove a construction theorem for symplectic grids:

Theorem 2.9. *Let I_3 be a totally ordered index set $\neq \emptyset$ with $1,2,3 \notin I_3$. Put $I_2 = \{3\} \cup I_3$, $I_1 = \{2\} \cup I_2$ and extend the total order to $I = \{1\} \cup I_1$ in the natural way. Assume*

i) *$\{f_{1i}; i \in I_1\}$ is a pure collinear system (of non-zero tripotents),*

ii) *there exists a non-zero tripotent f_{23} such that $f_{12} \top f_{23} \top f_{13}$, $\{f_{12}\, f_{23}\, f_{13}\} = 0$ and $f_{23} \perp f_{1i}$ for every $i \in I_3$.*

We define

for $i \in I_3$: $\quad f_{2i} := \{f_{23}\, f_{13}\, f_{1i}\} =: -f_{i2}$,

for $i,j \in I_2$, $i \neq j$: $\quad f_{ij} := \{f_{1i}\, f_{12}\, f_{2j}\}$

for $i \in I_1$: $\quad f_{i1} := -f_{1i}$

Then $f_{ij} = -f_{ji}$ for all $i,j \in I$, $i \neq j$ and $S = \{f_{ij}; i,j \in I, i < j\}$ is a symplectic grid.

Proof. We proceed in several steps:

1) $\{f_{1i}, f_{2i}; i \in I_2\}$ is a rectangular grid with $f_{12} \top f_{2i}$ for all $i \in I_2$. Indeed, the first assertion follows from Theorem 2.6 and $f_{12} \top f_{23}$ by ii). For $i \in I_3$ the definition of f_{2i} and formula (I.1.21) shows $f_{2i} \in V_1(f_{12})$. Because $f_{12} \in V_1(f_{23}) \cap V_1(f_{13}) \cap V_1(f_{1i})$ and $(f_{23}, f_{13}, f_{1i}, f_{2i})$ is a quadrangle we conclude $f_{12} \in V_1(f_{2i})$ from the Quadrangle Decomposition Theorem 2.2, whence $f_{12} \top f_{2i}$.

2) $\{f_{12}\} \cup \{f_{1i}, f_{2i}; i \in I_2\}$ is pure.
By i) resp. 1) purity is clear as long as all three tripotents involved lie in $\{f_{1i}, i \in I_1\}$ resp. $\{f_{1i}, f_{2i}; i \in I_2\}$. Thus we only need to consider products which contain f_{12} and at least one factor $f_{2i}, i \in I_2$. Keeping in mind that the order of the factors is irrelevant (Lemma I.3.7.a), we see that there are two types of such products:
$\{f_{12}\, f_{1i}\, f_{2i}\}$ and $\{f_{12}\, f_{2i}\, f_{2j}\}$ for $i, j \in I_2$, $i \neq j$.
Regarding the first we have $\{f_{12}\, f_{13}\, f_{23}\} = 0$ by assumption, and for $i \in I_3$ we get $\{f_{2i}\, f_{1i}\, f_{12}\} = \{f_{23}\, f_{13}\, f_{12}\} = 0$ by (I.2.3) and the fact that $(f_{23}, f_{13}, f_{1i}, f_{2i})$ is a quadrangle. Regarding the second we

similarly obtain $\{f_{2j}\ f_{2i} f_{12}\} = \{f_{1j}\ f_{1i}\ f_{12}\} = 0$ by i).

3) For $i,j \in I_2$, $i \neq j$, $(f_{1i}, f_{12}, f_{2j}, f_{ij})$ is a quadrangle, in particular $f_{ij} \neq 0$. This is an immediate consequence of 1) and the Quadrangle Criterion I.2.1.

4) $f_{ij} = -f_{ji}$ for $i,j \in I$, $i \neq j$.

This assertion holds by definition for i or j = 1,2. So we may assume $i,j \in I_2$. Then $f_{ij} = \{f_{2j}\ f_{12}\ f_{1i}\} =$ (by 1)) $\{f_{2j}\ f_{12}\{f_{2i}\ f_{2j}\ f_{1j}\}\} =$ $\{\{f_{2j}\ f_{12}\ f_{2i}\} f_{2j}\ f_{1j}\} - \{f_{2i}\ \{f_{12}\ f_{2j}\ f_{2j}\} f_{1j}\} +$ $\{f_{2i}\ f_{2j}\{f_{2j}\ f_{12}\ f_{1j}\}\} =$ (by 1) and 2)) $- \{f_{2i}\ f_{12}\ f_{1j}\} = -f_{ji}$.

5) $f_{ij} \top f_{ik}$ for $i,j,k \neq$.

Again, this is clear for i = 1 or i = 2. So we may assume $i \in I_2$. The case j or k $\in \{1,2\}$ follows from 1) and 3), so also $j,k \in I_2$, then $f_{ij} = \{f_{1i}\ f_{12}\ f_{2j}\} \in V_1(f_{1i}) \cap V_0(f_{12}) \cap V_0(f_{2k}) \subset V_1(f_{ik})$ using 3) and the Quadrangle Decomposition Theorem I.2.2. By symmetry $f_{ij} \top f_{ik}$.

6) $f_{ij} \perp f_{k\ell}$ for $\{i,j\} \cap \{k,\ell\} = \emptyset$.

This assertion follows by a straightforward case-by-case verification using the techniques of 5). The details are left to the reader.

7) For distinct $i,j,k,\ell \in I$ $(f_{ij}, f_{kj}, f_{k\ell}, f_{i\ell})$ is a quadrangle.

First we note that by 5) and 6) we have the correct relations: $f_{ij} \top f_{kj} \top f_{k\ell} \top f_{i\ell} \top f_{ij}$ and $f_{ij} \perp f_{k\ell}$, $f_{kj} \perp f_{i\ell}$, hence all we have to show is one of the multiplication rules, e.g. $\{f_{ij}\ f_{kj}\ f_{k\ell}\} = f_{i\ell}$. For this we distinguish two cases.

7.a) $1 \in \{i,j,k,\ell\}$: All possibilities can be reduced to the case i = 1. Then the subcase $\ell = 2$ follows from $\{f_{k2}\ f_{12}\ f_{1j}\} = f_{kj}$, similarly one handles j = 2 and k = 2, and $\ell,k,j \in I_2$ follows from $\{f_{1\ell}\ f_{1j}\ f_{kj}\} = \{f_{1\ell}\ f_{1j}\ \{f_{1k}\ f_{12}\ f_{2j}\}\} = \{f_{1k}\ f_{12}\ \{f_{1\ell}\ f_{1j}\ f_{2j}\}\} =$ $\{f_{1k}\ f_{12}\ f_{2\ell}\} = f_{k\ell}$.

7.b) $1 \notin \{i,j,k,\ell\}$: By 7.a) we already know, that $(f_{1\ell}, f_{1j}, f_{kj}, f_{k\ell})$ is a quadrangle, hence $\{f_{kj}\ f_{k\ell}\ f_{i\ell}\} = \{f_{1j}\ f_{1\ell}\ f_{i\ell}\}$ (by (I.2.3)) $= f_{ij}$, since $(f_{1j}, f_{1\ell}, f_{i\ell}, f_{ij})$ is also a quadrangle by 7.a).

8) $\mathcal{S}$ is pure.

By 4)-6) this follows if $\{f_{ij}\ f_{ik}\ f_{i\ell}\} = 0$ for $i,j,k,\ell \neq$. The case i = 1 or i = 2 was settled in 2). Hence $i \in I_2$ and without loss of generality also $j \in I_2$. Then there exists $r \in \{1,2\}$ such that $(f_{ij}, f_{ik}, f_{rk}, f_{rj})$ is a quadrangle, hence $\{f_{ij}\ f_{ik}\ f_{i\ell}\} = \{f_{rj}\ f_{rk}\ f_{i\ell}\}$ which vanishes by 6) if $r \neq \ell$ and by 2) if $r = \ell$. ∎

We now come back to the object of this chapter: the description of grids up to association. The results in the previous sections reduce the problem to the case where $\mathcal{E}$ is a connected ortho-collinear grid of rank $\mathcal{E} > 2$ and $\#\mathcal{E}_{12}(e_1,e_2) > 3$ for every orthogonal pair $(e_1,e_2) \subset \mathcal{E}$. The next result describes the possibilities for $\#\mathcal{E}_{12}(e_1,e_2)$:

Lemma 2.10. Let E be a closed ortho-collinear system and $(e_1,e_2) \subset E$ an orthogonal pair.

a) To every $f \in E_{12}(e_1,e_2)$ there exists exactly one $\tilde{f} \in E_{12}(e_1,e_2) \cap V_0(f)$. One has $\tilde{f} \approx \{e_1\ f\ e_2\}$.

b) $E_{12}(e_1,e_2)$ is the disjoint union of orthogonal pairs, such that

$$E_{12}(e_1,e_2) = \{f_1,f_2\} \dot{\cup} (E_{12}(e_1,e_2) \cap E_{12}(f_1,f_2))$$

for every orthogonal pair $(f_1,f_2) \subset E_{12}(e_1,e_2)$.

c) $\# E_{12}(e_1,e_2)$ is even or infinite.

Proof. a) By closedness there exists $\tilde{f} \in E_{12}(e_1,e_2) \cap V_0(f)$ with $\tilde{f} \approx \{e_1 f\ e_2\} = P(e_1+e_2)f$. For any $\tilde{f} \in E_{12}(e_1,e_2) \cap V_0(f)$ we have $f \approx \{e_1\ \tilde{f}\ e_2\} = P(e_1+e_2)\tilde{f}$, which proves uniqueness.

b) Let $E_{12}(e_1,e_2) \neq \emptyset$. By a) there exists an orthogonal pair $(f_1,f_2) \subset E_{12}(e_1,e_2)$ and any other $h \in E_{12}(e_1,e_2)$ is collinear to f_1 and f_2. Because $E_{12}(e_1,e_2)$ is again closed and ortho-collinear, we may substitute $E_{12}(e_1,e_2)$ for E and (f_1,f_2) for (e_1,e_2) and continue in this way if possible.

c) Immediately follows from b). ∎

We proceed with the discussion which was interrupted for Lemma 2.10. By this result we now can assume $\#E_{12}(e_1,e_2) > 4$ for every orthogonal pair. The next theorem handles the case $\#E_{12}(e_1,e_2) = 4$ for one orthogonal pair. We point out that by Theorem 2.4.c.2) this condition then holds for all orthogonal pairs.

Theorem 2.11. Let E be a connected ortho-collinear grid which contains an orthogonal system (e_1,e_2) such that $\#E_{12}(e_1,e_2) = 4$. Then E is associated to a symplectic grid.

Proof. By Lemma 2.10 we may write $E_{12}(e_1,e_2) = \{f_{13},f_{14},f_{23},\tilde{f}_{24}\}$ where the (non-zero) tripotents satisfy the relations $f_{13} \perp \tilde{f}_{24}$, $f_{14} \perp f_{23}$ and $f_{13} \top f_{14} \top \tilde{f}_{24} \top f_{23} \top f_{13}$. Let $h \in E_1(e_1) \cap E_{10}(f_{13},\tilde{f}_{24})$. Since $E_1(e_1)$ has rank 2 by Lemma 2.1.b and $f_{14} \perp f_{23}$, we have $f_{14} \top h$ or $f_{23} \top h$. After possibly renaming the elements we may assume $f_{14} \top h$. Because $f_{23} \approx \{f_{13}\ f_{14}\ \tilde{f}_{24}\} \in V_0(h)$ we then get $f_{23} \perp h$. Now let $g \in E_1(e_1) \cap E_{10}(f_{13},\tilde{f}_{24})$ be another tripotent and assume $g \perp f_{14}$. Then $f_{23} \top g$, and since $E_1(e_1) \cap V_0(\tilde{f}_{24})$ is a collinear system we obtain $g \top h$, hence $\{g\ h\ f_{14}\} \approx d \in E_1(e_1) \cap E_{12}(f_{13},\tilde{f}_{24}) \cap V_1(f_{23})$. But then $E_{12}(f_{13},\tilde{f}_{24}) \supset \{e_1,e_2,f_{14},f_{23},d\}$ has a bigger cardinality than 4 which gives a contradiction. Thus $g \top f_{14}$, and we proved so far that - after a possible change of indices which we assume for the following - we have

$E_1(e_1) \cap E_{10}(f_{13},\tilde{f}_{24}) \subset E_{10}(f_{14},f_{23})$. We define $f_{12} = e_1$ and

$$E_1(e_1) \cap E_{10}(f_{13},\tilde{f}_{24}) = \{f_{1i}; i \in \hat{I}\},$$

where $\hat{I}$ is a possibly empty index set with $1,2,3,4 \notin \hat{I}$. We put $I_3 = \{4\} \cup \hat{I}$ and see, using (I.4.8), that the assumptions of Theorem 2.9 are fulfilled. Let $S = \{f_{ij}; i,j \in I, i < j\}$ be the symplectic grid constructed there.

Clearly, we want to show $S \approx E$, i.e. every tripotent of S is associated to some tripotent of E and conversely. This is obvious for f_{1i}, $i \in I_1$, and f_{2i}, $i \in I_2$. For $i,j \in I_2$, $i < j$, let $f_{2j} \approx g \in E_0(f_{1i}) \cap V_1(f_{12})$ and put $\gamma = B(f_{1i}+f_{12},f_{1i}+f_{12}) \in \mathrm{Aut}\, V$. Then $f_{ij} = -\gamma f_{2j} \approx -\gamma g = \{f_{1i} f_{12} g\} \approx E$ by closedness. Conversely, since $E_1(e_1) = E_1$ is of rank 2 we have the decomposition

$$E_1 = \{f_{13},\tilde{f}_{24}\} \dot{\cup} (E_1)_{12} \dot{\cup} (E_1)_{10} \dot{\cup} (E_1)_{20}$$

where the Peirce spaces are formed relative to $(f_{13},\tilde{f}_{24}) \subset E_1$. Clearly, $\tilde{f}_{24} \approx f_{24}$, $(E_1)_{12} = \{f_{14},f_{23}\} \subset S$ and also $(E_1)_{10} \subset S$ by construction. Any $g \in (E_1)_{20}$ satisfies $\{f_{13}\ f_{23}\ g\} \approx f_{1j}$ for some j and hence $g \approx f_{2j}$. Finally, let $g \in E_0(e_1)$. By connectedness there exists $f \in E_{12}(e_1,g)$ and by Lemma 2.10.a there exists $\tilde{f} \in E_{12}(e_1,g)$ such that $f \perp \tilde{f}$ and hence $g \approx \{f\ e_1\ \tilde{f}\}$. We may assume $f = f_{1i}$ and $\tilde{f} \approx f_{2j}$. Then $e \approx \{f\ e_1\ \tilde{f}\} \approx \{f_{1i}\ f_{12}\ f_{2j}\} = f_{ij}$ and we are done. ∎

<u>2.5.</u> The last type of grid we consider in this section is that of an <u>even-dimensional quadratic form grid</u> which by definition is a family

$$\mathcal{Q}_e = \mathcal{Q}_e(I) = \{e_i^\varepsilon;\ i \in I,\ \varepsilon = \pm\}$$

of non-zero tripotents where I is an index set with $\# I > 2$ and $\mathcal{Q}_e$ has the following properties:

(2.8) For $i,j \in I$, $i \neq j$, and ε, $\mu = \pm$ arbitrary $(e_i^\varepsilon, e_j^\mu, e_i^{-\varepsilon}, -e_j^{-\mu})$ is a quadrangle;

(2.9) $\mathcal{Q}_e$ is pure.

We remark that (2.8) in particular implies

(2.10) $e_i^+ \perp e_i^-$, $e_i^\varepsilon \top e_j^\mu$ for $i \neq j$ and ε, μ arbitrary,

and that (2.9) is equivalent to

(2.9)' $\{e_i^\varepsilon\ e_j^\mu\ e_k^\delta\} = 0$ for $i,j,k \neq$ and ε,μ,δ arbitrary.

We also note that we already have come across an even-dimensional quadratic form grid: Whenever $\mathcal{Q}_o = \{e_o\} \cup \{e_i^\varepsilon; i \in I, \varepsilon = \pm\}$ is an odd-dimensional quadratic form grid (see § 1.1) the subfamily $\{e_i^\varepsilon; i \in I, \varepsilon = \pm\}$ is an even-dimensional quadratic form grid. Moreover, analogous to

$\mathcal{Q}_0$ an example of an even-dimensional quadratic form grid is provided by the standard basis $\{e_i^\varepsilon; i \in I, \varepsilon = \pm\}$ in the Jordan triple system $[K,{}^-,2I]$ - see Example I.1.6.b. Clearly, this example motivated our notation.

In case $\#I < \infty$ we also define

$$\mathcal{Q}(2m) = \mathcal{Q}_e(I) \text{ for } \#I = m < \infty,$$

in which case $\mathcal{Q}(2m)$ is called a 2m-dimensional quadratic form grid and is written in the form

$$\mathcal{Q}(2m) = (e_1^+, e_1^-, \dots, e_m^+, e_m^-).$$

The construction theorem for an even-dimensional quadratic form grid reads as follows:

Theorem 2.12. *Let I be an index set with $\#I > 2$ and $1 \in I$. Put $I_1 = I \setminus \{1\}$ and assume*

i) $\{e_i^+; i \in I\}$ is a pure collinear system (of non-zero tripotents) and

ii) there exists a non-zero tripotent e_1^- such that $e_1^+ \perp e_1^-$ and $\{e_1^-\} \cup \{e_i^+; i \in I_1\}$ is a pure collinear system.

Define $e_i^- = -\{e_1^+\, e_i^+\, e_1^-\}$ for $i \in I_1$. Then $\mathcal{Q}_e = \{e_i^\varepsilon; i \in I, \varepsilon = \pm\}$ is an even- dimensional quadratic form grid.

Proof. Let $i \in I$ and $j,k \in I_1$ be distinct. By definition $(e_1^+, e_j^+, e_1^-, -e_j^-)$ is a quadrangle, in particular $e_i^+ \perp e_i^-$ and $e_1^\varepsilon \top e_j^\mu$. Because $e_k^- = -\{e_1^+\, e_k^+\, e_1^-\} \in V_1(e_1^+) \cap V_1(e_j^+) \cap V_1(e_1^-) \subset V_1(e_j^-)$ we get by symmetry $e_k^\mu \top e_j^\varepsilon$. Now (2.8) follows from $\{e_k^+\, e_j^+\, e_k^-\} = -\{e_k^+\, e_j^+\, \{e_1^+ e_k^+ e_1^-\}\}$ = (using i) and ii)) $\{e_1^+\, \{e_j^+ e_k^+ e_k^+\}\, e_1^-\} = -e_j^-$.

Using the assumptions, Lemma I.3.7.a. and the automorphism $P(e_1^+ + e_1^-)|V_2(e_1^+ + e_1^-)$, which fixes e_1^ε and moves e_i^ε to $-e_i^{-\varepsilon}$, (2.9)' follows from the following 2 cases: $\{e_1^\varepsilon\, e_i^-\, e_j^+\}$ = (by (I.2.3)) - $\{e_i^+\, e_1^{-\varepsilon}\, e_j^+\} = 0$ by purity of $\{e_1^\varepsilon\} \cup \{e_i^+; i \in I_1\}$ and for $i,j,k \in I_1$ $\{e_i^+\, e_j^+\, e_k^-\} = -\{e_i^+\, e_j^+ \{e_1^+ e_k^+\, e_1^-\}\} = 0$ by the same reason. ∎

An application of this theorem is

Theorem 2.13. *Let E be a connected ortho-collinear grid and $(e_1, e_2) \subset E$ an orthogonal pair. Then $E_2(e_1+e_2) = \{e_1, e_2\} \cup E_{12}(e_1, e_2)$ is associated to an even-dimensional quadratic form grid.*

Proof. Let $\{e_i^+; i \in I_1\} \subset E_{12}(e_1, e_2)$ be a maximal collinear system, where I_1 is an index set with $1 \notin I_1$. By Corollary 2.2 $I_1 \neq \emptyset$. We put $I = \{1\} \cup I_1$, $e_1^+ = e_1$ and $e_2 = e_1^-$. Then $\{e_i^+; i \in I\}$ as well as

$\{e_1^-\} \cup \{e_i^+; i \in I_1\}$ are pure collinear systems. Let $\mathcal{Q}_e$ be the even-dimensional quadratic form grid constructed in Theorem 2.12. By closedness of E every element $\mathcal{Q}_e$ is clearly associated to one of $E_2(e_1+e_2)$.

Conversely, let $g \in E_{12}(e_1,e_2) \setminus \{e_i^+; i \in I_1\}$. By maximality there exists $i \in I_1$ with $g \perp e_1^+$, hence by Lemma 2.10 $g \approx \{e_1^+\ e_i^+\ e_1^-\} = -e_i^-$. This shows $\mathcal{Q}_e \approx E_2(e_1+e_2)$. ∎

The reader might wonder how this theorem corresponds to the grids already considered. For a rectangular grid $R_2(e_1+e_2)$ is a quadrangle which obviously is associated to a 4-dimensional quadratic form grid and for a symplectic grid S we may identify $S_2(e_1+e_2)$ with the symplectic grid $S(4) = \{f_{12},f_{13},f_{14},f_{23},f_{24},f_{34}\}$ which becomes a 6-dimensional quadratic form grid via the identification $(f_{12},f_{13},f_{14},f_{23},f_{24},f_{34}) = (e_1^+,e_2^+,e_3^+,e_3^-,-e_2^-,e_1^-)$. These considerations also explain why in the Résumé at the end of §3 we may restrict ourselves and consider symplectic grids $S(I)$ only for $\#I > 5$ and even-dimensional quadratic form grids $\mathcal{Q}_e(I)$ only for $\#I > 3$.

We explicitely mention the following corollary which pushes our investigation of connected grids one step further:

<u>Corollary 2.14.</u> <u>Let E be a connected ortho-collinear grid which contains an orthogonal system (e_1,e_2) such that $E_{00}(e_1,e_2) = E_{10}(e_1,e_2) = \emptyset$. Then E is associated to an even-dimensional quadratic form grid.</u>

<u>Proof.</u> By Theorem 2.4.c applied to (e_1,e_2) and (e_2,e_1) we also have $E_{20}(e_1,e_2) = \emptyset$, so $E = E_2(e_1+e_2)$. ∎

At the end of this section it is appropriate to realize the present situation in our study of grids: By the results proved so far it remains to investigate connected ortho-collinear grids E which contain an orthogonal pair (e_1,e_2) such that

i) $\#E_{12}(e_1,e_2) > 6$ and

ii) $E_2(e_1+e_2) \subsetneq E$, i.e. $E_{10}(e_1,e_2) \neq \emptyset$ or rank $E > 3$.

This will be done in the next section.

§3 Ortho-collinear grids II (the exceptional cases)

We introduce the two exceptional grids (Bi-Cayley grid and Albert grid) and describe the grids which are associated to them.

3.1. The grid, which we will study in this subsection, consists of $16 = 2 \cdot 8$ tripotents and can be realized in Mat(1,2;$\mathbb{O}$) where $\mathbb{O}$ is the split Cayley algebra over k. That is why we call it a Bi-Cayley grid. The details of the realization will be given in III§3.2.

Now to the definition: A Bi-Cayley grid is a family

$$\mathcal{B} = (e_j^\varepsilon;\ 1 \leq j \leq 8, \varepsilon = \pm)$$

of non-zero tripotents, built out of

(3.1) four 8-dimensional quadratic form grids, namely

a) $(e_1^+, e_1^-, \ldots, e_4^+, e_4^-)$,

b) $(e_5^+, e_5^-, \ldots, e_8^+, e_8^-)$,

c) $(e_1^\varepsilon, e_5^{-\varepsilon}, \ldots, e_4^\varepsilon, e_8^{-\varepsilon})$ for $\varepsilon = \pm$ and

(3.2) two rectangular grids $(c_{ij}^\varepsilon;\ 1 \leq i \leq 2, 1 \leq j \leq 4)$, $\varepsilon = \pm$, of size 2×4, where $c_{1j}^\varepsilon = e_j^\varepsilon$, $c_{2j}^\varepsilon = e_{j+4}^\varepsilon$,

such that in addition the following multiplication rules hold:

(3.3) $(e_j^-, e_k^+, e_{\ell+4}^-, \operatorname{sgn}\binom{1234}{jk\ell m} e_{m+4}^+)$ is a quadrangle, where $\operatorname{sgn}\binom{1234}{jk\ell m}$ denotes the signature of the permutation $\binom{1234}{jk\ell m}$, and

(3.4) $\mathcal{B}$ is pure.

By writing down what a quadratic form grid resp. a rectangular grid means we arrive at the following equivalent definition: A family $\mathcal{B} = (e_j^\varepsilon;\ 1 \leq j \leq 8, \varepsilon = \pm)$ of non-zero tripotents is a Bi-Cayley grid iff it satisfies (3.3) and

(3.5) for $1 \leq j,k \leq 4$, $j \neq k$ and ε, μ arbitrary the following families are quadrangles:

a) $(e_j^\varepsilon, e_k^\mu, e_j^{-\varepsilon}, -e_k^{-\mu})$,

b) $(e_{j+4}^\varepsilon, e_{k+4}^\mu, e_{j+4}^{-\varepsilon}, -e_{k+4}^{-\mu})$,

c) $(e_j^\varepsilon, e_k^\varepsilon, e_{j+4}^{-\varepsilon}, -e_{k+4}^{-\varepsilon})$ and

d) $(e_j^\varepsilon, e_k^\varepsilon, e_{k+4}^\varepsilon, e_{j+4}^\varepsilon)$,

(3.6) for ε, μ, δ arbitrary and $1 \leq j,k,\ell \leq 4$ with $j,k,\ell \neq$

$$\{e_j^\varepsilon\ e_k^\mu\ e_\ell^\delta\} = 0 = \{e_{j+4}^\varepsilon\ e_{k+4}^\mu\ e_{\ell+4}^\delta\},$$

$$\{e_j^\varepsilon\ e_k^\varepsilon\ e_{\ell+4}^{-\varepsilon}\} = 0 = \{e_{j+4}^\varepsilon\ e_{k+4}^\varepsilon\ e_\ell^{-\varepsilon}\},$$

$$\{e_j^\varepsilon\ e_k^{-\varepsilon}\ e_{k+4}^{-\varepsilon}\} = 0 = \{e_{j+4}^\varepsilon\ e_{k+4}^{-\varepsilon}\ e_k^{-\varepsilon}\} \text{ and}$$

$$\{e_j^{\varepsilon}\ e_k^{-\varepsilon}\ e_{j+4}^{\varepsilon}\} = 0 = \{e_{j+4}^{\varepsilon}\ e_{k+4}^{-\varepsilon}\ e_j^{\varepsilon}\}$$

We note that (3.4) follows from (3.6) together with Lemma I.3.7.a.

We record some of the immediate consequences of (3.1)-(3.6). First, (3.1) and (3.2) imply

(3.7) $\mathcal{B}$ is a connected ortho-collinear system, in particular we have for $1 \leq j,k \leq 4$, $j \neq k$

$e_j^{\varepsilon} \perp e_j^{-\varepsilon}$, $e_{j+4}^{\varepsilon} \perp e_{j+4}^{-\varepsilon} \perp e_j^{\varepsilon} \perp e_{k+4}^{\varepsilon}$ and

$e_j^{\varepsilon} \top e_k^{\mu}$, $e_{k+4}^{\mu} \top e_{j+4}^{\varepsilon} \top e_j^{\varepsilon} \top e_{k+4}^{-\varepsilon}$

By (3.4) and (3.7) the only non-zero triple products between three different elements of $\mathcal{B}$ are of quadrangle-type. It is easily checked:

(3.8) The only quadrangles in $\mathcal{B}$ are those of (3.3) and (3.5).

Therefore

(3.9) $\mathcal{B}$ is a grid.

As in the study of the other grids we now prove a construction theorem for a Bi-Cayley grid:

Theorem 3.1. *Let* $(e_1^+, e_2^+, e_3^+, e_4^+, e_1^-, e_5^-)$ *be non-zero tripotents such that*

i) $e_1^- \perp e_1^+ \perp e_5^-$

ii) $(e_1^+, e_2^+, e_3^+, e_4^+)$ *and* $(e_1^-, e_5^-, e_2^+, e_3^+, e_4^+)$ *are pure collinear systems.*

Define for $2 \leq j \leq 4$:

$$e_j^- = -\{e_1^+\ e_j^+\ e_1^-\}, \quad e_{j+4}^- = -\{e_1^+\ e_j^+\ e_5^-\}$$

$$e_5^+ = \{e_3^-\ e_2^+\ e_8^-\}, \quad e_{j+4}^+ = -\{e_5^+\ e_{j+4}^-\ e_5^-\}$$

Then $\mathcal{B} = (e_i^{\varepsilon};\ 1 \leq i \leq 8, \varepsilon = \pm)$ *is a Bi-Cayley grid.*

Proof. We proceed in several steps:

1) $(e_1^+, e_1^-, \dots, e_4^+, e_4^-)$ and $(e_1^+, e_5^-, \dots, e_4^+, e_8^-)$ are 8-dimensional quadratic form grids, in particular

$$e_{j+4}^- \perp e_j^+ \top e_{k+4}^- \text{ for } j \neq k.$$

This holds by Theorem 2.12 and the definition of e_j^- resp. e_{j+4}^-.

2) $e_{k+4}^- \perp e_j^- \top e_{j+4}^-$ for $j \neq k$.

For $j = 1$ we know $e_1^- \top e_5^-$ by assumption ii) and have

$e_{k+4}^- = -\{e_1^+\ e_k^+\ e_5^-\} \in V_{0-1+1}(e_1^-) = V_0(e_1^-)$, i.e. $e_{k+4}^- \perp e_1^-$. Similarly, for k = 1 we obtain $e_j^- = -\{e_1^+ e_j^+ e_1^-\} \in V_0(e_5^-)$. Hence we can assume $2 < j,k$. Then $e_{j+4}^- = -\{e_1^+ e_j^+ e_5^-\} \in V_1(e_1^+) \cap V_0(e_j^+) \cap V_0(e_1^-) \subset V_1(e_j^-)$ by the Quadrangle Decomposition Theorem I.2.2 and in the same way $e_j^- \in V_1(e_{j+4}^-)$, hence $e_j^- \top e_{j+4}^-$. Finally, $e_{k+4}^- \in V_1(e_1^+) \cap V_1(e_j^+) \cap V_0(e_1^-) \subset V_0(e_j^-)$ and thus $e_{k+4}^- \perp e_j^-$.

In the following analogous considerations will be referred to as the "method of 2)" and will be left to the reader.

3) $\{e_j^- e_k^+ e_{j+4}^-\} = 0$ for $j \neq k$.

By 1) $(e_j^-, e_k^+, e_j^+, -e_k^-)$ is a quadrangle, thus by (I.2.3) we have $\{e_j^- e_k^+ e_{j+4}^-\} = -\{e_k^- e_j^+ e_{j+4}^-\}$ which vanishes because $e_j^+ \perp e_{j+4}^-$ by 1).

4) For $1 < j,k,\ell < 4$ and $j,k,\ell \neq$ we have

$$\{e_j^- e_k^+ e_{\ell+4}^-\} = \{e_k^- e_\ell^+ e_{j+4}^-\} = -\{e_k^- e_j^+ e_{\ell+4}^-\},$$

in particular for $(jk\ell) \in \{(324),(243),(432)\}$

$$e_5^+ = \{e_j^- e_k^+ e_{\ell+4}^-\} = -\{e_k^- e_j^+ e_{\ell+4}^-\}.$$

Since $L(e_j^-, e_k^+) = -L(e_k^-, e_j^+)$ it is enough to show the first equation. We begin with $\ell=1$: $\{e_k^- e_1^+ e_{j+4}^-\} = -\{e_k^- e_1^+\{e_1^+ e_j^+ e_5^-\}\} = -\{e_k^- e_j^+ e_5^-\} = \{e_j^- e_k^+ e_5^-\}$. For $\ell > 1$ we obtain $\{e_j^- e_k^+ e_{\ell+4}^-\} = -\{\{e_j^- e_k^+ e_1^+\} e_\ell^+ e_5^-\} - \{e_1^+ e_\ell^+\{e_j^- e_k^+ e_5^-\}\}$ which for $k = 1$ gives $-\{e_j^- e_\ell^+ e_5^-\}$ = (by "ℓ=1") $- \{e_\ell^- e_1^+ e_{j+4}^-\} = \{e_1^- e_\ell^+ e_{j+4}^-\}$, for $j = 1$ we get $\{e_k^- e_\ell^+ e_5^-\}$ and for $j,k > 2$ we obtain $-\{e_1^+ e_\ell^+\{e_j^- e_k^+ e_5^-\}\}$ = (by "j=1") $- \{e_1^+ e_\ell^+\{e_1^- e_j^+ e_{k+4}^-\}\} = \{e_\ell^- e_j^+ e_{k+4}^-\}$ = (by repetition) $\{e_k^- e_\ell^+ e_{j+4}^-\}$. This proves 4).

5) $(e_5^+, e_5^-, \dots, e_8^+, e_8^-)$ is a quadratic form grid.

By Theorem 2.12 and the definition of e_{j+4}^+ this follows as soon as $(e_5^-, \dots, e_8^-)$ is a pure collinear system (which holds by (1)), $e_5^+ \perp e_5^-$ (which is trivial) and $(e_5^+, e_6^-, e_7^-, e_8^-)$ is a pure collinear system. Collinearity follows by the method of 2) and purity holds if $\{e_5^+ e_{j+4}^- e_{k+4}^-\} = 0$ for $2 < j,k$ and $j \neq k$, which follows from $e_5^+ = \pm\{e_k^- e_\ell^+ e_{j+4}^-\}$ by 4) and $\{e_{k+4}^- e_{j+4}^-\{e_k^- e_\ell^+ e_{j+4}^-\}\}$ = (since $e_{j+4}^- \perp e_k^-$ and $\{e_{j+4}^- e_{k+4}^- e_\ell^+\} = 0$ by 1)) $\{e_k^- e_\ell^+ e_{k+4}^-\} = 0$ by 3).

6) Put $c_{1j}^\varepsilon = e_j^\varepsilon$, $c_{2j}^\varepsilon = e_{j+4}^\varepsilon$. Then $(c_{ij}^\varepsilon;\ 1 < i < 2,\ 1 < j < 4)$ is a rectangular grid for $\varepsilon = \pm$.

First we consider $\varepsilon = -$. Since $(e_1^-, \dots, e_4^-)$ is a pure collinear system and $e_1^- \top e_5^- \perp e_j^-$ for $2<j<4$, 6) follows from Theorem 2.6 as soon as $\{e_5^-\ e_1^-\ e_j^-\} = e_{j+4}^- = -\{e_1^+\ e_j^+\ e_5^-\}$ which is valid because $(e_j^-,e_1^-,e_j^+,-e_1^+)$ is a quadrangle. In the same way $\varepsilon = +$ follows: $(e_1^+,\ldots,e_4^+)$ is a pure collinear system, $e_1^+ \top e_5^+ \perp e_j^+$ for $2<j$ by the method of 2) and $\{e_5^+\ e_1^+\ e_j^+\} = -\{e_5^+\ e_{j+4}^-\ e_5^-\} = e_{j+4}^+$ since $(e_j^+,e_1^+,e_{j+4}^-,-e_5^-)$ is a quadrangle by 1).

7) $(e_1^-,e_5^+,\ldots,e_4^-,e_8^+)$ is a quadratic form grid.

We note that $(e_1^-,e_2^-,e_3^-,e_4^-)$ is a pure collinear system, that $e_1^- \perp e_5^+$ and that $(e_5^+,e_2^-,e_3^-,e_4^-)$ is a collinear system by the method of 2), which is also pure: $\{e_j^-\ e_k^-\ e_5^+\} = 0$ is proven as in 5). Also, $\{e_1^-\ e_j^-\ e_5^+\} = \{e_5^-\ e_{j+4}^-\ e_5^+\} = -e_{j+4}^+$ because $(e_1^-,e_j^-,e_{j+4}^-,e_5^-)$ is a quadrangle by 6). Now Theorem 2.12 implies 7).

8) (3.3) holds.

We already know $e_j^- \top e_k^+ \top e_{\ell+4}^- \perp e_j^-$, thus it is to show

(*) $$\{e_j^-\ e_k^+\ e_{\ell+4}^-\} = \operatorname{sgn}\begin{pmatrix}1 & 2 & 3 & 4\\ j & k & \ell & m\end{pmatrix} e_{m+4}^-,$$

which by 4) holds for $m = 1$. Also by 4) a cyclic permutation of $(jk\ell)$ does not change the left side of (*) and an anticyclic permutation of $(jk\ell)$ gives a minus sign. Since the same holds for the right side of (*), it is enough to prove for fixed m one of the 6 formulas (*). By 4) $(e_3^-,e_4^+,e_6^-,-e_5^+)$ is a quadrangle, thus $\{e_3^-\ e_4^+\ e_5^-\} = -\{e_5^+\ e_6^-\ e_5^-\} = e_6^+$ which shows (3.3) for $m = 2$. Similarly, $m = 3$ and $m = 4$ follow.

9) (3.4) holds, i.e. B is pure.

By what we already have shown, (3.4) follows from (3.6), which in turn holds, as soon as for $j \neq k$

$$\{e_j^\varepsilon\ e_k^{-\varepsilon}\ e_{k+4}^{-\varepsilon}\} = 0 = \{e_{j+4}^\varepsilon\ e_{k+4}^{-\varepsilon}\ e_k^{-\varepsilon}\},$$

$$\{e_j^\varepsilon\ e_k^{-\varepsilon}\ e_{j+4}^\varepsilon\} = 0 = \{e_{j+4}^\varepsilon\ e_{k+4}^{-\varepsilon}\ e_j^\varepsilon\}$$

is valid. But indeed, $\{e_j^\varepsilon\ e_k^{-\varepsilon}\ e_{k+4}^{-\varepsilon}\} = -\{e_k^\varepsilon\ e_j^{-\varepsilon}\ e_{k+4}^{-\varepsilon}\}$ (by (I.2.3)) $= 0$ because $e_j^{-\varepsilon} \perp e_{k+4}^{-\varepsilon}$. In the same way the other equations follow. ∎

We again take up the investigation of grids. The next step is to characterize those grids which are associated to a Bi-Cayley grid. We will need the following lemma:

<u>Lemma 3.2.</u> <u>Let</u> E <u>be a connected closed ortho-collinear system and</u> $(e_1,e_2) \subset E$ <u>an orthogonal pair. Then</u> $E_{00}(e_1,e_2) \neq \emptyset$ <u>iff there are</u>

orthogonal tripotents $f_{i0}, \bar{f}_{i0} \in E_{i0}(e_1,e_2)$ for $i = 1,2$.

Proof. Let $(i,j) \in \{(1,2),(2,1)\}$ and assume $e_3 \in E_{00}(e_1,e_2)$. Since E is connected there exists $f_{i0} \in E$ such that $e_i \top f_{i0} \top e_3$. From Theorem I.1.10.b follows $f_{i0} \in E_0(e_j)$, thus $f_{i0} \in E_{i0}(e_1,e_2)$. Let $\bar{f}_{i0} \in E$ be associated to $\{e_i\ f_{i0}\ e_3\} \in V_{i0}(e_1,e_2)$. Then $(f_{i0},\bar{f}_{i0}) \subset E_{i0}(e_1,e_2)$ is an orthogonal pair.

Conversely, $\{f_{i0}\ e_i\ \bar{f}_{i0}\} \in V_{00}(e_1,e_2)$ and thus $c \in E$ with $c \approx \{f_{i0} e_i\ \bar{f}_{i0}\}$ lies in $E_{00}(e_1,e_2)$. ∎

Theorem 3.3. Let E be a connected ortho-collinear grid which contains an orthogonal system (e_1,e_2) such that

i) $\# E_{12}(e_1,e_2) > 6$ and

ii) $E_{00}(e_1,e_2) = \emptyset$, but $E_{10}(e_1,e_2) \neq \emptyset$.

Then $\# E_{12}(e_1,e_2) = 6$, $\# E_{i0}(e_1,e_2) = 4$ and E is associated to a Bi-Cayley grid.

Proof. We put $E_{ij} = E_{ij}(e_1,e_2)$ and first prove

1) $\#(E_{12} \cap V_0(f_{20})) = \#(E_{12} \cap V_1(f_{20})) = \#(E_{10} \cap V_1(f_{20})) > 3$ for any $f_{20} \in E_{20}$.

The map $P(e_1,e_2)|V_{12}$ is an automorphism of order 2 which maps $V_{12} \cap V_j(f_{20})$ onto $V_{12} \cap V_{1-j}(f_{20})$ for $j = 0,1$. Hence, assigning to every $e \in E_{12} \cap V_j(f_{20})$ the tripotent $\bar{e} \in E_{12} \cap V_{1-j}(f_{20})$ satisfying $\bar{e} \approx \{e_1\ e\ e_2\}$ induces a bijection between $E_{12} \cap V_j(f_{20})$ and $E_{12} \cap V_{1-j}(f_{20})$. In an analogous way the second equation follows: $E_{12}(e_1,f_{20}) = (E_{12} \cap V_1(f_{20})) \cup (E_{10} \cap V_1(f_{20}))$ and $P(e_1,f_{20})$ induces a bijection between the two components.

2) $\operatorname{rank}(E_{10} \cup E_{20}) = 2$.

Assume the converse. Then by Theorem 2.4.c), for all $f_{10} \in E_{10}$ we know that $E_{10}(f_{10},e_2) \cup E_{20}(f_{10},e_2)$ is a collinear system. In terms of our fixed Peirce decomposition these Peirce spaces are $E_{10}(f_{10},e_2) = \{e_1\} \cup (E_{10} \setminus \{f_{10}\})$ and $E_{20}(f_{10},e_2) = E_{12} \cap E_0(f_{10})$, in particular $E_{10} \setminus \{f_{10}\} \cup (E_{12} \cap E_0(f_{10}))$ is a collinear system and $E_{12} \cap E_0(f_{10}) \subset E_{12} \cap E_1(g_{10})$ for every $g_{10} \in E_{10} \setminus \{f_{10}\}$. It follows $P(e_1,e_2)(E_{12} \cap V_0(f_{10})) \approx E_{12} \cap V_1(f_{10}) \subset E_{12} \cap V_0(g_{10})$, hence by symmetry $E_{12} \cap V_1(f_{10}) = E_{12} \cap V_0(g_{10})$. But this equation gives a contradiction if one uses $\# E_{10} > 3$ which follows from 1).

By Lemma 3.2 we have $\operatorname{rank} E_{10} = 1 = \operatorname{rank} E_{20}$. Therefore, by 2), there exists an orthogonal pair (f_{10},f_{20}) with $f_{i0} \in E_{i0}$, which will be fixed in the following.

3) $E_{12} = (E_{12} \cap V_1(f_{10}) \cap V_0(f_{20})) \cup (E_{12} \cap V_0(f_{10}) \cap V_1(f_{20}))$ and both components have rank 1.

Let $e_{12} \in E_{12} \cap E_{12}(f_{10},f_{20})$. Then $P(e_1,e_2)e_{12} \approx c \in E_{12} \cap E_{00}(f_{10},f_{20}) = \emptyset$ because rank $E = 2$. Now 3) easily follows.

4) $\# E_{12} = 6$, i.e. by (1) $\#(E_{12} \cap V_j(f_{20})) = 3$, $j = 0,1$.
We already know from 1) and 3) that there exist at least three distinct $c_j, c_k, c_\ell \in E_{12} \cap V_0(f_{20}) \cap V_1(f_{10})$. Let $\{c_j\ e_2\ f_{20}\} \approx d \in E$. Then $\{c_j\ e_2\ f_{20}\}$, $d \in V_1(f_{10}) \cap V_0(c_k)$ because $c_j \top c_k$ by 3). Applying the automorphism $-B(f_{10}+c_k, f_{10}+c_k)$ to $\{c_j\ e_2\ f_{20}\} \approx d$ gives - according to Theorem I.1.13 - the relation $h = \{\{c_j\ e_2\ f_{20}\} f_{10}\ c_k\} \approx \{d\ f_{10}\ c_k\}$. There exists $g \in E$ with $\{d\ f_{10}\ c_k\} \approx g$. Because $h \in V_{12} \cap V_0(c_\ell)$ we obtain $g \in E_{12} \cap V_0(c_\ell)$. By construction, h is independent of ℓ, so for any fourth $c_m \in E_{12} \cap V_0(f_{20}) \cap V_1(f_{10})$ we also have $g \in E_{12} \cap V_0(c_m)$. But this contradicts Lemma 2.10.

5) $\# E_{10} = \# E_{20} = 4$, $\# E = 16$
We already know $\#(E_{10} \cap V_1(f_{20})) = 3$, so it is enough to show $E_{10} \cap V_0(f_{20}) = \{f_{10}\}$. Assume $E_{10} \cap V_0(f_{20}) \supset \{f_{10}, g_{10}\}$. Then $f_{10} \top g_{10}$ and $E_{12} \cap V_0(f_{20}) = E_{12} \cap V_1(f_{10}) = E_{12} \cap V_1(g_{10})$ by 3), but the non-zero tripotent g constructed in 4) lies in $E_0(f_{10})$ as well as in $E_1(g_{10})$, contradiction. Thus $\# E_{10} = 4$ and $\# E_{20} = 4$ follows using Theorem 2.4.c. Because $E = \{e_1,e_2\} \cup E_{12} \cup E_{10} \cup E_{20}$, the last assertion is clear.

We now put $e_1^+ = e_1$, $e_1^- = e_2$, $e_5^- = f_{20}$ and $E_{12} \cap V_1(f_{20}) = \{e_2^+, e_3^+, e_4^+\}$. Using 3) we see that $(e_1^+,e_2^+,e_3^+,e_4^+)$ and $(e_1^-,e_5^-,e_2^+,e_3^+,e_4^+)$ are collinear systems, which are pure since E is a grid. Hence the assumptions of Theorem 3.1 are fulfilled. So we obtain a Bi-Cayley grid B and we claim $B \approx E$. By closedness of E we clearly have $e_j^- \approx E$ and $e_{j+4}^- \approx E$ for $2 \leq j \leq 4$. The argument for e_k^+, $5 \leq k \leq 8$, is the same as the one used in 4) to prove $h \approx g$. Thus B is associated to a subfamily of E, but since both families consist of 16 tripotents we actually have $E \approx B$. ∎

3.2. The last type of a grid we consider has 27 tripotents and can be realized in the exceptional Jordan triple system $H_3(\mathbb{O},k,',*)$ (see Example I.1.5) where $\mathbb{O}$ is the split octonian algebra over k, and ' resp. * is the standard resp. nonstandard involution. For the precise details of the realization the reader is refered to Lemma III.3.4. Since $H_3(\mathbb{O},k,',*)$ is closely related to the exceptional Jordan algebra

$H_3(\mathbb{O})$, the Albert algebra, we call this last type of grid an Albert grid.

By definition, an <u>Albert grid</u> is a family A of 27 non-zero tripotents, named in the following form (whose reason will become apparent in III§3.3.)

$$[i],\quad i = 1,2,3, \text{ and}$$
$$_r^{\varepsilon}[ij],\quad \varepsilon = \pm,\ 1 \le r \le 4,\ 1 \le i < j \le 3$$

such that putting

(3.10) $_1^{\varepsilon}[ij] = {}_1^{-\varepsilon}[ji]$ and $_r^{\varepsilon}[ij] = -{}_r^{\varepsilon}[ji]$ for $r \geq 2$, $i \neq j$

the extended family

$$\hat{A} = ([i];\ 1 \le i \le 3) \cup ({}_r^{\varepsilon}[ij],\ \varepsilon = \pm,\ 1 \le r \le 4, 1 \le i,j \le 3,\ i \neq j)$$

is built out of

(3.11) three Bi-Cayley grids: For each $i \in \{1,2,3\}$ the tripotents
$e_r^{\varepsilon} = {}_r^{\varepsilon}[ij]$, $e_{r+4}^{\varepsilon} = {}_r^{\varepsilon}[ik]$, $1 \le r \le 4$, $\varepsilon = \pm$, form a Bi-Cayley grid,

(3.12) three ten-dimensional quadratic form grids, namely
$({}_1^{+}[ij], {}_1^{-}[ij], \ldots, {}_4^{+}[ij], {}_4^{-}[ij], [i], -[j])$,

such that in addition the following families are quadrangles,

(3.13) $({}_1^{\varepsilon}[ij], {}_r^{\varepsilon}[kj], {}_r^{\varepsilon}[ki], [i])$, $1 \le r \le 4$,

(3.14) $({}_r^{\varepsilon}[ij], {}_1^{\varepsilon}[kj], {}_r^{-\varepsilon}[ki], -[i])$, $2 \le r \le 4$,

(3.15) $({}_r^{\varepsilon}[ij], {}_s^{-\varepsilon}[kj], {}_t^{\varepsilon}[ki], \operatorname{sgn}\binom{234}{rst}[i])$
where $\operatorname{sgn}\binom{234}{rst}$ is the signature of the permutation $\binom{234}{rst}$,

and moreover

(3.16) A is pure.

We want to state some of the immediate consequences of (3.11)-(3.16). First, we look at the relations between two tripotents in an Albert grid. They are:

(3.17) For $\{i,j,k\} = \{1,2,3\}$, $\varepsilon,\mu = \pm$ and $r \neq s$ we have

a) $[i] \perp [j]$,

b) $_r^{\varepsilon}[ij] \top [i] \perp {}_r^{\varepsilon}[jk]$,

c) $_r^{\varepsilon}[ij] \top {}_s^{\mu}[ij] \perp {}_s^{-\mu}[ij]$,

d) $_s^{-\varepsilon}[ij] \top {}_r^{\varepsilon}[ik] \top {}_r^{\varepsilon}[ij]$,

e) $_s^{\varepsilon}[ij] \perp {}_r^{\varepsilon}[ik] \perp {}_r^{-\varepsilon}[ij]$.

With (3.17) it is obvious that A is a connected ortho-collinear system. That it is also a grid follows from the following remark which is - by again using (3.17) - straightforward to check.

(3.18) All possible quadrangles in A belong to (3.11)-(3.15).

For later use we explicitly write down what (3.16) means: If A fulfills (3.11) and (3.12) (which already imply (3.17)), then A is pure as soon as (3.16)' the following collinear systems are pure ($1 \leq r,s \leq 4$), $r \neq s$, $\varepsilon = \pm$)

a) $([i], {}^{\varepsilon}_{r}[ij], {}^{-\varepsilon}_{s}[ik])$

b) $([i], {}^{\varepsilon}_{r}[ij], {}^{\varepsilon}_{r}[ik])$

c) $({}^{\varepsilon}_{1}[ij], {}^{\varepsilon}_{1}[ik], {}^{\varepsilon}_{r}[kj])$, $r \geq 2$

d) $({}^{\varepsilon}_{r}[ij], {}^{\varepsilon}_{r}[ik], {}^{\varepsilon}_{r}[kj])$, $r \geq 2$

e) $({}^{\varepsilon}_{r}[ij], {}^{\varepsilon}_{r}[ik], {}^{-\varepsilon}_{s}[kj])$, $r,s \geq 2$

f) $({}^{\varepsilon}_{1}[ij], {}^{-\varepsilon}_{r}[ik], {}^{\varepsilon}_{s}[jk])$, $r,s \geq 2$.

The construction theorem for the Albert grid is

<u>Theorem 3.4.</u> <u>Let</u> $(c, e_1^+, e_2^+, e_3^+, e_4^+, e_1^-, e_5^-)$ <u>be non-zero tripotents such that</u>

i) $e_1^- \perp e_1^+ \perp e_5^-$ <u>and</u>

ii) $(c, e_1^+, e_2^+, e_3^+, e_4^+)$ <u>and</u> $(c, e_1^-, e_5^-, e_2^+, e_3^+, e_4^+)$ <u>are pure collinear systems</u>.

<u>Using</u> (3.10), <u>we put for</u> $r \geq 1$, $s \geq 2$

$${}^{+}_{r}[12] = e_r^+, \quad {}^{-}_{1}[12] = e_1^-, \quad {}^{-}_{s}[12] = -\{e_1^+ \, e_s^+ \, e_1^-\},$$

$${}^{-}_{1}[13] = e_5^-, \quad {}^{-}_{s}[13] = -\{e_1^+ \, e_s^+ \, e_5^-\},$$

$${}^{+}_{1}[13] = \{{}^{-}_{3}[12] \; {}^{+}_{2}[12] \; {}^{-}_{4}[13]\}, \quad {}^{+}_{s}[13] = -\{{}^{+}_{1}[13] \; {}^{-}_{s}[13] \; {}^{-}_{1}[13]\},$$

$$[1] = c, \quad [2] = \{e_1^+ \, c \, e_1^-\}, \quad [3] = \{{}^{+}_{1}[13] \; c \; {}^{-}_{1}[13]\},$$

$${}^{\varepsilon}_{r}[32] = \{{}^{\varepsilon}_{1}[12] \; [1] \; {}^{\varepsilon}_{r}[31]\}.$$

<u>Then</u> $A = ([i];\ 1 \leq i \leq 3) \cup ({}^{\varepsilon}_{r}[ij];\ \varepsilon = \pm,\ 1 \leq r \leq 4,\ 1 \leq i < j \leq 3)$ <u>is an Albert grid</u>.

<u>Proof</u>. Applying Theorem 3.1 resp. Theorem 2.12 immediately shows the following two steps

1) $({}^{\varepsilon}_{r}[12], {}^{\varepsilon}_{r}[13];\ \varepsilon = \pm,\ 1 \leq r \leq 4)$ is a Bi-Cayley grid.

2) $({}^{+}_{1}[12], {}^{-}_{1}[12], \ldots, {}^{+}_{4}[12], {}^{-}_{4}[12], [1], -[2])$ is a quadratic form grid.

3) $[i] \perp {}^{\varepsilon}_{r}[jk]$

Indeed, $[1] \perp {}^{\varepsilon}_{r}[32]$, because $({}^{\varepsilon}_{1}[12], [1], {}^{\varepsilon}_{r}[31], {}^{\varepsilon}_{r}[32])$ forms a quadrangle, and $[i] \perp {}^{\varepsilon}_{r}[1k]$ for $i = 2,3$ since $[i] = \{{}^{+}_{1}[1i] \; [1] \; {}^{-}_{1}[1i]\} \in V_0({}^{\varepsilon}_{r}[1k])$.

4) $\{[i] \; {}^{\varepsilon}_{r}[ij] \; {}^{-\varepsilon}_{s}[ik]\} = 0 = \{[i] \; {}^{\varepsilon}_{r}[ij] \; {}^{\varepsilon}_{r}[ik]\}$ for $i = 1,2$.

By Lemma I.3.7.a it is enough to prove

$$\{[1]\ {}^{\varepsilon}_{r}[12]\ {}^{-\varepsilon}_{s}[13]\} = 0 = \{[1]\ {}^{\varepsilon}_{r}[12]\ {}^{\varepsilon}_{r}[13]\} \text{ and}$$

$$\{[2]\ {}^{\varepsilon}_{r}[21]\ {}^{-\varepsilon}_{s}[23]\} = 0 = \{[2]\ {}^{\varepsilon}_{r}[21]\ {}^{\varepsilon}_{r}[23]\}$$

These equations are easy consequences of 3) and the formula (I.2.3) for the quadrangle $([1],\ {}^{\varepsilon}_{r}[12],\ [2],\ {}^{-\varepsilon}_{r}[12])$.

5) $({}^{+}_{1}[13],\ {}^{-}_{1}[13],\ \ldots,\ {}^{+}_{4}[13],\ {}^{-}_{4}[13],\ [1],\ -[3])$ is a quadratic form grid.

By Theorem I.1.13 and 1)-4) the automorphism $B({}^{+}_{1}[12]+{}^{+}_{1}[13],\ {}^{+}_{1}[12]+{}^{+}_{1}[13])$ maps $\vec{x} = ({}^{+}_{1}[13],\ {}^{+}_{2}[13],\ \ldots,\ {}^{+}_{4}[13],\ [1])$ to $({}^{+}_{1}[12],\ {}^{+}_{2}[12],\ \ldots,\ {}^{+}_{4}[12],\ -[1])$ and $\vec{y} = ({}^{-}_{1}[13],\ {}^{+}_{2}[13],\ \ldots,\ {}^{+}_{4}[13],\ [1])$ to $({}^{-}_{1}[12],\ {}^{+}_{2}[12],\ \ldots,\ {}^{+}_{4}[12],\ -[1])$, hence by assumption ii) $\vec{x}$ and $\vec{y}$ are pure collinear systems. Applying now Theorem 2.12 proves 5).

6) The equations of 4) also hold for $i = 3$.

Indeed, knowing 5) the proof of 4) now also works for $i = 3$.

7) $({}^{\varepsilon}_{r}[21],\ {}^{\varepsilon}_{r}[23];\ \varepsilon=\pm,\ 1\leq r\leq 4)$ is a Bi-Cayley grid.

8) $({}^{+}_{1}[23],\ {}^{-}_{1}[23],\ \ldots,\ {}^{+}_{4}[23],\ {}^{-}_{4}[23],\ [2],\ -[3])$ is a quadratic form grid.

This will follow by applying two automorphisms. The first one is $\alpha = B([1]+{}^{+}_{1}[12],\ [1]+{}^{+}_{1}[12])$. Using 1)-4) and Theorem I.1.13 it is straightforward to compute that α maps

$$\vec{v} = ([1],\ {}^{+}_{1}[12],\ {}^{-}_{1}[12],\ {}^{\varepsilon}_{r}[12],\ {}^{+}_{1}[13],\ {}^{-}_{1}[13],\ {}^{+}_{r}[13],\ {}^{-}_{r}[13],\ [2],$$
$$ {}^{+}_{1}[32],\ {}^{-}_{1}[32],\ {}^{+}_{r}[32],\ {}^{-}_{r}[32],\ [3])$$

to $({}^{+}_{1}[12],\ [1],\ -[2],\ -{}^{\varepsilon}_{r}[12],\ -{}^{+}_{1}[13],\ -{}^{+}_{1}[32],\ {}^{+}_{r}[32],\ -{}^{-}_{r}[13],\ -{}^{-}_{1}[12],$
$\quad -{}^{-}_{1}[13],\ {}^{-}_{1}[32],\ {}^{+}_{r}[13],\ {}^{-}_{r}[32],\ [3])$.

In the same way it follows that $\beta = B({}^{+}_{1}[12]+[2],\ {}^{+}_{1}[12]+[2])$ maps $\vec{v}$ to $(-{}^{-}_{1}[12],\ [2],\ -[1],\ -{}^{\varepsilon}_{r}[12],\ -{}^{-}_{1}[32],\ {}^{-}_{1}[13],\ {}^{+}_{r}[13],\ {}^{-}_{r}[32],\ {}^{+}_{1}[12],$
$\quad -{}^{-}_{1}[23],\ -{}^{+}_{1}[13],\ -{}^{+}_{r}[32],\ -{}^{-}_{r}[13],\ [3])$.

So $\beta\circ\alpha$ maps the Bi-Cayley grid $({}^{\varepsilon}_{r}[12],\ {}^{\varepsilon}_{r}[13])$ to $(-{}^{\varepsilon}_{r}[21],\ {}^{\varepsilon}_{r}[23])$, which implies (7), and the quadratic form grid $({}^{+}_{1}[13],\ {}^{-}_{1}[13],\ \ldots,\ [1],\ -[3])$ to $({}^{+}_{1}[23],\ {}^{-}_{1}[23],\ \ldots,\ [2],\ -[3])$.

9) $$ {}^{\varepsilon}_{r}[32] = \{{}^{\varepsilon}_{1}[12]\ [1]\ {}^{\varepsilon}_{r}[31]\} = \{[2]\ {}^{-\varepsilon}_{1}[12]\ {}^{\varepsilon}_{r}[31]\}$$
$$\text{(for } r > 2) = \{[2]\ {}^{-\varepsilon}_{r}[12]\ {}^{\varepsilon}_{1}[13]\} = \{{}^{\varepsilon}_{r}[12]\ [1]\ {}^{\varepsilon}_{r}[13]\}.$$

The second equality follows from the first by (I.2.3), and for $r > 2$ we have in the same way $\{[2]\ {}^{-\varepsilon}_{1}[12]\ {}^{\varepsilon}_{r}[31]\} = -\{{}^{\varepsilon}_{r}[13]\ {}^{-\varepsilon}_{1}[12]\ [2]\} = \{{}^{\varepsilon}_{1}[13]\ {}^{-\varepsilon}_{r}[12]\ [2]\} = \{{}^{\varepsilon}_{1}[13]\ [1]\ {}^{\varepsilon}_{r}[12]\}$.

10) $({}^{\varepsilon}_{r}[31],\ {}^{\varepsilon}_{r}[32];\ \varepsilon=\pm,\ 1<r<4)$ is a Bi-Cayley grid.

We first show that $({}^{+}_{r}[31],\ 1<r<4) \cup ({}^{-}_{1}[31],\ {}^{-}_{1}[32])$ fulfills the assumptions of Theorem 3.1. We already know ${}^{-}_{1}[31] \perp {}^{+}_{1}[31]$ and that $({}^{+}_{r}[31],\ 1<r<4)$ is a pure collinear system. The relations ${}^{+}_{1}[31] \perp {}^{-}_{1}[32]$ and ${}^{-}_{1}[31] \top {}^{-}_{1}[32] \top {}^{+}_{r}[31]$ for $r > 2$ are proved by the usual methods (compare 2) in the proof of Theorem 3.1). Therefore $({}^{+}_{r}[31];\ 2<r<4) \cup ({}^{-}_{1}[31],\ {}^{-}_{1}[32])$ is a collinear system. It is pure as soon as $\{{}^{+}_{1}[13]\ {}^{+}_{r}[13]\ {}^{+}_{1}[23]\} = 0 = \{{}^{+}_{r}[13]\ {}^{+}_{s}[13]\ {}^{+}_{1}[23]\}$ for $r,s > 2$, $r \neq 2$. But these equations follow from 7) and $L({}^{+}_{1}[13],\ {}^{+}_{r}[13]) = L({}^{+}_{1}[12],\ {}^{+}_{r}[12])$ resp. $L({}^{+}_{r}[13],\ {}^{+}_{s}[13]) = L({}^{+}_{r}[12],\ {}^{+}_{s}[12])$. Thus the assumptions of Theorem 3.1 are fulfilled and 10) follows, if for $r > 2$

α) ${}^{-}_{r}[31] = -\{{}^{+}_{1}[31]\ {}^{+}_{r}[31]\ {}^{-}_{1}[31]\}$,

β) ${}^{-}_{r}[32] = -\{{}^{+}_{1}[31]\ {}^{+}_{r}[31]\ {}^{-}_{1}[32]\}$,

γ) ${}^{+}_{1}[32] = \{{}^{-}_{3}[31]\ {}^{+}_{2}[31]\ {}^{-}_{4}[32]\}$ and

δ) ${}^{+}_{r}[32] = -\{{}^{+}_{1}[32]\ {}^{-}_{r}[32]\ {}^{-}_{1}[32]\}$.

Here α) holds by 1), and β)-δ) follow from (I.1.14) and the substitutions ${}^{-}_{1}[32] = \{{}^{-}_{1}[12]\ [1]\ {}^{-}_{1}[31]\}$, ${}^{-}_{4}[32] = \{{}^{-}_{4}[12]\ [1]\ {}^{-}_{1}[13]\}$ by 9), ${}^{-}_{r}[32] = \{[2]\ {}^{+}_{r}[32]\ [3]\}$.

We pause to note that so far we have shown (3.11), (3.12) and a), b) of (3.16)'. So it remains to prove (3.13)-(3.15) and (3.16)'c)-f).

11) $({}^{\varepsilon}_{1}[ij],\ {}^{\varepsilon}_{r}[kj],\ {}^{\varepsilon}_{r}[ki],\ [i])$ is a quadrangle, i.e. (3.13) holds.

For $r > 2$ we know that $({}^{\varepsilon}_{r}[kj],\ {}^{\varepsilon}_{r}[ki],\ {}^{\varepsilon}_{1}[ki],\ {}^{\varepsilon}_{1}[kj])$ is a quadrangle, hence $L({}^{\varepsilon}_{r}[kj],\ {}^{\varepsilon}_{r}[ki]) = L({}^{\varepsilon}_{1}[kj],\ {}^{\varepsilon}_{1}[ki])$ and all we have to show is the case $r = 1$. But $({}^{\varepsilon}_{1}[ij],\ {}^{\varepsilon}_{1}[kj],\ {}^{\varepsilon}_{1}[ki],\ [i])$ is a quadrangle iff $({}^{\varepsilon}_{1}[ij],\ {}^{\varepsilon}_{1}[kj],\ [k],\ {}^{-\varepsilon}_{1}[ki])$ is a quadrangle iff (by shifting) $({}^{-\varepsilon}_{1}[ki],\ {}^{-\varepsilon}_{1}[ji],\ -{}^{\varepsilon}_{j}[jk],\ [k])$ is a quadrangle. So it is enough to do "i=1". By symmetry the problem is reduced to the family $({}^{\varepsilon}_{1}[12],{}^{\varepsilon}_{1}[32],\ {}^{\varepsilon}_{1}[31],\ [i])$ which, by definition of ${}^{\varepsilon}_{1}[32]$, clearly is a quadrangle.

12) $({}^{\varepsilon}_{r}[ij],\ {}^{\varepsilon}_{1}[kj],\ {}^{-\varepsilon}_{r}[ki],\ -[i])$, $r > 2$, is a quadrangle, i.e. (3.14) holds.

By (3.11) we know that $({}^{-\varepsilon}_{1}[ji],\ -{}^{\varepsilon}_{r}[jk],\ {}^{\varepsilon}_{1}[jk],\ {}^{-\varepsilon}_{r}[ji])$ is a quadrangle, hence $L({}^{\varepsilon}_{1}[ij],\ {}^{\varepsilon}_{r}[kj]) = -L({}^{-\varepsilon}_{r}[ij],\ {}^{-\varepsilon}_{1}[kj])$ and 11) implies 12).

13) (3.15) holds: $({}^{\varepsilon}_{r}[ij],\ {}^{-\varepsilon}_{s}[kj],\ {}^{\varepsilon}_{t}[ki],\ \mathrm{sgn}\binom{234}{rst}[i])$ is a quadrangle. By (3.11) we know that $L({}^{-\varepsilon}_{s}[kj],\ {}^{\varepsilon}_{t}[ki]) = \mathrm{sgn}\binom{234}{rst}L({}^{\varepsilon}_{r}[kj],\ {}^{-\varepsilon}_{1}[ki])$, so $\{{}^{-\varepsilon}_{s}[kj]\ {}^{\varepsilon}_{t}[ki]\ [i]\} = \mathrm{sgn}\binom{234}{rst}\{{}^{\varepsilon}_{r}[kj]\ {}^{\varepsilon}_{1}[ik]\ [i]\} = \mathrm{sgn}\binom{234}{rst}\ {}^{\varepsilon}_{r}[ij]$.

14) (3.16)'c)-f) hold.
For c) we have $\{{}^{\varepsilon}_{1}[kj]\ {}^{\varepsilon}_{1}[ki]\ {}^{\varepsilon}_{r}[ij]\} = \{{}^{\varepsilon}_{1}[ij]\ [i]\ {}^{\varepsilon}_{r}[ij]\} = 0$ by 11) and 6), and in the same way d)-f) follow. ∎

With the next theorem we undertake the final step in our classification of grids:

Theorem 3.5. *Let E be a connected ortho-collinear grid which contains an orthogonal system (e_1,e_2) such that*
i) $\# E_{12}(e_1,e_2) > 6$ *and*
ii) $E_{00}(e_1,e_2) \neq \emptyset$.
Then E is associated to an Albert grid.

Proof. We put $E_{ij} = E_{ij}(e_1,e_2)$ and first show

1) $E_1(e_1) = E_{12} \cup E_{10}$ is a connected ortho-collinear grid.
The only non-obvious statement being connectedness, assume $f,g \in E_1(e_1)$ such that $f \perp g$. Let $\{f\ e_1\ g\} \approx d \in E_0(e_1)$. Then $f,g \in E_{12}(e_1,d)$ and $\# E_{12}(e_1,d) > 6$ by Theorem 2.4.c). Now Lemma 2.10.b shows the existence of $h \in E_{12}(e_1,d) \cap E_{12}(f,g)$, in particular $f \top h \top g$.

By Lemma 3.2 we may choose an orthogonal pair $(f_{10},\bar{f}_{10}) \subset E_{10}$. To every $c \in E_{12} \cap E_j(f_{10})$, $j = 1,0$ there exists a unique $\bar{c} \in E_{12} \cap E_{1-j}(f_{10})$ with $\bar{c} \approx \{e_1\ c\ e_2\}$, and similarly for $\bar{f}_{10}$. This in particular implies $\#(E_{12} \cap E_{12}(f_{10},\bar{f}_{10})) = \#(E_{12} \cap E_{00}(f_{10},\bar{f}_{10}))$ but the latter set is empty by Lemma 2.1. These arguments together with Lemma 3.2 applied to $E_1(e_1)$ show the first part of

2) $E_{12} = (E_{12} \cap E_{10}(f_{10},\bar{f}_{10})) \cup (E_{12} \cap E_{20}(f_{10},\bar{f}_{10}))$ and both components have rank 1. Moreover $\#(E_{12} \cap E_{i0}(f_{10},\bar{f}_{10})) = 4$. Since $\# E_{12} > 6$ there exist $c_i \in E_{12} \cap E_{10}(f_{10},\bar{f}_{10}))$, $i = 1,2,3$, and $\bar{c}_i \in E_{12} \cap E_{20}(f_{10},\bar{f}_{10}))$ such that $\bar{c}_i \approx \{e_1\ c_i\ e_2\}$. It follows that $h = \{\{c_1\ \bar{c}_2\ \bar{f}_{10}\} f_{10}\ c_3\}$ is a non-zero tripotent lying in $V_{12} \cap V_1(c_i) \cap V_1(\bar{f}_{10})$. Thus there exists a fourth element $\bar{c}_4 \in E_{12} \cap E_{20}(f_{10},\bar{f}_{10})$ such that $\bar{c}_4 \approx h$. For every $g \in E_{12} \cap E_{10}(f_{10},\bar{f}_{10})$, $g \neq c_i$, we have

$g \perp \bar{c}_4$. Therefore the uniqueness assertion of Lemma 2.10 implies $\#(E_{12} \cap E_{i0}(f_{10},\bar{f}_{10})) = 4$.

3) rank $E = 3$, $\# E = 27$.
For the orthogonal pair $(c_1,\bar{c}_1) \subset E_{12}$ constructed in 2) we have $\#[E_1(e_1)]_{12}(c_1,\bar{c}_1) \geq 6$, $[E_1(e_1)]_{00}(c_1,\bar{c}_1) = \emptyset$ and $f_{10} \in [E_1(e_1)]_{10}(c_1,\bar{c}_1)$. Thus the assumptions of Theorem 3.3 are fulfilled for $E_1(e_1)$ in place of E and thus $\# E_1(e_1) = 16$, i.e. $\# E_{10} = 8$. On the other hand, let $(e_1,e_2,\dots,e_r)$ be an orthogonal system. We know $r \geq 3$, $E_{10} \supset \dot{\bigcup}_{i\geq 3} E_{1i}(e_1,\dots,e_r)$ and $\# E_{i1}(e_1,\dots,e_r) = 8$ by $\# E_{12} = 8$ and Theorem 2.4.c. Therefore $r = 3$ and $\# E = 3+3\cdot 8 = 27$ follows.

We now want to apply Theorem 3.4 and put $c = e_1$, $e_i^+ = c_i$ for $i = 1,\dots,4$, $e_1^- = \bar{c}_1$ and $e_5^- = \bar{f}_{10}$. Because E is a grid the assumptions of Theorem 3.4 are fulfilled. Let A be the Albert grid constructed there. The construction of the elements of A shows that A is associated to a subfamily of E. But because $\# A = 27 = \# E$ we actually have $A \approx E$. ∎

We have arrived at the end of the grid classification. The remark at the end of §2 and Theorems 3.3 and 3.5 imply the following résumé:

<u>Grid Classification Theorem.</u> <u>Every connected grid E is associated to exactly one of the following seven standard grids:</u>

I) <u>rectangular grid</u> $R(I,J)$, $\# I \geq 1$, $\# J \geq 1$,
II) <u>symplectic grid</u> $S(I)$, $\# I \geq 5$,
III) <u>hermitian grid</u> $H(I)$, $\# I \geq 2$,
IV.a) <u>odd-dimensional quadratic form grid</u> $\mathcal{Q}_o(I)$, $\# I \geq 2$,
b) <u>even-dimensional quadratic form grid</u> $\mathcal{Q}_e(I)$, $\# I \geq 3$,
V) <u>Bi-Caley grid</u> B,
VI) <u>Albert grid</u> A.

We remark that the numeration of the grids reflects the usual numeration of simple Jordan pairs ([31]12.12). More details about this can be found in IV §1.

§4 Construction of Peirce-dense atomic grids with minimal tripotents

In Corollary I.5.11 we considered a special class of Jordan triple systems (namely those with porperties (4.1) - (4.3) below) and proved the existence of a Peirce-dense atomic grid with minimal tripotents. In this section we will improve this result by explicitly constructing such a grid which has better density-properties. These grids will be used in the classification of Hilbert triples and JBW*-triples.

We investigate Jordan triple systems V having the following properties:

(4.1) If $e \in V$ is a minimal tripotent and $0 \neq h \in V_1(e)$ is another tripotent, then $e \top h$ or $e \dashv h$.

(4.2) Each two minimal collinear tripotents of V are rigidly imbedded.

(4.3) V contains an orthogonal system $\mathcal{O} = (e_i)_{i \in I}$ of minimal tripotents such that

a) $V_{(0)}(\mathcal{O}) = \bigcap_{i \in I} V_0(e_i) = 0$ and

b) if $\mathcal{E} \supset \mathcal{O}$ is an atomic closed cog with minimal tripotents, then $V_J(\mathcal{E})$ for $J \neq (0)$ contains a non-zero tripotent or vanishes.

We already came across such triple systems: In Corollary I.5.11 we proved that V contains a Peirce-dense atomic grid $\mathcal{E}^* \supset \mathcal{O}$ with minimal tripotents. However, Peirce-densedness does not seem to be sufficient in topological situations like Hilbert triples (see IV §2) or JBW*-triple (see IV §3) where one wants to prove the existence of a topologically dense grid or at least the existence of a grid with a topologically dense cover. In order to show this we will construct in this section a maximal $\mathcal{E}^* \supset \mathcal{O}$ with better density-properties.

To begin with, we first note that (4.3.b) is equivalent to the easier to handle condition (4.3'.b) where one does not assume closedness of $\mathcal{E}$:

(4.3'.b) If $\mathcal{E} \supset \mathcal{O}$ is an atomic cog with minimal tripotents and if $V_J(\mathcal{E}) \neq 0$ for $J \neq (0)$, then $V_J(\mathcal{E})$ contains a non-zero tripotent.

Indeed, by Theorem I.4.11 $\mathcal{E}$ is imbedded in a closed cog $\mathcal{E}_c$ with minimal tripotents (Lemma I.5.1), which is atomic (Lemma I.5.6) and has the same Peirce spaces as $\mathcal{E}$, whence (4.3.b) applied to $\mathcal{E}_c$, shows (4.3.b) $\rightarrow$ (4.3'.b).

Because we are working in an algebraic setting we will only be able to say something about the Peirce sum $PS_V(\mathcal{O})$. (The study of the imbedding $PS_V(\mathcal{O}) \to V$ will require additional structure, like topology.) The next step therefore is to show that $PS_V(\mathcal{O})$ inherits the properties (4.1)-(4.3) from V. Throughout this section we denote the Peirce spaces of $\mathcal{O}$ in the following familiar way ($i,j \in I$):

$$V_{ii} = V_2(e_i)\ ,\quad V_{00} = \bigcap_i V_0(e_i)$$

(4.4) $$V_{ij} = V_1(e_i) \cap V_1(e_j)$$

$$V_{0i} = V_1(e_i) \cap \bigcap_{j \neq i} V_0(e_j).$$

Thus the Peirce sum of $\mathcal{O}$ is given by

(4.5) $$PS_V(0) = \sum_{i,j} V_{ij}$$

where i,j varies in $I \cup \{0\}$ (we assume of course $0 \notin I$).

<u>Lemma 4.1.</u> <u>In any Jordan triple system is the Peirce sum of an orthogonal system an inner ideal.</u>

<u>Proof.</u> We want to show $P(PS_V(\mathcal{O}))V \subset PS_V(\mathcal{O})$ which will follow from $P(V_{\alpha\beta})V + P(V_{\alpha\beta},V_{\gamma\delta})V \subset PS(\mathcal{O})$ for all possible choices of $\alpha, \ldots, \delta \in I \cup \{0\}$. For all finite subsets $F \subset I$ the element $e_F = \sum_{i \in F} e_i$ is a tripotent with $V_2(e_F) = \sum_{i,j \in F} V_{ij}$ (apply Theorem I.1.10, noting that V_{ij} for $i,j \in F$ is also a Peirce space relative to $(e_i)_{i \in F}$). Because $V_2(e_F)$ is always an inner ideal ($P(x_2)V = P(P(e_F)\bar{x}_2)V = P(e_F)P(\bar{x}_2)P(e_F)V \subset P(e_F)V = V_2(e_F)$), we only need to consider products $P(V_{\alpha\beta})V$ and $P(V_{\alpha\beta},V_{\gamma\delta})V$ involving at most one facter from a $V_2(e_F)$. These are all of the following type ($i,j,k \in I, i,j,k,\neq$): $P(x_{i0}),P(x_{00})$ and $P(x_{i0},x_{00})$, $P(x_{ii},x_{i0})$, $P(x_{ii},x_{j0})$, $P(x_{ii},x_{00})$, $P(x_{ij},x_{i0})$, $P(x_{ij},x_{k0})$, $P(x_{ij},x_{00})$. But all these operators map V into $PS(\mathcal{O})$, since $Z = \bigcap_{\ell \neq i,j,k} V_0(e_\ell)$ is an inner ideal (every $V_0(e_\ell)$ is inner) and the Peirce decomposition of Z relative to (e_i,e_j,e_k) shows $Z = \sum_{\alpha,\beta = i,j,k,0} V_{\alpha\beta}$. ∎

<u>Decomposition Lemma 4.2.</u> <u>Assume</u> V <u>satisfies</u> (4.1) - (4.3). <u>Then</u> $PS_V(\mathcal{O})$ <u>also satisfies</u> (4.1) - (4.3). <u>Moreover,</u> $PS_V(\mathcal{O})$ <u>is a direct sum of ideals of</u> $PS_V(\mathcal{O})$:

(4.6) $$PS_V(\mathcal{O}) = \oplus_\alpha V_\alpha\ ,\qquad V_\alpha \lhd PS_V(\mathcal{O})$$

<u>where each</u> V_α <u>satisfies</u> (4.1) - (4.3) <u>and</u> (4.3.a) <u>in the following stronger form</u> (<u>for</u> V_α, $\mathcal{O}_\alpha = \mathcal{O} \cap V_\alpha$ in place of $V, \mathcal{O}$):

(4.3'.a) V <u>contains an orthogonal system</u> $\mathcal{O} = (e_i)_{i \in I}$ <u>of minimal tripotents such that</u> $V_{00}(\mathcal{O}) = 0$, $V_{ij}(\mathcal{O}) = V_1(e_i) \cap V_1(e_j) \neq 0$ ($i,j \in I, i \neq j$) <u>and</u> $V = PS_V(\mathcal{O})$.

<u>Every ideal</u> W <u>of</u> $PS_V(\mathcal{O})$ <u>respects the decompositions</u> (4.5) <u>and</u> (4.6):

(4.7) $$W = \sum_{i,j}(W \cap V_{ij}) = \sum_\alpha (W \cap V_\alpha) \qquad (\text{for } W \lhd PS_V(\mathcal{O}))$$

<u>For every proper ideal</u> W_α <u>of</u> V_α <u>and every</u> $V_{ij} \subset V_\alpha$ <u>the space</u> $W \cap V_{ij}$

<u>does not contain non-zero tripotents. The ideal V_α is simple iff V_α is nondegenerate and every $V_{ii} \subset V_\alpha$ is simple.</u>

<u>Proof.</u> I) Since $U = PS_V(\mathcal{O})$ is an inner ideal, every tripotent $c \in U$ satisfies $V_2(c) = P(c)V \subset U$, hence $V_2(c) = U_2(c)$ and in particular every minimal tripotent of U stays minimal in V. By this remark, (4.1) and (4.2) hold for U. Obviously, (4.3.a) holds too, and the validity of (4.3.b) follows from the observation that not only any cog $E \supset \mathcal{O}$ already lies in U, but also all Peirce spaces $V_J(E) \subset U$ and that therefore any atomic closed cog of U with minimal tripotents remains such a cog in V. If we have proven that $U = \oplus V_\alpha$ is a direct sum of ideals, it is obvious that every V_α inherits (4.1) - (4.3) from U. Since (4.3'.a) will be clear from our construction of the ideals V_α, we are left with proving the decomposition (4.6).

II) To this end we first show that any Peirce space $V_{ij} \neq 0$, $i,j \in I$, $i \neq j$, always contains a tripotent $h_{ij} \in V_{ij}$ such that $e_i \dashv h_{ij} \vdash e_j$. Indeed, by (4.3.b) there exists a tripotent $0 \neq h_{ij} \in V_{ij}$ and by (4.1) we have $e_i \dashv h_{ij}$ or $e_i \top h_{ij}$. Assume the latter, then also $e_j \top h_{ij}$ (since $e_i \top h_{ij}$ forces h_{ij} to be minimal, so $e_j \dashv h_{ij}$ gives a contradiction), thus $(e_i, h_{ij}, e_j, \{e_i h_{ij} e_j\})$ is a quadrangle and $e_i \dashv (h_{ij} + \{e_i h_{ij} e_j\}) \vdash e_j$. Now claim II) follows.

III) Next we prove for $i,j,k \in I$, $i,j,k \neq$

$$V_{ij} \neq 0 \text{ and } V_{jk} \neq 0 \rightarrow V_{ik} \neq 0$$

For, by II) we may choose $h_{ij} \in V_{ij}$ and $h_{jk} \in V_{jk}$ satisfying $e_i \dashv h_{ij} \vdash e_j \dashv h_{jk} \vdash e_k$. Since $h_{jk} \in V_1(e_i + e_j) = V_1(h_{ij})$ we have $h_{ij} \top h_{jk}$ and $h_{ik} = \{h_{ij}\ e_j\ h_{jk}\} \in V_{ik}$ is a non-zero tripotent by the Diamond Criterion I.2.7.

IV) By III), $i \sim j \leftrightarrow V_{ij} \neq 0$ defines an equivalence relation on I, and for any equivalence class α we put

$$V_\alpha = \oplus_{i \in \alpha} (V_{i0} \oplus \oplus_{j \in \alpha} V_{ij})$$

Then $U = \oplus_\alpha V_\alpha$, and it remains to show $V_\alpha \lhd U$, i.e.

$$\{U\ U\ V_\alpha\} + P(U)V_\alpha + P(V_\alpha)U \subset V_\alpha,$$

which is an exercise in Peirce multiplication rules.

$\{U\ U\ V_\alpha\} \subset V_\alpha$: By linearity we only have to consider $\{V_{rs} V_{si} V_{ij}\} \subset V_{rj}$ for $i,j \in \alpha_0 = \alpha \cup \{0\}$, $(i,j) \neq (0,0)$. If $i = 0$ we get $j \in \alpha$, so we may assume $r \notin \alpha_0$. Since $i = 0$ we have $s \neq 0$, so $V_{rs} \neq 0$ implies $V_{sj} = 0$, in particular V_{s0}, $V_{sr} \subset V_0(e_j)$, thus $\{V_{rs} V_{s0} V_{0j}\} =$ (by (I.1.24)) $\{V_{rs}\{e_j V_{j0} V_{s0}\} e_j\} = 0$ because $\{e_j V_{j0} V_{0s}\} \subset V_{js} = 0$. If $j = 0$ we have $i \in \alpha$, $s \in \alpha_0$. We may then assume $r \notin \alpha$. In case $s = 0$ we get $\{V_{r0} V_{0i} V_{i0}\} = \{\{V_{r0} V_{0i} e_i\} e_i V_{i0}\} = 0$ since $\{V_{r0} V_{0i} V_{ii}\} \subset V_{ri} = 0$ and in case $s \neq 0$ we have $s \in \alpha$, $r \in \alpha_0$, whence $r = 0$ and the product becomes $\{V_{0s} V_{si} V_{i0}\} \subset V_{00} = 0$. Finally, in case $i,j \in \alpha$ we are done if $r \in \alpha_0$.

Otherwise, $V_{si} \neq 0$ and $V_{rs} \neq 0$ forces $s = 0$: $\{V_{r0}V_{0i}V_{ij}\} = \{e_r\{e_rV_{r0}V_{0i}\}V_{ij}\} = 0$ because $\{e_rV_{r0}V_{0i}\} \subset V_{ri} = 0$.

The second condition $P(U)V_\alpha \subset V_\alpha$ splits into 2 cases: $P(V_{ri})V_{ij} \subset V_{rs}$ for $ri = js$ and $\{V_{ri}V_{ij}V_{js}\} \subset V_{rs}$, and for the third condition $P(V_\alpha)U \subset V_\alpha$ it remains to consider $P(V_{ij})V_{jm} \subset V_{in}$ for $ij = mn$, and always $V_{ij} \subset V_\alpha$. All these cases are treated using the same technique as in the verification of the first condition.

V) For an ideal W of U every $w \in W$ decomposes according to (4.5): $w = \sum_{ij} w_{ij}$. Because of $w_{ii} = P(e_i)^2 w \in W$ and for $i,j \in I$, $i \neq j$ $w_{ij} = \{e_i e_i \{e_j e_j w\}\} \in W$ we also have $w_{i0} = \{e_i e_i w\} - 2w_{ii} - \sum_{i,j \in I,\ i \neq j} w_{ij} \in W$ which implies (4.7). Now let W_α be a proper ideal of V_α, so $W_\alpha = \sum W_\alpha \cap V_{ij}$ by what we proved so far. We will show: if some $W_\alpha \cap V_{ij}$ contains a non-zero tripotent, then $W_\alpha = V_\alpha$. According to the definition of V_{ij} there are various cases to consider: If $c \in W_\alpha \cap V_{i0}$ is a non-zero tripotent, then $P(c)e_i \in V_{00} = 0$ implies $e_i \top c$ and therefore, by (4.2), $V_{ii} = \{ccV_{ii}\} \subset W_\alpha$. If $c \in W_\alpha \cap V_{ij}$, $i,j \in I$, $i \neq j$, is a non-zero tripotent, then either $e_i \top c$ or $e_i \dashv c$ and in both cases $V_{ii} \subset W_\alpha$ follows. A similar argument applies if $c \in W_\alpha \cap V_{ii}$. Hence we can always assume $V_{ii} \subset W_\alpha$. But then for every other $e_j \in V_\alpha$ we have $V_{jj} = P(h_{ij})V_{ii} \subset W_\alpha$, and therefore all $U_1(e_i) \subset W_\alpha$ implying $W_\alpha = V_\alpha$.

The implication "V_α simple → $V_{ii} \subset V_\alpha$ simple" is the main result of [39], and the implication "V_α simple → V_α nondegenerate" is clear: Every absolute zero divisor lies in the Jacobson radical which vanishes by simplicity and the existence of tripotents. Conversely, for every proper ideal W of V_α the space $W \cap V_{ii}$ is an ideal of $V_{ii} \subset V_\alpha$ whence $W \cap V_{ii} = 0$. For $x_{ij} \in W \cap V_{ij}$ for $i \neq j$ we therefore have $P(x_{ij})y_{ij}$ = (by(I.1.28)) $\{\{x_{ij}y_{ij}e_i\}e_i x_{ij}\} - \{e_i y_{ij} P(x_{ij})e_i\} = 0$, implying $P(x_{ij}) = 0$, so $x_{ij} = 0$, and $W = 0$ by (4.7). ∎

What follows is a series of lemmata in which we construct standard grids for the triple systems satisfying (4.1), (4.2), (4.3') = (4.3'.a) + (4.3'.b) and an additional condition, specific for each grid. In the end we will show that we considered all possibilities.

Hermitian Lemma 4.3. Assume V satisfies (4.1)-(4.3') with respect to $\mathcal{O} = (e_i)_{i \in I}$, $\#I > 2$, and in addition assume that

(4.8) there exists $m, n \in I$, $m \neq n$, such that V_{mn} does not contain minimal tripotents of V .

Then V is covered by an atomic hermitian grid $H(I)$ with minimal tripotents.

Proof. As a preliminary result we show for V satisfying (4.1) - (4.3'):

(I) $(4.8) \to V_{m0} = 0.$

Otherwise V_{m0} contains a non-zero tripotent e_{m0} satisfying $e_{m0} \top e_m$ since $P(e_{m0})e_m \in V_{00} = 0$. Relative to e_{m0}, V_{mn} decomposes into a Peirce 0- and a Peirce-1-part which are interchanged by $P(e_m,e_n)$. Any non-zero tripotent $c \in V_{mn} \cap V_1(e_{m0})$ satisfies $e_{m0} \top c$ since e_{m0} is minimal and $e_{m0} \dashv c$ is impossible because $V_2(c) \cap V_{m0} = 0$. Therefore $c \in V_{mn}$ is minimal, contradiction.

II. Fix $i \in I$. From step II in the proof of Lemma 4.2 we know there exist tripotents h_{ij} satisfying $e_i \dashv h_{ij} \vdash e_j$. Thus, for $j \neq k$, $h_{jk} \in V_1(e_j + e_k) = V_1(h_{ik})$, so by symmetry $h_{ij} \top h_{ik}$ and by Theorem 1.4 $h_{ii} = e_i$, h_{ij} for $i \neq j$ generate a hermitian grid $\mathcal{H} = \mathcal{H}(I)$. By construction $\{h_{jj};\ j \in I\} \approx \mathcal{O}$ and every off-diagonal Peirce space V_{jk}, $j,k \in I$, $j \neq k$ contains a $h_{jk} \in \mathcal{H}(I)$, which is minimal in V_{jk} because there exists an automorphism of V mapping V_{jk} onto V_{mn} and h_{jk} onto h_{mn} (apply Theorem I.3.11 to $\mathcal{H}^{(2)}$), and h_{mn} is minimal in V_{mn} by (4.8): Every non-zero tripotent $c \in V_{mn}$ satisfies $e_m \dashv c \vdash e_n$ whence $c \approx h_{mn}$. This also shows that (4.8) holds for all Peirce spaces V_{ij}, $i \neq j$, $i,j \in I$, thus $V_{i0} = 0$ for all $i \in I$, and $\mathcal{H}(I)$ covers V. ∎

Quadratic Form Lemma 4.4. *Assume V satisfies (4.1) - (4.3') with respect to an orthogonal system $\mathcal{O} = (e_1,e_2)$ such that*

(4.8) $$V = V_{11} \oplus V_{12} \oplus V_{22}$$

Then (e_1,e_2) can be imbedded in a Peirce-dense atomic quadratic form grid $\mathcal{Q}_e(J)$ or $\mathcal{Q}_o(J)$ with minimal tripotents such that $\{e_j^+;\ j \in J\}$ is a maximal collinear system in V.

Proof. Let $C \subset V_{11} \oplus V_{12}$ be a maximal collinear system containing e_1 (note $C = \{e_1\}$ is possible, in which case V is covered by a triangle $\mathcal{Q}_0(1)$). By (4.2) C is pure and by Theorem 2.12 $C \cup \{e_2\}$ generates an even-dimensional quadratic form grid $\mathcal{Q}_e(I) = \{e_i^\varepsilon; i \in I, \varepsilon = \pm\}$ consisting of minimal tripotents. Since $V_2(e_i^\varepsilon) = V_0(e_i^{-\varepsilon}) \subset V_1(e_j^\delta)$, $i \neq j$, either $\mathcal{Q}_e(I)$ is Peirce-dense or $V_1 = \bigcap_{i \in I} (V_1(e_i^+) \cap V_1(e_i^-)) \neq 0$, hence, by maximality of C, V_1 contains a V_1-minimal tripotent e_o satisfying $e_o \vdash e_i^\varepsilon$. Then $\mathcal{Q}_e(I) \cup \{e_o\}$ is associated to a Peirce-dense atomic odd-dimensional quadratic form grid. ∎

Rectangular Lemma 4.5. *Assume V satisfies (4.1) - (4.3') with respect to $\mathcal{O} = (e_i)_{i \in I}$, $\#I > 2$, and in addition*

some V_{ij}, $i,j \in I$, $i \neq j$, is covered by two orthogonal

(4.10) *minimal tripotents*

$$V_{ij} = V_2(e_{ij}) \oplus V_2(e_{ji}),\ e_{ij} \perp e_{ji}.$$

Then there exists a Peirce-dense atomic grid $R(I,J)$ with minimal tripotents such that $I \subset J$ and

a) $\mathcal{O} \approx \{e_{kk} \in R(I,J); k \in I\}$

b) the Peirce spaces of $R(I,J)$ lying in V_{kl}, $k,l \in I$, $k \neq l$, are $V_2(e_{kl})$ and $V_2(e_{lk})$, in particular $V_2(\mathcal{O}) = \sum_{k,l \in I} V_{kl}$ is covered by $R(I,I) \subset R(I,J)$.

c) the Peirce spaces of $R(I,J)$ lying in V_{k0} $(k \in I)$ are $V_2(e_{kl}), l \in J \setminus I$,

d) $R_{k0} = V_{k0} \cap R(I,J)$ is a maximal collinear family in V_{k0}.

Proof. In the following we fix $i,j \in I$.

I) Because $P(e_i + e_j)e_{ij} \in V_2(e_{ji})$ we might as well assume $e_{ji} = \{e_i e_{ij} e_j\}$, thus $(e_i, e_{ij}, e_j, e_{ji})$ is a quadrangle. Denoting its Peirce spaces by $V_{(mnpq)}$ we have $V_{i0} = (V_{i0} \cap V_{(1100)}) \oplus (V_{i0} \cap V_{(1001)})$. We want to show that only one of these summands can be non-zero, so assume otherwise. Then there exist non-zero tripotents $f \in V_{i0} \cap V_{(1100)}$ and $g \in V_{i0} \cap V_{(1001)}$. Because every non-zero tripotent in V_{i0} is collinear to e_i, we have $V_{i0} = V_{i0} \cap V_2(f) \oplus V_{i0} \cap V_1(f)$. Since $P(f)g \in V_{-1}(e_{ji}) = 0$, $g \in V_1(f)$ and by symmetry $g \top f$. Then $\{gfe_{ij}\} = -T_{g,f}(e_{ij}) \neq 0$ by Theorem I.1.13, but $\{gfe_{ij}\} \in V_1(e_{ij}) \cap V_1(e_{ji}) \cap V_{ij} = 0$, contradiction. So we may assume $V_{i0} \subset V_1(e_{ij})$.

II) Let $C_{i0} \subset V_{i0}$ be a maximal collinear system. Then $(e_i, e_{ij}) \top C_{i0}$ and we can extend $(e_i, e_{ij}) \cup C_{i0}$ to a maximal collinear system $C \subset \cup_{k \in I \cup \{0\}} V_{ik}$. We claim $\#(C \cap V_{ik}) = 1$ for every $k \in I$. This is clear for $k = j$, so let $k \neq i,j$ and assume $C \cap V_{ik} = \emptyset$. As in I) V_{ik} splits relative to $(e_i, e_{ij}, e_j, e_{ji})$: $V_{ik} = V_{ik} \cap V_{(1100)} \oplus V_{ik} \cap V_{(1001)}$. The automorphism $P(e_i + e_j + e_k)|V_2(e_i + e_j + e_k)$ fixes e_i, e_j and exchanges e_{ij} and e_{ji}, hence exchanges the 2 summands of V_{ik}, thus in particular $V_{ik} \cap V_{(1100)} \neq 0$ and by (4.3'.b) there exists a tripotent $e_{ik} \in V_{ik} \cap V_{(1100)}$, $e_{ik} \neq 0$. We want to show $e_{ik} \top C$ (then we can extend C, so by maximality of C, our assumption $C \cap V_{ik} = \emptyset$ was wrong). By the Peirce multiplication rules $P(e_{ik})e_{ij} = 0$ and therefore $e_{ij} \top e_{ik}$ by (4.1), in particular e_{ik} is minimal. Using (4.1) again, it is enough to show $e_{ik} \in V_1(c)$ for the remaining $c \in C$. We may assume $c \neq e_i$, say $c = e_{i\ell} \in V_{i\ell}$, $\ell \neq i,j,k$. Then $e_{i\ell} \top e_{ij} \top e_j \perp e_{i\ell}$ and we get a quadrangle $(e_{i\ell}, e_{ij}, e_j, e_{j\ell} = \{e_{i\ell} e_{ij} e_j\})$ with $e_{ik} \in V_{(100)}$ relative to $(e_{ij}, e_j, e_{j\ell})$, so $e_{ik} \in V_1(e_{i\ell})$ by (I.2.4), and, as noted before, $\#(C \cap V_{ik}) \geq 1$.

We now show $\#(C \cap V_{ik}) = 1$: We know there exist tripotents h_{ij}, h_{ik} satisfying $e_j \dashv h_{ij} \vdash e_i \dashv h_{ik} \vdash e_k$, $h_{ij} \top h_{ik}$. The exchange automorphism between h_{ij} and h_{ik}, restricted to $V_1(e_i)$, exchanges V_{ij}

and V_{ik}, so we have $V_{ik} = V_2(e_{ik}) \oplus V_2(e_{ki})$ for a tripotent $e_{ki} \perp e_{ik}$, in particular $(V_{ik})_1(e_{ik}) = 0$, so $\#(C \cap V_{ik}) = 1$.

III) We may now write $C = \{e_{ik}; k \in J\}$ ($e_i = e_{ii}$) for an index set $J \supset I$ such that $e_{ik} \in V_{ik}$ for $k \in I$. As shown in II) we have $V_{ik} = V_2(e_{ik}) \oplus V_2(e_{ki})$ for $e_{ki} \perp e_{ik}$. We claim that $\{e_{ki};\ k \in I\}$ is another collinear family satisfying $e_{i\ell} \perp e_{ki}$ for $\ell,k,i \neq$: Indeed, we have $e_{i\ell} \in V_1(e_i + e_k) = V_1(e_{ik} + e_{ki})$ and $e_{i\ell} \in V_1(e_{ik})$, so $e_{i\ell} \perp e_{ki}$. Similarly $e_{ki} \in V_1(e_{i\ell} + e_{\ell i}) \cap V_0(e_{i\ell})$, whence $e_{ki} \in V_1(e_{i\ell})$ and by symmetry $e_{ki} \top e_{\ell i}$. Since both collinear families $\{e_{ik};k \in J\}$ and $\{e_{ki};k \in I\}$ are pure, they generate, by Theorem 2.6, a rectangular grid $R = R(I,J)$ consisting of minimal tripotents.

IV) R satisfies a) and b), the latter because of (4.2): Minimal collinear tripotents are rigid-imbedded. This also implies part of c): Every $V_2(e_{k\ell})$, $\ell \in J \setminus I$, is a Peirce space of R. Conversely, assume the Peirce space $V_K(R) \subset V_{k0}$ does not contain any $e_{k\ell} \in R$. Because $V_{k0} = V_2(e_{k\ell}) \oplus V_{k0} \cap V_1(e_{k\ell})$ we have $V_K(R) \subset \cap_{\ell \in J\setminus I} V_1(e_{k\ell})$ and any non-zero tripotent in $V_K(R)$ would be collinear to R_{k0}. Thus c) follows from d).But d) is clear for $k = i$ by our choice of C_{i0}. For $k \neq i$ the product of the two exchange automorphisms determined by the collinear pairs (e_{ii},e_{ik}) and (e_{ik},e_{kk}) exchanges e_{ii} and e_{kk} and fixes $e_{\ell\ell}$, $\ell \neq i,k$, so it maps V_{i0} onto V_{k0}. Because $e_{i\ell}$, $\ell \notin I$, is mapped onto $e_{k\ell}$, d) follows. The remaining claims, Peirce-denseness and atomicity, follow from a) - c). ⊞

<u>Symplectic Lemma 4.6.</u> <u>Assume</u> V <u>satisfies</u> (4.1) - (4.3') <u>with respect to</u> $O = (e_i)_{i \in I}$, $\#I > 2$, <u>and in addition</u>
(4.11) <u>some</u> $V_{ij}, i,j \in I$, $i \neq j$, <u>is covered by a quadrangle</u>.
<u>Then</u> V <u>contains a covering symplectic grid</u> S <u>of minimal tripotents such that</u> O <u>is associated to a subfamily of</u> S.

<u>Proof.</u> It is suggestive to enumerate part of the grid and the Peirce spaces by natural numbers. So we assume that V_{12} is covered by the quadrangle S_{12}.

I) We claim that in case $V_{10} \neq 0$ there are 2 minimal collinear tripotents $f,g \in V_{10}$ such that $V_{10} = V_2(f) \oplus V_2(g)$ and $f \in V_{(1100)}(S_{12})$, $g \in V_{(0011)}(S_{12})$ for a suitable ordering of S_{12}. Indeed, V_{10} splits relative to S_{12}, and since $V_{(1111)}(S_{12}) = V_{11} \oplus V_{22}$, these Peirce spaces are of type $V_{(ijkl)}$ with $i+j+k+l = 2$. Therefore we may assume that there exists a non-zero minimal tripotent $f \in V_{10} \cap V_{(1100)}(S_{12})$ for $S_{12} = (f_{13},f_{14},f_{24},f_{23})$. Then $g = \{f_{23}f_{13}f\} \in V_{10} \cap V_{(0011)}(S_{12}) \cap V_1(f)$, whence (f,g) is a rigid-collinear pair. Thus $V_{10} = V_2(f) \oplus V_2(g) \oplus X$ with $X = V_{10} \cap V_{(11)}(f,g)$. The space X splits relative to the quadrangles $Q = (f_{13}, f, g, f_{23})$ and $S_{12} = (f_{13}, f_{14}, f_{24}, f_{23})$ where $X_{(i11\ell)}(Q) = X_{(ijk\ell)}(S_{12})$. By the Quadrangle Decomposition Theorem I.2.2 for Q we know $(i\ell) = (00)$ or $= (11)$. In the first case $(jk) = (11)$ and $L(f_{13}, f)$ maps $X_{(0110)}(S_{12}) = X_{(0110)}(Q)$ injectively into $V_{12} \cap V_{(1111)}(S_{12}) = 0$, in the second case $(jk) = (00)$ and $L(f_{14}, f)$ maps $X_{(1001)}(S_{12})$ injectively into $V_{12} \cap V_{(1111)}(S_{12}) = 0$. Thus $X = 0$ and $V_{10} = V_2(f) \oplus V_2(g)$. In case $\#I = 2$ the family $(f_{12} = e_1, f_{13}, f_{14}, f = f_{15}, f_{23})$ fulfills the assumptions of Theorem 2.9, thus generates a symplectic grid which covers V. Therefore, in what follows we may assume $\#I \geq 3$.

II) Similiar as in I) we decompose $U = V_{13}$ relative to S_{12} : $U = U_{(1100)} \oplus U_{(0011)} \oplus U_{(1001)} \oplus U_{(0110)}$. Since at least one of these spaces does not vanish we may rename S_{12} so that there exists a non-zero tripotent $h \in U_{(1100)}$ for $S_{12} = (f_{13}, f_{14}, f_{24}, f_{23})$. Then $f_{13} \top h \top f_{14}$ and $0 \neq \bar{h} = P(e_1 + e_3)h = \{e_1 h e_3\} \in U_{(0011)} \cap U_0(h)$. Therefore $(f_{13} + f_{24}) \top (h + \bar{h})$ and $\phi = L(h + \bar{h}, f_{13} + f_{24}) : V_{12} \to V_{13}$ is a Jordan triple isomorphism by Theorem I.1.13. In particular, we get $\phi f_{13} = \{h\, f_{13}\, f_{13}\} = h$, $\phi f_{24} = \bar{h}$ and $\phi f_{14} = \{\bar{h}\, f_{24}\, f_{14}\}$ because $\{h\, f_{13}\, f_{14}\} \in V_{13} \cap V_2(f_{14}) = 0$. Thus $\phi f_{14} \in V_1(f_{13}) \cap V_1(f_{14}) \cap V_{13} = U_{(1100)}$ is a non-zero tripotent such that $h \top \phi f_{14}$. Now $(f_{12} = e_1, f_{13}, f_{14}, f_{15} = h, f_{16} = \phi f_{14})$ is a pure collinear family and generates together with f_{23} a symplectic grid $\{f_{ij}; 1 \leq i < j \leq 6\}$ (by Theorem 2.9) We note that all f_{ij} lie in Peirce spaces relative to $\mathcal{O}$, e.g. $e_2 \approx f_{34}$, $e_3 \approx f_{56}$.

III) We assume that we are given a symplectic grid base $C = \{f_{23}\} \cup \{f_{1i};\ i \geq 2\}$ i.e. a family satisfying the assumptions of Thoerem 2.9, such that there are at least 5 tripotents f_{1i} and all f_{1i} and f_{23} lie in Peirce spaces V_{1j} relative to $\mathcal{O}$. We will show that we can enlarge this grid base in case $\#\,(C \cap V_{1k}) \leq 1$ for some $k \in I$ or if $C \cap V_{10} = \emptyset$.

We may assume $C \cap V_{ik} = \emptyset$. Because C generates a sympletic grid, for any two collinear tripotents $f, g \in \tilde{C} = C \setminus \{f_{23}, f_{12}\}$ there exists a tripotent h satisfying $f \perp h \perp g$ and $V_{1k} \subset V_1(f+h) = V_1(g+h)$ whence $L(f,f) = L(g,g)$ on V_{1k}, i.e. $V_{1k} = K_1 \oplus K_0$ where $K_1 = V_1(f_{1i})$, $K_0 = V_0(f_{1i})$ are independent of i. If $k \in I$ we have $e_k \perp f_{1i}$ whence $P(e_1 + e_k)$ interchanges K_1 and K_0 and both spaces are non-zero. Thus there exists a non-zero tripotent $c \in K_1$. As in II) we can now show that K_1 actually contains a collinear pair (c,d). Then $\{f_{1i}; i \geq 2\} \cup \{c,d\}$ is a pure collinear family, and $c \perp f_{23} \perp d$. Therefore $C \cup \{c,d\}$ generates a symplectic grid with all tripotents lying in Peirce spaces of $\mathcal{O}$.

If $k = 0$ we may clearly assume $V_{10} \neq 0$. Then there exists a non-zero tripotent $c \in K_i$, $i = 1$ or 0. If $i = 0$ we have $f_{23} \top c \perp f_{13}$, $d = \{f_{13} f_{23} c\} \neq 0$ is a tripotent in K_1 and $C \cup \{d\}$, generates a symplectic grid. If $i = 1$ we take $C \cup \{c\}$.

IV) Now let $\mathcal{C}$ be maximal among all the families considered in III) and let $\mathcal{S}$ be the symplectic grid generated by $\mathcal{C}$. As shown in III) $\#(\mathcal{C} \cap V_{1i}) = 2$ for all $i \in I$, $i \neq 1$, and $\#(\mathcal{C} \cap V_{10}) = 1$. It is then straightforward to show that $\mathcal{S} \cap V_{ij}, i,j \in I, i \neq j$, is a quadrangle covering V_{ij}, $\mathcal{O}$ is associated to a subfamily of $\mathcal{S}$ and $\mathcal{S} \cap V_{i0}$ (if $V_{i0} \neq 0$) is a collinear pair covering V_{i0}. ∎

We can now put together the various lemmata and prove the

Main Construction Theorem 4.7. *Let V be a Jordan triple system satisfying the conditions (4.1) - (4.3) with respect to the orthogonal system $\mathcal{O}$. Then V contains Peirce-dense atomic grid $\mathcal{E}$ with minimal tripotents, such that $\mathcal{O}$ is associated to a subfamily of $\mathcal{O}$, and the Peirce sum of $\mathcal{O}$ is a direct sum of ideals*

$$PS_V(\mathcal{O}) = \oplus_\alpha V_\alpha \quad , \quad V_\alpha \lhd PS_V(\mathcal{O})$$

where

$$\mathcal{E} = \cup_\alpha \mathcal{E}_\alpha \quad , \quad \mathcal{E}_\alpha = \mathcal{E} \cap V_\alpha$$

and each $\mathcal{E}_\alpha$ is a standard grid which in the case of a symplectic, hermitian, Bi-Cayley or Albert grid covers V_α, whereas in the case of a quadratic form grid or rectangular grid it has the density properties described in Lemma 4.4 resp. Lemma 4.5, in particular is Peirce-dense.

Proof. First we apply the Decomposition Lemma 4.2, so $PS_V(\mathcal{O}) = \oplus_\alpha V_\alpha$ for $V_\alpha \lhd PS_V(\mathcal{O})$, and note that $\mathcal{E} = \cup \mathcal{E}_\alpha$ will be a Peirce-dense atomic grid with minimal tripotents as soon as each $\mathcal{E}_\alpha$ has these properties. We therefore can restrict our attention to an ideal V_α, henceforth called V, which satisfies (4.1) - (4.3').

If $\mathcal{O} = (e)$, i.e. $\#I=1$, we choose a maximal collinear system $\mathcal{C} = (e_j)_{j \in J}$ containing e. Then $\mathcal{C}$ consists of minimal tripotents which are rigid-imbedded by (4.2). Therefore $\mathcal{C}$ is pure, i.e. a rectangular grid of type $R(1,J)$. It is also Perice-dense by (4.3.b) and maximality of $\mathcal{C}$. Hence, all the conclusions of the Rectangular Lemma 4.5 hold mutatis mutandis for $\mathcal{C}$.

We can now assume $\mathcal{O} = (e_i)_{i \in I}$, $\#I \geq 2$. Then the Hermitian Lemma 4.3 implies that we may also suppose

(α) all Peirce spaces V_{ij}, $i,j \in I$ with $i \neq j$, contain minimal tripotents of V.

Moreover, by the Quadratic Form Lemma 4.4 we can restrict our attention to the case when V in addition satisfies

(β) $\#I=2$ and $V_{i0} \neq 0$ for some i , or $\#I \geq 3$.

We fix $i,j \in I$, $i \neq j$, and if $\#I=2$, $V_{i0} \neq 0$ we choose a non-zero tripotent $c \in V_{i0}$. Such a c is minimal, whence $e_i \top c \perp e_j$. If $\#I \geq 3$

we choose a minimal tripotent $c \in V_{ik}$ for $k \in I$, $i,j,k \neq$ (possible because of (α)), so again $e_i \top c \perp e_j$. In both cases V_{ij} decomposes relative to c into a Peirce-0- and a Peirce-1-part which are exchanged by $P(e_i,e_j)$. Any non-zero tripotent $d \in V_{ij} \cap V_1(c)$ satisfies $c \top d$, so in particular is minimal and collinear to e_i and e_j. Let $\mathcal{C}$ be a maximal collinear system in $V_{ij} \cap V_1(c)$. Then, by Theorem 2.12, $\mathcal{C} \cup \{e_i,e_j\}$ generates an even-dimensional quadratic form grid $\mathcal{Q}_e \subset V_2(e_i+e_j)$, and $\mathcal{Q}_e \cup \{c\}$ is a connected ortho-collinear cog which, by Theorem I.4.11, is imbedded in a connected ortho-collinear closed cog $\mathcal{E}$. By Theorem I.4.12 $\mathcal{E}$ is a grid. Therefore, by the classification of grids, $\mathcal{E}$ is a rectangular, symplectic, Bi-Cayley or Albert grid yielding $\mathcal{Q}_e = \mathcal{Q}_e(m)$ for m = 1,2,3 or 4. By maximality of $\mathcal{C}$ and rigidity, $\mathcal{Q}_e$ covers V_{ij}. Thus the Rectangular Lemma 4.5 and the Symplectic Lemma 4.6 take care of m = 1 and m = 2 respectively. In the remaining cases $\mathcal{E}$ is a Bi-Cayley or Albert grid consisting of minimal tripotents and by finiteness of $\mathcal{E}$ we have $V = \oplus V_K(\mathcal{E})$. If there would be a non-zero Peirce space $V_K(\mathcal{E})$, $K \neq 0$, with $V_K(\mathcal{E}) \cap \mathcal{E} = \emptyset$ we could choose a tripotent $0 \neq d \in V_K(\mathcal{E})$ such that $d \top f$ or $d \perp f$ for all $f \in \mathcal{E}$ and $d \top f$ for at least one $f \in \mathcal{E}$. Then as before $\mathcal{E} \cup \{d\}$ would generate a bigger connected grid containing $\mathcal{E}$ which is impossible. Hence $\mathcal{E}$ covers V. ∎

§5 The 27 lines upon a cubic surface and the Albert grid.

In this section we show that the configuration of the 27 lines upon a smooth cubic surface in $\mathbb{P}^3$ is the same as the configuration of the Albert grid if one identifies intersecting lines with orthogonal tripotents. This connection is then used to give Jordan theoretic interpretations and proofs of some well-known theorems about the geometry of the 27 lines.

We will use the term <u>cubic surface</u> to denote the zero set of a polynomial F in projective space $\mathbb{P}^3 = \mathbb{P}^3(K)$, K an algebraically closed field, where F is homogeneous of degree 3 in 4 variables. We will always assume that the surface is smooth. The following famous classical result goes back to Cayley and Salmon:

<u>Theorem 5.1.</u> <u>There are exactly 27 straight lines upon a cubic surface. For each such surface the configuration of the 27 lines is the same.</u>

For the history of this theorem and related topics the reader is referred to [13], historical summary. A proof using "enumerative geometry" can be found in [14] §24, 25. Within the framework of algebraic geometry the theorem is for example proven in [45] §8D, and also in [12]V Theorem 4.9.

As a consequence of the theorem the configuration of the 27 lines can be found by looking at a special example. A particularly nice one is

<u>Example 5.2.</u> (The Fermat cubic surface, [12]V Ex. 4.16, [45] p.176). This is the surface given by the equation $x_0^3 + x_1^3 + x_2^3 + x_3^3 = 0$. Let $\eta \in K$ be a primitive 3rd root of unity. Then the 27 lines are $(k,\ell = 1,2,3)$:

$$\left[{1 \atop k\ell}\right] = \{(x_0:x_1:x_2:x_3) \in \mathbb{P}^3;\ x_0 + \eta^k x_1 = 0,\ x_2 + \eta^\ell x_3 = 0\}$$

$$\left[{2 \atop k\ell}\right] = \{(x_0:x_1:x_2:x_3) \in \mathbb{P}^3;\ x_0 + \eta^k x_2 = 0,\ x_1 + \eta^\ell x_3 = 0\}$$

$$\left[{3 \atop k\ell}\right] = \{(x_0:x_1:x_2:x_3) \in \mathbb{P}^3;\ x_0 + \eta^k x_3 = 0,\ x_1 + \eta^\ell x_2 = 0\}$$

The configuration of these lines follows from the following rules $(1 \leq i,k,\ell,p,q \leq 3)$:

$$\left[{i \atop k\ell}\right] \cap \left[{i \atop pq}\right] \neq \emptyset \quad \leftrightarrow \quad k = p \text{ or } \ell = q$$

$$\left[{1 \atop k\ell}\right] \cap \left[{2 \atop pq}\right] \neq \emptyset \quad \leftrightarrow \quad k - \ell \equiv m - n \pmod 3$$

$$\left[{1 \atop k\ell}\right] \cap \left[{3 \atop pq}\right] \neq \emptyset \quad \leftrightarrow \quad k + \ell \equiv m - n \pmod 3$$

$[^{2}_{k\ell}] \cap [^{3}_{pq}] \neq \emptyset \quad \leftrightarrow \quad k + \ell \equiv m + n \pmod 3$

It is worked out in the following table where a dot stands for "intersecting". The order of the lines from left to right and top to bottom is $([^{1}_{11}], [^{1}_{22}], [^{1}_{33}], [^{2}_{12}], [^{2}_{23}], [^{2}_{31}], [^{1}_{23}], [^{1}_{31}], [^{1}_{12}], [^{2}_{33}], [^{2}_{11}], [^{2}_{22}], [^{1}_{21}], [^{1}_{13}], [^{3}_{12}], [^{3}_{23}], [^{3}_{31}], [^{1}_{32}], [^{3}_{21}], [^{3}_{32}], [^{3}_{13}], [^{3}_{33}], [^{3}_{11}], [^{3}_{22}], [^{2}_{13}], [^{2}_{32}], [^{2}_{21}]\}$. This order is chosen such that the table coincides with [13] Intersection Table.

							•	•	•	•	•	•	•	•	•	•										
						•		•	•	•	•	•					•	•	•	•						
						•	•		•	•	•		•				•				•	•	•			
						•	•	•		•	•			•				•			•			•	•	
						•	•	•	•		•				•				•			•		•		•
						•	•	•	•	•						•				•			•		•	•
	•	•	•	•	•							•	•	•	•	•										
•		•	•	•	•							•					•	•	•	•						
•	•		•	•	•								•				•				•	•	•			
•	•	•		•	•									•				•			•			•	•	
•	•	•	•		•										•				•			•		•		•
•	•	•	•	•												•				•			•		•	•
•	•					•	•															•	•	•	•	•	•
•		•				•		•										•	•	•				•	•	•
•			•			•			•								•		•	•		•	•			•
•				•		•				•							•	•		•	•		•		•	
•					•	•					•						•	•	•		•	•		•		
	•	•					•	•						•	•	•								•	•	•
	•		•				•		•				•		•	•						•	•			•
	•			•			•			•			•	•		•					•		•		•	
	•				•		•				•		•	•	•						•	•		•		
		•	•					•	•			•			•	•			•	•						•
		•		•				•		•		•		•		•		•		•					•	
		•			•			•			•	•		•	•			•	•					•		
			•	•					•	•		•	•			•	•			•			•			
			•		•				•		•	•	•		•		•		•			•				
				•	•					•	•	•	•	•			•	•			•					

TABLE 5.3.

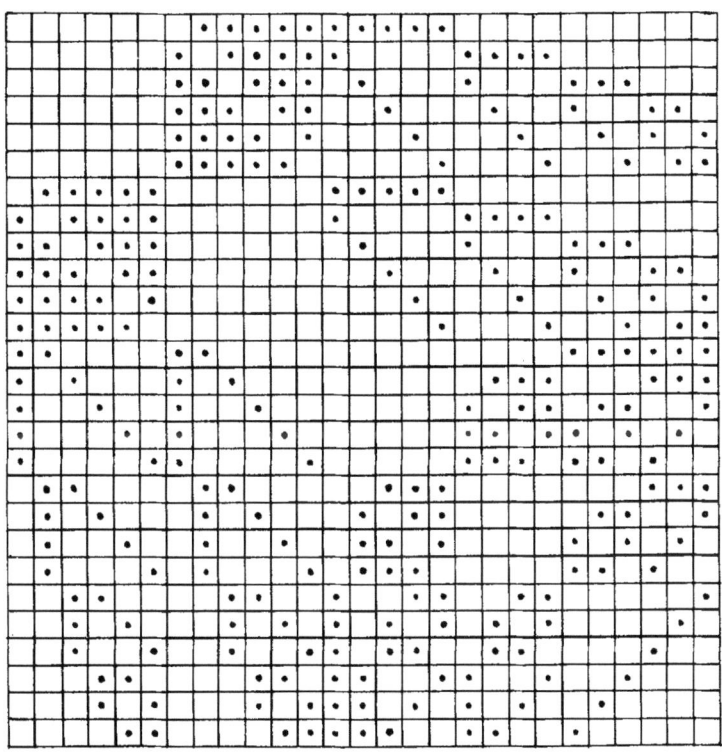

Remark 5.4. Every cubic surface can be obtained from $\mathbb{P}^2$ by blowing up six properly chosen points $P_1,\dots,P_6$ (see e.g. [12]). In this way one obtains a (non-unique) division of the 27 lines into

6 exceptional curves $E_1,\dots,E_6$,

6 strict transforms G_j, $1 \leq j \leq 6$, of the conic in $\mathbb{P}^2$ containing the five points P_i, $i \neq j$,

15 strict transforms F_{ij}, $1 \leq i < j \leq 6$, of the line in $\mathbb{P}^2$ containing P_i and P_j.

If one orders these lines in a natural way as $(E_1,E_2,E_3,E_4,E_5,E_6,G_1,G_2,G_3,G_4,G_5,G_6,F_{12},F_{13},F_{14},F_{15},F_{16},F_{23},F_{24},F_{25},F_{26},F_{34},F_{35},F_{36},F_{45},F_{46},F_{56})$ the intersection table of the lines becomes exactly the same as table 5.3.

Theorem 5.5. There exists a bijection between the 27 lines on a cubic surface and the 27 tripotents of an Albert grid (see §3) such that lines intersect iff the corresponding tripotents are orthogonal. An example for such a map is the map sending the lines $([{}^{1}_{11}], [{}^{1}_{22}], \dots, [{}^{2}_{21}])$, ordered as in table 5.3, onto the 27 tripotents ordered in the following way

$([1], {}^{+}_{1}[12], {}^{+}_{2}[12], {}^{+}_{3}[12], {}^{-}_{4}[12], {}^{-}_{4}[13], {}^{+}_{4}[13], {}^{+}_{4}[23], {}^{-}_{3}[23], {}^{-}_{2}[23],$
${}^{+}_{1}[23], [3], {}^{-}_{4}[23], {}^{+}_{3}[23], {}^{+}_{2}[23], {}^{-}_{1}[23], [2], {}^{+}_{3}[13], {}^{+}_{2}[13], {}^{-}_{1}[13],$
${}^{-}_{1}[12], {}^{+}_{1}[13], {}^{-}_{2}[13], {}^{-}_{2}[12], {}^{-}_{3}[13], {}^{-}_{3}[12], {}^{+}_{4}[12])$.

Proof. It is straightforward to compute the configuration of the ordered family of tripotents using (3.17). One obtains the same distribution of orthogonal signs as the distribution of dots in table 5.3. ∎

Remark 5.6. a) The fact, that the incidence relation on the Albert grid is given by orthogonality of the tripotents, is not surprising if one recalls the Jordan theoretic construction of the real projective Cayley plane due to Jordan and Freudenthal (see [19]IX.7): Points and lines of the plane are identified with the minimal idempotents in the exceptional formally real Jordan algebra and a point $P = c$ is incident with a line $\ell = d$ iff the idempotents c,d are orthogonal.

b) That there is a connection between the 27 lines on a cubic surface and exceptional Jordan structures was a widely believed conjecture. The author gratefully acknowledges discussions about this topic with J. Faulkner, J. Ferrar and O. Loos who each had their own point of view supporting the conjecture. In a recent preprint [22] K. Johnson and W. Lichtenstein interpreted this connection from a representation theoretic point of view.

The common ground for all efforts to understand the connection between the 27 lines on a cubic surface and the 27-dimensional exceptional Jordan structures (algebras, pairs and triples) seems to be that both are related to the exceptional group resp. Lie algebra of type E_6 (see [35]IV).

By Theorem 5.5 one can interpret results referring to the configuration of the 27 lines as results about the Albert grid. In the following we will give some examples for this procedure. Throughout, let A be an Albert grid which, by Theorem III.3.5, we may assume to be realized in an Albert triple V. We start out with:

(5.1) Each line is intersected by exactly 10 other lines, more precisely by exactly 5 pairs of intersecting lines.

This means for A:

(5.1') For each $c \in A$ the family $A_0(c) = \{d \in A;\ d \perp c\}$ consists of five orthogonal pairs. (We know even more: $A_0(c)$ is a grid of type IV_{10}.)

Both statements can be proven by examining table 5.3. A simplification is obtained by checking (5.1') for $c = [1]$ and then mapping any other $d \in A$ onto c by an automorphism of V leaving A invariant up to association. The existence of such an automorphism follows from Theorem 2.4.c . As a consequence, we have 27.10 = 270 orthogonal pairs of tripotents in A, taking order into account. This shows:

(5.2) There are - up to a change of order - 135 pairs of intersecting lines, in other words there are 135 pairs of orthogonal tripotents in A.

Another application of Theorem 2.4.c is:

(5.3) For each pair of intersecting lines there exists exactly one more line intersecting both, or: two orthogonal tripotents in A are pairwise orthogonal to exactly one other tripotent.

By (5.3) every maximal orthogonal system in A consists of 3 tripotents. Such a system is called a <u>frame</u>. Interpreted geometrically a frame consists of 3 pairwise intersecting lines on the surface, which determine a plane in $\mathbb{P}^3$ whose intersection with the surface are the given 3 lines. In [13] p.11 such a plane is called a <u>triple tangent plane</u>.

Lemma 5.7. There are exactly 45 triple tangent planes for a cubic surface, or: there are exactly 45 frames in A. They are (up to a change of order)

a) $([1], [2], [3])$

b) $([i], {}^{\varepsilon}_{r}[jk], {}^{-\varepsilon}_{r}[jk])$, $i,j,k \neq$ (12 types)

c) $({}^{\varepsilon}_{1}[12], {}^{-\varepsilon}_{1}[13], {}^{\varepsilon}_{1}[23])$ (2 types)

d) $({}^{\varepsilon}_{1}[ij], {}^{\varepsilon}_{r}[ik], {}^{-\varepsilon}_{r}[kj])$, $r > 2, i,j,k \neq$ (18 types)

e) $({}^{\varepsilon}_{r}[12], {}^{\varepsilon}_{s}[13], {}^{\varepsilon}_{t}[23])$, $r,s,t > 2$ and $r,s,t \neq$, (12 types)

Proof. The number of triple tangent planes = frames is easily computed: By (5.2) we have 2.135 = 6.45 orthogonal systems (c_1,c_2) each of which uniquely determines a frame by (5.3).

The only family containing an orthogonal pair $\{[i],[j]\}$ is $\{[1],[2],[3]\}$. Since $A_0([i]) = \{[j],[k],{}^{\varepsilon}_{r}[jk]\}$ the only families with exactly one $[i]$ are of type b). Thus in the sequel we only look at families which do not contain $[i]$ or a pair of type $\{{}^{\varepsilon}_{r}[ij],{}^{-\varepsilon}_{r}[ij]\}$. Assume the family in question contains ${}^{\varepsilon}_{1}[ij]$. Either it contains another tripotent of type ${}^{\mu}_{\varepsilon}[jk]$ (and then is of type c)) or we have type d). In the remaining cases only tripotents ${}^{\varepsilon}_{r}[ij]$ with $r > 2$ occur. These lead to type e). ∎

A more complicated geometric figure are triheders which by definition consist of 3 nonintersecting triple tangent planes (e_1,e_2,e_3), (c_1,c_2,c_3) and (f_1,f_2,f_3) such that $(e_i,c_i,f_i), i = 1,2,3,$ are again triple tangent planes. A geometric proof of the existence of triheders is given in [13] p.26, an algebraic one in

Lemma 5.8. Let (e_1,e_2,e_3) and (c_1,c_2,c_3) be two orthogonal systems in A such that $\{e_1,e_2,e_3\} \cap \{c_1,c_2,c_3\} = \emptyset$.

a) There exists a unique $c_{\sigma(i)}$ with $e_i \perp c_{\sigma(i)}$.

b) Let f_i be the unique element in A such that $(e_i,c_{\sigma(i)},f_i)$ is an orthogonal system. Then (f_1,f_2,f_3) is an orthogonal system, too.

Proof. a) Denoting by V_{ij} the Peirce spaces of V relative to (e_1,e_2,e_3) every c_i lies in a Peirce space V_{jk}, which, by assumption, is off-diagonal. Assume $c_1,c_2 \in V_{ij}$. Then $c_1 + c_2$ and $e_i + e_j$ are associated, hence have the same Peirce spaces. In particular, $c_3 \in V_0(c_1+c_2) = V_0(e_i+e_j) = V_2(e_k)$ for $i,j,k \neq$, whence $c_3 = e_k$, a contradiction. Thus each off-diagonal Peirce space contains exactly one c_i, which implies a).

b) For simpler notation assume $\sigma = \mathrm{Id}$. Then $c_i \in V_{jk}$ $(i,j,k \neq)$ for all i and $V_{jj} \oplus V_{kk} \subset V_1(c_i)$. Hence $f_i \in V_0(e_i) \cap V_0(c_i) \subset V_{jk}$, and we can apply a) to the two orthogonal systems (e_i,c_i,f_i), (e_j,c_j,f_j), $j \neq k$. Since $e_i \perp e_j$, $c_i \perp c_j$ we have $f_i \perp f_j$ and b) is proven. ∎

Finally we look at the famous configuration of a <u>double six</u> which is a family of 12 lines in $\mathbb{P}^3$, written as $(f_1^+,\dots,f_6^+;f_1^-,\dots,f_6^-)$, such that f_j^ε intersects exactly with $f_k^{-\varepsilon}$, $k \neq j$. An example of a double six is given by the first 12 lines in table 5.3. In fact, the notion of a double six was introduced by Schläfli in 1854 while studying the 27 lines. It leads to a very useful notation for the 27 lines: Starting out with a double six $(f_1^+,\dots,f_6^-)$ among the 27 lines, one obtains for all distinct i,j a pair of planes, (f_i^+,f_j^-) and (f_i^-,f_j^+), whose intersection line g_{ij} meets all 4 lines $f_i^\pm$, $f_j^\pm$ and hence lies entirely on the surface. In this way one obtains the remaining 15 lines.

<u>Schläfli's Double Six Theorem 5.9.</u> <u>Given 5 pairwise non-intersecting lines</u> $f_1^+,\dots,f_5^+$ <u>in</u> $\mathbb{P}^3$ <u>and a sixth line</u> f_6^- <u>intersecting them</u>. <u>Then</u> $(f_1^+,\dots,f_5^+,f_6^-)$ <u>can be completed to a double six</u>.

This is of course a theorem in projective geometry and can be proven in that context. Moreover, Schläfli's proof and subsequently given proofs (see [13]6 for the history of the theorem) made use of the fact that $f_1^+,\dots,f_5^+$, f_6^- can be imbedded in a cubic surface: Choose 4 points on f_6^- and 3 points on each f_i^+, in total 19 points. Since a general cubic homogeneous polynomial has 20 coefficients, there exists a cubic surface X through these 19 points. Then f_6^-, containing at least 4 points of X, entirely lies on X, whence also $f_1^+,\dots,f_5^+$. In this way the proof is reduced to showing the theorem for the 27 lines on a cubic surface which we will do using Jordan theory. We need the following

<u>Lemma 5.10.</u> <u>Let</u> $(f_1,f_2,f_3,f_4,h) \subset A$ <u>such that</u> $(f_1,\dots,f_4)$ <u>is a collinear family in</u> $V_0(h)$. <u>Then there exists a unique</u> $g \in A_1(h) \cap \bigcap_{k=1,\dots,4} V_0(f_k)$.

<u>Proof</u>. Since $V_2(f_j) \subset V_0(h)$ we see that $W = V_1(h)$ decomposes relative to $(f_1,\dots,f_4)$ as $W = \oplus W_I$ where $W_I = W_{(ijk\ell)} = W \cap V_i(f_1) \cap V_j(f_2) \cap V_k(f_3) \cap V_\ell(f_4)$ with $i,j,k,\ell \in \{0,1\}$. There are 16 possible W_I's. Since $\#(A\cap W) = 16$ it is more than enough to show that each W_I contains the same number of elements of A. This is done in 2 steps:

Let $I = (i_1,i_2,i_3,i_4) \in \{0,1\}^4$ and $J = (j_1,j_2,j_3,j_4) \in \{0,1\}^4$ such that $i_1 + i_2 + i_3 + i_4 = j_1 + j_2 + j_3 + j_4$. Then there exists a suitable product γ of exchange automorphisms lying in $G(f_1,f_2,f_3,f_4) \subset G(A) \subset \mathrm{Aut}\, V$ (see Theorem 2.4) such that $\gamma W_I = W_J$, in particular $\#(A \cap W_I) = \#(A \cap W_J)$.

By (5.1') $A_0(h)$ is a 10-dimensional quadratic form grid $(f_i^\varepsilon;\ \varepsilon = \pm 1,\ 1 \leq i \leq 4)$. We may assume $f_i = f_i^+$. The automorphism $B(f_5^+ + f_1^-, f_5^+ + f_1^-) \circ B(f_1^+ + f_5^+,\ f_1^+ + f_5^+) \in G(A)$ maps $(f_1^+,f_2^+,f_3^+,f_4^+,f_1^-,h)$ onto $(f_1^-,f_2^+,f_3^+,f_4^+,f_1^+,h)$, therefore $W_{(ijk\ell)}$ onto $W_{(1-i,j,k,\ell)}$ (note $W = V_1(f_1^+ + f_1^-)$, thus $W \cap V_i(f_1^+) = W \cap V_{1-i}(f_1^-)$) and hence $\#(A \cap W_{(ijk\ell)}) = \#(A \cap W_{(1-i,j,k,\ell)})$. This implies that each W_I contains the same number of elements of A. ∎

<u>Jordan theoretic proof of Schläfli's Double Six Theorem</u>. We interpret f_i^ε as tripotents in the Albert grid $A \subset V$, thus $(f_1^+,\ldots,f_5^+)$ is a collinear system in $V_0(f_6^-)$. We have to show the existence of $f_1^-,\ldots,f_5^-,\ f_6^+$ such that

a) $(f_1^\varepsilon,f_2^\varepsilon,\ldots,f_6^\varepsilon)$ is a collinear system,

b) $f_i^- \perp f_j^+ \top f_j^-$ for $i \neq j$.

By Lemma 5.10 we may choose the unique $f_j^- \in A_1(f_6^-) \cap \bigcap_{k \neq j} V_0(f_k^+)$, $1 \leq j \leq 5$. Since A is ortho-collinear we have $f_j^- \top f_6^-$. If $f_i^- \perp f_j^-$ for $1 \leq i < j \leq 5$ then (f_i^-,f_j^-,f_m^+) for $i,j,m \neq$ is an orthogonal system contradicting uniqueness in (5.3), whence $(f_1^-,\ldots,f_6^-)$ is a collinear system. Next, suppose $f_j^+ \perp f_j^-$ and let $\ell \neq j$, $\ell \leq 5$, so $f_\ell^-,\ f_j^- \in A_1(f_6^-) \cap \bigcap_{m \neq \ell} V_0(f_m^+)$ contradicting uniqueness in Lemma 5.10. Thus $f_j^+ \top f_j^-$ and a),b) hold for all $f_j^\varepsilon \neq f_6^+$, which we are going to choose now: Let f_6^+ be the unique element in $A_1(f_5^+) \cap \bigcap_{k=1,\ldots,4} V_0(f_k^-)$. Since also $\{f_1^+\ f_1^-\ f_6^-\} \in V_1(f_5^+) \cap \bigcap_{k=1,\ldots,4} V_0(f_k^-) \cap (\pm A)$ we have $f_6^+ = \pm\{f_1^+ f_1^- f_6^-\}$, however by the already known rules $\{f_1^+\ f_1^-\ f_6^-\} \in V_1(f_6^-) \cap V_1(f_j^+) \cap V_0(f_j^-)$ for $1 \leq j \leq 5$, which implies the remaining statements in a), b). ∎

CHAPTER III. COORDINATIZATION THEOREMS

The aim of this chapter is to find out the structure of a Jordan triple system V which is covered by a grid E. This will be done by introducing "coordinates" and proving that these together with E completely determine V ("coordinatization theorems").

Before we start proving these theorems we want to remark:
(1) If V is not covered by E, then under additional assumptions (see Theorem I.4.14) the cover $C_V(E)$ of E is a subsystem of V. The results of this chapter then apply to $C_V(E)$ instead of V. Thus, henceforth we will always assume $V = C_V(E)$.
(2) If $E = \dot{\bigcup}_{j\in J} E_j$ where the E_j are the connected components of E then V is a triple system direct sum of the subsystems $C_V(E_j)$:

$$V = \bigoplus_{j\in J} C_V(E_j).$$

Since E_j covers $C_V(E_j)$ it is enough to consider the case of a connected E.
(3) Because associated cogs have the same Peirce spaces we may finally assume that E is one of the 7 standard grids, defined and studied in II§ 1-3. We will consider each type individually and correspondingly have 7 coordinatization theorems. We will find in §1 and §2 that the cover of a rectangular grid is a rectangular matrix system, the cover of a symplectic grid is a symplectic matrix system, the cover of a hermitian grid is - under mild additional assumptions - a hermitian matrix system and the cover of an even- or odd-dimensional quadratic form grid is an even- or odd-dimensional quadratic form triple. In §3 we will see that the cover of a Bi-Cayley grid is a certain 1×2 rectangular matrix system over a split octonion algebra $\mathbb{O}$, to be called Bi-Cayley triple, and similarly, the cover of a Albert grid is a certain 3×3 hermitian matrix system over $\mathbb{O}$, a so-called Albert triple. Using the term <u>standard example</u> for a rectangular matrix system or .... Albert triple, the results of this chapter can be summarized as

<u>Cover Classification Theorem (short form)</u>. <u>The cover of a standard grid is a standard example, except for the hermitian grid where one needs additional hypotheses</u>.

The additional assumptions needed for a hermitian grid $H(I)$ are stated in Theorem 1.9. for $\#I \geq 3$ and in [43] for $\#I = 2$. Of course, for $\#I = 1$ nothing more can be said than that V is a mutation of a Jordan algebra - as every Peirce-2-space is of this type.

§1 Coordinatization Theorems for rectangular, symplectic and hermitian grids

Coordinatization theorems for Jordan triple systems covered by rectangular, symplectic or hermitian grids were proven in [42] for the case of finite grids. In this section these results are generalized to the case of arbitrary index sets. As in [42] we first prove symmetries theorems. Once these are known, the corresponding coordinatization theorems are straightforward generalizations of the finite case.

1.1. We recall from II.2.3 that a rectangular grid $R = R(I,J) = \{e_{ij};\ i \in I,\ j \in J\}$, where I and J are arbitrary index sets, is a family of non-zero tripotents satisfying

(1.1) $\quad (e_{ij}, e_{i\ell}, e_{k\ell}, e_{kj})$ for $i \neq k$, $j \neq \ell$, is a quadrangle,

(1.2) $\quad \{e_{ij}\ e_{k\ell}\ e_{rs}\} = 0$ if $(k,\ell) \neq (r,j)$ or (i,s).

In this subsection we are looking at a Jordan triple system V over k which is covered by a rectangular grid $R(I,J)$. We will coordinatize V so that it becomes isomorphic to $\mathrm{Mat}(I,J;D,\bar{\ })$, which was defined in Example I.1.3 and is our standard example for a Jordan triple system covered by a rectangular grid.

Since V is covered by $R(I,J)$, we have a simultaneous Peirce decomposition

(1.3)
$$V = \bigoplus_{i \in I}\ \bigoplus_{j \in J} V_{ij}, \text{ where}$$
$$V_{ij} = V_2(e_{ij}) \subset V_1(e_{i\ell}) \cap V_1(e_{kj}) \cap V_0(e_{k\ell}),\ i \neq k,\ j \neq \ell$$

(The Peirce spaces here should not be confused with the Peirce spaces of an orthogonal system). Because of (I.1.29) the product of Peirce spaces lies again in a Peirce space, the following multiplication rules are therefore easily checked:

(1.4) $\quad P(V_{ij})V_{ij} \subset V_{ij}$,

(1.5) $\quad \{V_{ij}\ V_{ij}\ V_{i\ell}\} \subset V_{i\ell}$, $\{V_{ij}\ V_{ij}\ V_{kj}\} \subset V_{kj}$

(1.6) $\quad \{V_{ij}\ V_{i\ell}\ V_{k\ell}\} \subset V_{kj}$ for $i \neq k$, $j \neq \ell$,

(1.7) $\quad$ all other types of products between the spaces V_{ij} vanish.

The example $\mathrm{Mat}(I,J;D,\bar{\ })$ has a lot of automorphisms. Indeed, let ΣI resp. ΣJ denote the group of all permutations (= bijections) of I resp. J. Then for $\rho \in \Sigma I$, $\sigma \in \Sigma J$ the map

(1.8)
$$T_{\rho,\sigma}: \mathrm{Mat}(I,J;D,\bar{\ }) \to \mathrm{Mat}(I,J;D,\bar{\ })$$
$$\sum x_{ij} E_{ij} \to \sum x_{ij} E_{\rho(i)\sigma(j)}$$

is an automorphism of $\mathrm{Mat}(I,J;D,\bar{\ })$ which follows from the multiplication rules given in Example I.1.3.b. The next theorem shows that $T_{\rho,\sigma}$ also exists in general, i.e. for V satisfying (1.3). Decisive for this is that the "symmetries" which exchange the ith and k^{th} row resp. the jth and ℓ^{th} column are automorphisms. They are constructed using the exchange automorphism $T_{e,f} = B(e+f,e+f)$ induced by two collinear tripotents (see Theorem I.1.13) and an adjusting Peirce reflection $S_{(e,f)}$ (see I.1.4.). The precise result, which we will use, is the following

Lemma 1.1. (= [42]1.6) *Let e,f be collinear tripotents in a Jordan triple system V. If $V_{(22)} = 0 = V_{(10)}$ or if e,f imbed in a quadrangle (e,f,g,h) satisfying $V_{(2200)} = 0 = V_{(1111)}$, then the map $S_{(e,f)} T_{e,f}$ is an automorphism of order 2 having the following actions*

$$L(f,e) \text{ on } V_{(21)} + V_{(10)}, \quad L(e,f) \text{ on } V_{(12)} + V_{(01)}$$
$$\mathrm{Id} \text{ on } V_{(11)} + V_{(00)}.$$

As a generalization of [42] 3.3. we prove

Rectangular Symmetries Theorem 1.2. *Let V be covered by a rectangular grid $\mathcal{R}(I,J)$. For $\rho \in \Sigma I$, $\sigma \in \Sigma J$ define $T_{\rho,1}$, $T_{1,\sigma}$ and $T_{\rho,\sigma}$ by*

$$(1.9) \qquad T_{\rho,1}\, x_{ij} = \begin{cases} x_{ij}, & \text{if } \rho(i) = i \\ \{e_{\rho(i)j}\, e_{ij}\, x_{ij}\}, & \text{if } \rho(i) \neq i \end{cases}$$

$$(1.10) \qquad T_{1,\sigma}\, x_{ij} = \begin{cases} x_{ij}, & \text{if } \sigma(j) = j \\ \{e_{i\sigma(j)}\, e_{ij}\, x_{ij}\}, & \text{if } \sigma(j) \neq j \end{cases}$$

$$(1.11) \qquad T_{\rho,\sigma} = T_{\rho,1}\, T_{1,\sigma} = T_{1,\sigma}\, T_{1,\rho}$$

Then

$$T: \ \Sigma I \times \Sigma J \to \mathrm{Aut}\, V: \ (\rho,\sigma) \to T_{\rho,\sigma}$$

is a group monomorphism such that $T_{\rho,\sigma}$ permutes the "rows" $R_i = \bigoplus_{j \in J} V_{ij}$ according to ρ and the "columns" $C_j = \bigoplus_{i \in I} V_{ij}$ according to σ:

$$(1.12) \qquad T_{\rho,\sigma} V_{ij} \subset V_{\rho(i),\sigma(j)}, \quad T_{\rho,\sigma}\, e_{ij} = e_{\rho(i)\sigma(j)}$$

$$(1.13) \qquad T_{\rho,\sigma} | V_{ij} = T_{\pi,\tau} | V_{ij}, \ \text{if } \rho(i) = \pi(i),\ \sigma(j) = \tau(j).$$

The symmetries exchanging the ith and kth row resp. the jth and ℓth column are given by

$$(1.14) \qquad T_{(ik),1} = S_{(e_{ij},e_{kj})}\, T_{e_{ij},e_{kj}} \quad (\text{independent of } j \in J)$$

$$(1.15) \qquad T_{1,(j\ell)} = S_{(e_{ij},e_{i\ell})}\, T_{e_{ij},e_{i\ell}} \quad (\text{independent of } i \in I)$$

Proof. We start out by showing

(a) $\quad T_{\rho,1}\, T_{\pi,1} = T_{\rho\pi,1}$ for $\rho, \pi \in \Sigma I$.

We check this equation by evaluating both sides on V_{ij}. The cases $\pi i = i$ or $\rho\pi i = \pi i$ are obvious, so we assume $\pi i \neq i$ and $\rho\pi i \neq \pi i$. Then $T_{\rho,1}T_{\pi,1}x_{ij} = \{e_{\rho\pi(i)j}\, e_{\pi(i)j}\, \{e_{\pi(i)j}\, e_{ij}\, x_{ij}\}\} =$ (by (I.1.14)) $\{e_{\rho\pi(i)j}\, e_{ij}\, x_{ij}\} - \{e_{\pi(i)j}\, \{e_{\pi(i)j}\, e_{\rho\pi(i)j}\, e_{ij}\}\, x_{ij}\}$, since $\{e_{\rho\pi(i)j}\, e_{\pi(i)j}\, x_{ij}\} = 0$ by (1.4) - (1.7). If $\rho\pi(i) = i$ we get $2x_{ij} - \{e_{\pi(i)j}\, e_{\pi(i)j}\, x_{ij}\} = x_{ij} = T_{\rho\pi,1}x_{ij}$ and if $\rho\pi(i) \neq i$ we get $\{e_{\rho\pi(i)}\, e_{ij}\, x_{ij}\} = T_{\rho\pi,1}x_{ij}$. This shows (a). Similarly one proves

(b) $\quad T_{1,\sigma}T_{1,\tau} = T_{1,\sigma\tau}$ for $\sigma,\tau \in \Sigma J$.

Next, we claim

(c) $\quad T_{\rho,1}\, T_{1,\sigma} = T_{1,\sigma}\, T_{\rho,1}$ for $\rho \in \Sigma I$, $\sigma \in \Sigma J$.

Again, we check this equation on V_{ij}. It is obvious if $\rho i = i$ or $\sigma j = j$. So we assume $\rho i \neq i$ and $\sigma j \neq j$. Then $(e_{ij}, e_{i\sigma(j)}, e_{\rho(i)\sigma(j)}, e_{\rho(i)j})$ is a quadrangle, hence $L(e_{\rho(i)\sigma(j)}, e_{i\sigma(j)}) = L(e_{\rho(i)j}, e_{ij})$ by (I.2.3). Thus $T_{\rho,1}T_{1,\sigma}\, x_{ij} = \{e_{\rho(i)\sigma(j)}\, e_{i\sigma(j)}\{e_{i\sigma(j)}\, e_{ij}\, x_{ij}\}\} =$ $\{e_{\rho(i)j}\, e_{ij}\{e_{i\sigma(j)}\, e_{ij}\, x_{ij}\}\} =$ (by (I.1.14)) $\{\{e_{\rho(i)j}\, e_{ij}\, e_{i\sigma(j)}\}e_{ij}\, x_{ij}\} - 0 + L(e_{i\sigma(j)}, e_{ij})\{e_{\rho(i)j}\, e_{ij}\, x_{ij}\} =$ (by (1.1) and (I.2.3)) $\{e_{\rho(i)\sigma(j)}\, e_{ij}\, x_{ij}\} + L(e_{\rho(i)\sigma(j)}, e_{\rho(i)j})\, T_{\rho,1}x_{ij} =$ (by (1.1)) $T_{1,\sigma}\, T_{\rho,1}\, x_{ij}$.

Clearly, (a)-(c) imply that $T: \Sigma I \times \Sigma J \to GL(V)$ is a group homomorphism. (1.12) and (1.13) immediately follow from the definitions. The second equation of (1.12) in particular implies that T is injective.

Now we show (1.14): By Lemma 1.1 the map on the right side operates as $L(e_{kj}, e_{ij})$ on $V_{(21)} + V_{(10)} = \sum_\ell V_{i\ell}$, $L(e_{ij}, e_{kj})$ on $V_{(12)} + V_{(01)} = \sum_\ell V_{k\ell}$, Id on $V_{(11)} + V_{(00)} = \sum_{r\neq i,k} \sum_s V_{rs}$. On the other side, the action of $T_{(ik),1}$ is $L(e_{k\ell}, e_{i\ell})$ on $V_{i\ell}$, $L(e_{i\ell}, e_{k\ell})$ on $V_{k\ell}$, Id on $\sum_{r\neq i,k} \sum_s V_{rs}$. So both actions clearly are the same on V_{ij}. They also coincide on $V_{i\ell}$, $\ell\neq j$, by (I.2.3) applied to the quadrangle $(e_{kj}, e_{ij}, e_{i\ell}, e_{k\ell})$. Similarly, equality is shown on $V_{k\ell}$, whence (1.14). In the same way one proves (1.15).

It remains to show, that $T_{\rho,1}$ and $T_{1,\sigma}$ are automorphisms, which we will only do for $T_{\rho,1}$, since $T_{1,\sigma}$ can be treated analogously. Because of (1.12) it is enough to check that $T_{\rho,1}$ is a homomorphism relative to the products (1.4)-(1.6). This will follow, as soon as $T_{\rho,1}: U \to V$ for $U = V_{ij} + V_{i\ell} + V_{kj} + V_{k\ell}$ is a homomorphism. We now distinguish several cases: If $\rho i = i$ then (1.13) implies $T_{\rho,1}|U = T_{(k\rho(k)),1}|U$ or $Id|U$. For $\rho k = k$ we have $T_{\rho,1}|U = T_{(i\rho(i)),1}|U$ or $Id|U$. If $\rho i \neq i$, $\rho k \neq k$ and $i \neq \rho k$ resp. $k \neq \rho i$, then $T_\rho|U = T_{(i\rho(i)),1}\, T_{(k\rho(k)),1}|U$ resp. =

$T_{(k\rho(k)),1}\, T_{(i\rho(i)),1}|U$. Finally, for $\rho i = k$, $\rho k = i$ we have $T_{\rho,1}|U = T_{(ik),1}|U$. In all cases we represented $T_{\rho,1}|U$ as a product of maps $T_{(mn),1}$, which are automorphisms by (1.14) and Lemma 1.1. ∎

All coordinatization theorems make essential use of the 1×2 coordinatization which we quote here for later reference:

1×2 Coordinatization Theorem 1.3. (=[42]2.2). *If e, f are collinear tripotents in a Jordan triple system V, then the subsystem $V_{(21)} \oplus V_{(12)}$ is isomorphic to $Mat(1,2;D,^-)$ for an alternative algebra D with involution $a \to \bar{a}$. We may take $D = V_{(21)}$ with product, unit and involution*

(1.16) $\quad a.b = \{\{aef\}fb\} = \{T(a)fb\} = \{ef\{abf\}\} = T\{abf\},\ T = T_{e,f}$

(1.17) $\quad 1 = e$,

(1.18) $\quad \bar{a} = P(e)a.$

The isomorphism $F\colon V_{(21)} \oplus V_{(12)} \to Mat(1,2;D)$ *is given by* $x_{21} \oplus y_{12} \to (x_{21}, \{efy_{12}\})$ *and sends* $T = T_{e,f}$ *to the canonical exchange automorphism* $T'\colon\ (a,b) \to (b,a)$,

(1.19) $\quad FT(x) = T'F(x)$,

(1.20) $\quad P(a)b = (a.\bar{b}).a = a.(\bar{b}.a)$,

(1.21) $\quad \{a\ b\ Tc\} = T(a.(\bar{b}.c))$.

Before we come to the general rectangular coordinatization we remark that there is no harm in assuming $\#I \le \#J$. Indeed, if $R(I,J) = \{e_{ij}\}$ is a rectangular grid of size I x J, then $\{c_{ji}\}$ with $c_{ji} = e_{ij}$ is a rectangular grid of size J x I. It is also reasonable to assume $\#I + \#J > 3$, since otherwise $\#I = \#J = 1$ and V is covered by a single tripotent: $V = V_2(e)$, and V is a mutation of the Jordan algebra $V_2(e)^{(e)}$.

Using the Rectangular Symmetries Theorem 1.2. instead of [42]3.3 the proof of the Rectangular Coordinatization Theorem [42]3.4 for $\#R < \infty$ carries over nearly word by word to yield the

Rectangular Coordinatization Theorem 1.4. *A Jordan triple system V is covered by a rectangular grid $R(I,J)$ with $\#I + \#J > 3$ iff V is isomorphic to a rectangular matrix system.*

More precisely, let V be covered by $R(I,J) = \{e_{ij}\}$ and assume $2 \le \#J$. Choose different pairs $(i,j), (i,\ell) \in I \times J$. Define a unital alternative algebra D with involution on V_{ij} as in Theorem 1.2 :

$$a.b = \{T_{1,(j\ell)}(a)\ e_{i\ell}\ b\},\ 1 = e_{ij},\ \bar{a} = P(e_{ij})a.$$

Then

$$F\colon V \to Mat(I,J;D,^-)\colon x_{kh}\ (\in V_{kh}) \to T_{(ki),(jh)}(x_{kh})E_{kh}$$

<u>is an isomorphism. The algebra D is associative as soon as #I + #J ≥ 4.</u>

We point out that the construction of D depends on our choices. Clearly, $\mathrm{Mat}(I,J;D) \approx \mathrm{Mat}(I,J;D')$ for $D \approx D'$ but note also $\mathrm{Mat}(I,J;D,^{-}) \cong \mathrm{Mat}(I,J;D^{op},^{-})$.

<u>1.2.</u> A symplectic grid as defined in II.2.4 is a family $\mathcal{S} = \mathcal{S}(I) = \{f_{ij}; i,j \in I, i < j\}$ of non-zero tripotents which is indexed by a well-ordered set I with #I ≥ 4 and which satisfies

(1.22) $(f_{ij}, f_{kj}, f_{k\ell}, f_{i\ell})$ for $i,j,k,\ell \neq$ is a quadrangle, where we put $f_{ji} = -f_{ij}$ for $i < j$,

(1.23) $\mathcal{S}$ is pure.

In this subsection we consider a Jordan triple system V which is covered by $\mathcal{S}(I)$:

$$V = \bigoplus_{i<j} V_{ij}, \text{ where for } i,j,k,\ell \neq \tag{1.24}$$
$$V_{ij} = V_{ji} = V_2(f_{ij}) \subset V_1(f_{i\ell}) \cap V_1(f_{kj}) \cap V_0(f_{k\ell})$$

Our aim is to coordinatize such a V to become isomorphic to a symplectic matrix system A(I;K), which was introduced in Example I.1.4. We recall that the symplectic matrix units $F_{ij} = E_{ij} - E_{ji}$ form a symplectic grid in A(I;K).

The properties (1.22) and (1.23) of a symplectic grid immediately imply the following multiplication rules between the Peirce spaces V_{ij}

(1.25) $P(V_{ij})V_{ij} \subset V_{ij}$,

(1.26) $\{V_{ij}\, V_{ij}\, V_{i\ell}\} \subset V_{i\ell}$,

(1.27) $\{V_{ij}\, V_{kj}\, V_{k\ell}\} \subset V_{i\ell}$ for $i,j,k,\ell \neq$,

(1.28) all other types of products vanish.

As with the rectangular coordinatization we first prove that V has a lot of automorphisms. In the example A(I;K) the maps T_σ, $\sigma \in \Sigma I$

$$T_\sigma\colon A(I;K) \to A(I;K)\colon \sum x_{ij} F_{ij} \to \sum x_{ij} F_{\sigma(i)\sigma(j)}$$

are automorphisms, which is an easy consequence of the multiplication rules stated in Example I.1.4. To construct these maps in general we use

<u>Lemma 1.5.</u> (=[42]1.7) <u>Let</u> (e,f,g,h) <u>be a quadrangle in</u> V <u>and</u> d <u>a tripotent collinear with</u> e,f,g,h <u>such that</u> $V_{(2200)} + V_{(0011)} \subset V_0(d)$ <u>and</u> $V_{(1111)} \subset V_2(d) + V_0(d)$. <u>Then</u> $S_{(e,f,d)}T_{e,f}$ <u>is an automorphism of</u> V <u>of order</u> 2 <u>acting as</u>

$$L(f,e) \text{ on } V_{(21)} + V_{(11)}, \quad L(e,f) \text{ on } V_{(12)} + V_{(01)}$$

$P(e)P(f)$ on $V_{(02)}$, $P(f)P(e)$ on $V_{(20)}$
$L(e,f)L(f,e) - \mathrm{Id}$ on $V_{(11)} \cap V_2(d)$
Id on $V_{(22)} + V_{(00)} + (V_{(11)} \cap V_0(d))$

The following theorem generalizes [42]4.3.

Symplectic Symmetries Theorem 1.6. Let V be covered by a symplectic grid $S(I)$. For $\sigma \in \Sigma I$ define T_σ by the following rules ($i \neq j$, $x_{ij} \in V_{ij}$):

$$T_\sigma(x_{ij}) = \begin{cases} x_{ij}, & \text{if } \sigma(i) = i,\ \sigma(j) = j \\ \{f_{\sigma(i)\sigma(j)}\ f_{ij}\ x_{ij}\}, & \text{if } \sigma(i)=i,\ \sigma(j) \neq j \text{ or } \sigma(i) \neq i,\ \sigma(j)=j, \\ -x_{ij}, & \text{if } \sigma(i) = j,\ \sigma(j) = i \\ \{f_{\sigma(i)\sigma(j)}\ f_{mn}\{f_{mn}\ f_{ij}\ x_{ij}\}\}, & \text{if } \sigma(i) \neq i,\ \sigma(j) \neq j \text{ and } \\ & (m,n) = (i,\sigma(j)) \text{ for } i \neq \sigma(j) \\ & \text{resp.} = (\sigma(i),j) \text{ for } \sigma(i) \neq j. \end{cases}$$

Then

$$T:\quad \Sigma I \to \mathrm{Aut}\, V: \sigma \to T_\sigma$$

is a group monomorphism such that

(1.29) $\quad T_\sigma V_{ij} \subset V_{\sigma(i)\sigma(j)}, \quad T_\sigma f_{ij} = f_{\sigma(i)\sigma(j)}$,

(1.30) $\quad T_\rho|V_{ij} = T_\sigma|V_{ij}$, if $\rho(i) = \sigma(i)$ and $\rho(j) = \sigma(j)$,

(1.31) $\quad T_\rho|V_{ij} = -T_\sigma|V_{ij}$, if $\rho(i) = \sigma(j)$ and $\rho(j) = \sigma(i)$.

For a transposition (ij) the symmetry $T_{(ij)}$ has the form

(1.32) $\quad T_{(ij)} = S_{(f_{ik},f_{jk},f_{ij})}T_{f_{ik},f_{ik}}$ (independent of $k \neq i,j$).

Proof. First of all we note that (1.29) and (1.30) immediately follow from the definition of T_σ.

The first major step in this proof is to show (1.32): We can apply Lemma 1.5 and see that the map on the right side is an automorphism of order 2 with the following actions: $L(f_{jk},f_{ik})$ on $V_{i\ell}$, $\ell \neq j$, $L(f_{ik},f_{jk})$ on $V_{j\ell}$, $\ell \neq i$, $-\mathrm{Id}$ on V_{ij} and Id on $V_{k\ell}$, $\{k,\ell\} \cap \{i,j\} = \emptyset$. On the other hand $T_{(ij)}$ operates in the following way: $L(f_{j\ell},f_{i\ell})$ on $V_{i\ell}$, $\ell \neq j$, $L(f_{i\ell},f_{j\ell})$ on $V_{j\ell}$, $\ell \neq i$, $-\mathrm{Id}$ on V_{ij} and Id on $V_{k\ell}$, $\{k,\ell\} \cap \{i,j\} = \emptyset$. We consider both actions on $V_{i\ell}$ and $V_{j\ell}$, $\ell \neq i,j$. They agree if $k = \ell$. Otherwise $k,\ell,i,j \neq$ and $(f_{jk},f_{ik},f_{i\ell},f_{j\ell})$ is a quadrangle by (1.22). Hence $L(f_{jk},f_{ik}) = L(f_{j\ell},f_{i\ell})$ and $L(f_{ik},f_{jk}) = L(f_{i\ell},f_{j\ell})$, and (1.32) follows including the independance of k: $T_{(ij)}$ is defined without reference to k.

Next we claim $T_{(ij)}T_{(jk)}T_{(ij)} = T_{(ik)}$ for $i,j,k \neq$. Indeed, let $\tau = T_{(ij)}$ and $T_{(jk)} = S_{(e,f,g)}T_{f,g}$. Then $T_{(ij)}T_{(jk)}T_{(ij)} = (\tau S_{(e,f,g)}\tau^{-1})(\tau T_{f,g}\tau^{-1}) = S_{(\tau e,\tau f,\tau g)}T_{\tau f,\tau g}$. Because $\tau f = T_{(ij)}f_{ji} = f_{ij}$, $\tau g =$

$T_{(ij)}f_{ki} = f_{kj}$ and $\tau e = T_{(ij)}f_{jk} = f_{ik}$ by (1.29), the asserted equation is proven. Now let $\Sigma_0 I$ be the subgroup of ΣI consisting of all permutations which move only a finite number of points in I. It is generated by the transpositions (ij) and its defining relations are $(ij)^2 = 1$, $(ij)(jk)(ij) = (ik)$ for $i,j,k \neq$. Therefore we get a group homomorphism $\mathcal{T}: \Sigma_0 I \to \mathrm{Aut}\, V$ induced by $(ij) \to T_{(ij)}$. We claim:

(a) $$\mathcal{T}_\sigma = T_\sigma \quad \text{for} \quad \sigma \in \Sigma_0 I$$

We will check (a) on V_{ij}. By (1.30) $T_\sigma = T_\rho$ on V_{ij} for any $\rho \in \Sigma_0 I$ with $\rho|\{i,j\} = \sigma|\{i,j\}$. But for such a ρ we also have $\mathcal{T}_\sigma = \mathcal{T}_\rho$ on V_{ij}, since $T_{(k,\ell)} = \mathrm{Id}$ on V_{ij} if $\{k,\ell\} \cap \{m,n\} = \emptyset$. Depending on σ we now make the following choices for ρ: 1) $\sigma(i) = i$, $\sigma(j) = j$: $\rho = \mathrm{Id}$, 2) $\sigma(i) = i$, $\sigma(j) \neq j$: $\rho = (j\sigma(j))$, 3) $\sigma(i) \neq i$, $\sigma(j) = j$: $\rho = (i\sigma(i))$, 4) $\sigma(i) = j$, $\sigma(j) = i$: $\sigma = (ij)$, 5) $\sigma(i) \neq i$, $\sigma(j) \neq j$, $\sigma(j) \neq i$: $\rho = (i\sigma(i))(j\sigma(j))$, 6) $\sigma(i) \neq i$, $\sigma(j) \neq j$, $\sigma(i) \neq j$: $\rho = (j\sigma(j))(i\sigma(i))$. By definition of $\mathcal{T}$ we have $\mathcal{T}_\rho = T_\rho$ in the cases 1)-4). In 5) we obtain $\mathcal{T}_\rho = T_{(i\sigma(i))}T_{(j\sigma(j))}$ which on V_{ij} equals $T_\sigma|V_{ij}$ by definition of T_σ. By symmetry 6) follows and therefore (a) holds.

We are now in the position to prove our claims about T, the first being $T_\sigma \in \mathrm{Aut}\, V$. Since V satisfies the multiplication rules (1.25)-(1.28), it is enough to show that $T_\sigma: U \to V$, $U = V_{ij} + V_{i\ell} + V_{kj} + V_{k\ell}$, $i,j,k,\ell \neq$, is a homomorphism. But $T_\sigma|U = T_\rho|U$ for some $\rho \in \Sigma_0 I$ by (1.30) and T_ρ is an automorphism by (a).

Our next claim is $T_\sigma T_\rho = T_{\sigma\rho}$ for $\sigma,\rho \in \Sigma I$. It is enough to check this equation on V_{ij}. We choose $\pi,\omega \in \Sigma_0 I$ with $\rho|\{i,j\} = \pi|\{i,j\}$ and $\sigma|\{\rho(i),\rho(j)\} = \omega|\{\rho(i),\rho(j)\}$. Then $T_\sigma T_\rho|V_{ij} = T_\omega T_\pi|V_{ij} =$ (by (a)) $T_{\omega\pi}|V_{ij} = T_{\sigma\rho}|V_{ij}$ since $\omega\pi|\{i,j\} = \sigma\rho|\{i,j\}$.

Finally, T is a monomorphism by (1.29), and (1.31) holds since $T_\sigma^{-1}\, T_\rho|V_{ij} = T_{(ij)}|V_{ij} = -\mathrm{Id}|V_{ij}$. ∎

Using now the Symplectic Symmetries Theorem instead of [42]4.3 the proof of the Symplectic Coordinatization Theorem for $\#I < \infty$, given in [42]4.4, carries over to the general case:

Symplectic Coordinatization Theorem 1.7. A Jordan triple system V is covered by a symplectic grid $\mathcal{S}(I)$ with $\#I \geq 4$ iff V is isomorphic to a symplectic matrix system $A(I;K)$ with $\#I \geq 4$.

More precisely, let V be covered by $\mathcal{S}(I)$. Choose different $i,j,k \in I$. Then an extension K of k is defined on V_{ij} by
$$a \cdot b = \{T_{(jk)}(a)\ f_{ik}\ b\}.$$

<u>It has the unit</u> f_{ij} <u>and an involution</u> $a \to \bar{a} = P(f_{ij})a$. <u>An isomorphism between</u> V <u>and</u> A(I;K) <u>is given by</u>

$$V_{rs} \ni x_{rs} \to T_\rho(x_{rs})F_{rs}\,, \quad r,s \in I \text{ with } r < s$$

<u>where</u> $\rho \in \Sigma I$ <u>satisfies</u> $\rho(r) = i$, $\rho(s) = j$.

<u>1.3.</u> In this subsection we want to coordinatize a Jordan triple system V which is covered by a hermitian grid $H = H(I) = \{h_{ij} = h_{ji};\ i,j \in I\}$ with #I ⩾ 3 (see II.1.2.). Hence we have

$$(1.33) \qquad V = \bigoplus_{i \leq j} V_{ij}$$

where for i,j,k,ℓ ≠

$$V_{ij} = V_1(h_{ii}) \cap V_1(h_{jj}) \subset V_2(h_{ij}) \cap V_1(h_{i\ell}) \cap V_1(h_{jk}) \cap V_0(h_{k\ell})$$

$$V_{ii} = V_2(h_{ii}) \subset V_2(h_{ij}) \cap V_0(h_{jj}) \cap V_0(h_{k\ell})$$

satisfying the following multiplication rules for arbitrary $r,s,t,u \in I$:

$$(1.34) \qquad P(V_{rs})V_{rs} \subset V_{rs},\quad P(V_{rs})V_{rr} \subset V_{ss},$$

$$(1.35) \qquad \{V_{rs}\ V_{st}\ V_{tu}\} \subset V_{ru},$$

(1.36) all other types of products vanish.

We remark, that (1.33) is the Peirce decomposition relative to the orthogonal system $\{h_{ii}; i \in I\}$ and that (1.34)-(1.36) are the usual Peirce multiplication rules (see Theorem I.1.10).

As in the previous subsections we will first construct automorphisms of V. In the example of a hermitian Jordan triple system $H_I(D,D_0,\pi,^-)$, which was defined in Example I.1.5 and satisfies our assumptions relative to the hermitian grid $H(I) = \{H_{ij}\}$, $H_{ii} = E_{ii}$, $H_{ij} = E_{ij} + E_{ji}$ for $i \neq j$, the automorphisms we are looking for are given by $T_\sigma(a[rs]) = a[\sigma(r)\sigma(s)]$ for $\sigma \in \Sigma I$ (recall that the elements $a[rs] = aE_{ij} + a^\pi E_{ji}$ span $H_I(D,D_0,\pi,^-)$. In general, we have the

<u>Hermitian Symmetries Theorem 1.8.</u> <u>Let</u> V <u>be covered by a hermitian grid</u> $H(I)$ <u>with</u> #I ⩾ 3. <u>For</u> $\sigma \in \Sigma I$ <u>define</u> T_σ <u>by the following rules</u>

$$T_\sigma x_{ii} = P(h_{i\sigma(i)})P(h_{ii})x_{ii}, \qquad \text{and for } i \neq j$$

$$T_\sigma(x_{ij}) = \begin{cases} x_{ij}\,, & \text{if } \sigma(i) = i,\ \sigma(j) = j \\ L(h_{\sigma(i)\sigma(j)},h_{ij})x_{ij}, & \text{if } \sigma(i)=i,\ \sigma(j)\neq j \text{ or } \sigma(i)\neq i,\ \sigma(j)=j, \\ P(h_{ij})P(h_{ii},h_{jj})x_{ij}\,, & \text{if } \sigma(i) = j,\ \sigma(j) = i \\ L(h_{\sigma(i)\sigma(j)},h_{mn})L(h_{mn},h_{ij})x_{ij}, & \text{if } \sigma(i) \neq i,\ \sigma(j) \neq i \text{ and } (m,n) = (i,\sigma(j)) \text{ for } i \neq \sigma(j) \text{ resp.} = (\sigma(i),j) \text{ if } \sigma(i) \neq j \end{cases}$$

<u>Then</u>

$$T: \Sigma I \to \mathrm{Aut}\, V: \quad \sigma \to T_\sigma$$

is a group monomorphism such that

(1.37) $T_\sigma V_{rs} \subset V_{\sigma(r)\sigma(s)}$, $T_\sigma h_{rs} = h_{\sigma(r)\sigma(s)}$

(1.38) $T_\sigma | V_{rs} = T_\rho | V_{rs}$, if $\sigma(r) = \rho(r)$ and $\sigma(s) = \rho(s)$.

For a transposition (ij) the symmetry $T_{(ij)}$ has the form

(1.39) $T_{(ij)} = S_{(h_{ik}, h_{jk}, h_{ij})} T_{h_{ik}, h_{jk}}$ (independent of $k \neq i,j$).

With this theorem at hand one now can coordinatize V as in the finite case, which was done in [42]5.6. But unlike the previous 2 cases covering is not enough to ensure that V is isomorphic to a hermitian matrix system. We need an additional condition which forces the extreme radical of V to vanish - a condition which is fulfilled for hermitian matrix systems.

Hermitian Coordinatization Theorem 1.9. A Jordan triple system V is covered by a hermitian grid $H(I)$, $\#I \geq 3$ and satisfies for $r,s \in I$, $r \neq s$:

(1.40) $L(h_{rs}, h_{ss}): V_{ss} \to V_{rs}$ is injective,

iff V is isomorphic to a hermitian matrix system $H_I(D, D_0, \pi, {}^-)$.

More precisely, let V be covered by $H(I)$ and satisfy (1.40). Choose different $i,j,k \in I$ and define an algebra D on V_{ij} by

(1.41) $a \cdot b = \{T_{(jk)}(a)\ h_{ik}\ b\}$ $(a,b \in V_{ij})$.

Then D is alternative, even associative for $\#I \geq 4$, has h_{ij} as unit, $a \to a^\pi = P(h_{ij})P(h_{ii}, h_{jj})a$ as first involution and $a \to \bar{a} = P(h_{ij})a$ as second involution commuting with π and leaving the ample subspace $D_0 = L(h_{ij}, h_{ii})V_{ii}$ invariant.

There exists an isomorphism mapping $H(I)$ onto the canonical hermitian grid $\{H_{ij} = H_{ji}; i,j \in I\}$.

The case of a Jordan triple system covered by a hermitian grid $H(I)$, $\#I = 2$, is treated in [43].

We remark that another method to extend the coordinatization theorems, proven in [42] for finite index sets, to the case of arbitrary index sets, is to take direct limits of the pieces of V covered by finite subgrids. The reasons for the procedure chosen here are that for later applications we want to have the symmetries theorems at our disposal, and that moreover we want to emphasize that the proofs given in [42] for finite grids also work in the arbitrary cases as soon as one has the corresponding symmetries theorems.

§2 Coordinatization Theorems for quadratic form grids

In this section we prove symmetries and coordinatization theorems for Jordan triple systems covered by even- and odd-dimensional quadratic form grids. Both cases are treated simultaneously.

2.1. We first recall from section II.2.5 the definition of an even-dimensional quadratic from grid $\mathcal{Q}_e = \mathcal{Q}_e(I) = \{e_i^\varepsilon; i \in I, \varepsilon = \pm\}$. The tripotents in $\mathcal{Q}_e(I)$ satisfy:

(2.1) $e_i^+ \perp e_i^-$ and $e_i^\varepsilon \top e_j^\mu$ for $i \neq j$ and ε,μ arbitrary,

(2.2) $\{e_i^+ \, e_j^\mu \, e_i^-\} = -e_j^{-\mu}$ for $i \neq j$ and

(2.3) $\{e_i^\varepsilon \, e_j^\mu \, e_k^\delta\} = 0$ for $i,j,k \neq$ and ε,μ,δ arbitrary.

Our only example of a Jordan triple system V covered by an even-dimensional quadratic form grid is $[K,\bar{\ },2I]$, defined in Example I.1.6.a; it is therefore not surprising that we will prove $V \cong [K,\bar{\ },2I]$ for an extension K of k.

Next we recall from section II.1.1. the definition of an odd-dimensional quadratic form grid $\mathcal{Q}_o = \mathcal{Q}_o(I) = \{e_o\} \dot\cup \{e_i^\varepsilon; i \in I, \varepsilon = \pm\}$. The tripotents in $\mathcal{Q}_o$ satisfy:

(2.4) $\mathcal{Q}_e(I) = \{e_i^\varepsilon;\ i \in I,\ \varepsilon = \pm\}$ is an even-dimensional quadratic form grid,

(2.5) for all $i \in I$ $e_o \vdash e_i^\varepsilon$ and $P(e_o)e_i^\varepsilon = e_i^{-\varepsilon}$, $\{e_i^+ \, e_o \, e_i^-\} = e_o$, i.e. (e_o, e_i^+, e_i^-) is a triangle,

(2.6) $\{e_o \, e_i^\varepsilon \, e_j^\mu\} = 0 = \{e_i^\varepsilon \, e_o \, e_j^\mu\}$ for $i \neq j$, ε and μ arbitrary.

The most general example of a Jordan triple system covered by a grid $\mathcal{Q}_o(I)$ is an odd-dimensional quadratic form triple, $[K,\bar{\ },X \oplus 2I]$ defined in Example I.1.6.b, where X contains a tripotent e_o satisfying $q_X(e_o) = 1$, $S_X e_o = e_o$. We will prove that this example already is the general case.

To treat both cases simultaneously we consider a Jordan triple system V over k which contains a grid $\mathcal{Q}_e(I)$ inducing a Peirce decomposition

(2.7) $V = C_V(\mathcal{Q}_e(I)) \oplus X$, $X = \bigcap_{f \in \mathcal{Q}_e(I)} V_1(f)$.

Clearly, if V is covered by an even- resp. odd-dimensional quadratic form grid, then $X = 0$ resp. $e_o \in X$. To simplify notation we make the convention that all assertions involving $e_0 \in X$ are void if there is no $e_0 \in X$ such that $\mathcal{Q}_0 = \{e_0\} \cup \mathcal{Q}_e$. We put

$$V_i^\varepsilon = V_2(e_i^\varepsilon).$$

Then (2.7) implies

(2.8) $\quad V_i^\varepsilon = V_0(e_i^{-\varepsilon}) \subset V_1(e_j^\mu), \quad i \neq j$, and

(2.9) $\quad V = \underset{i \in I}{\oplus}(V_i^+ \oplus V_i^-) \oplus X = V_2(e_0)$

In particular, all the spaces V_i^ε and X are Peirce spaces relative to $\mathcal{Q}_e(I)$ (and of course also relative to $\mathcal{Q}_0(I)$). By (I.1.29) a triple product between these spaces is contained in one of them. The precise relations are straightforward to check. They are for $i \neq j$ and ε,μ arbitrary

(2.10) $\quad P(V_i^\varepsilon)V_i^\varepsilon \subset V_i^\varepsilon$

(2.11) $\quad \{V_i^\varepsilon\ V_i^\varepsilon\ V_j^\mu\} \subset V_j^\mu, \qquad \{V_i^\varepsilon\ V_i^\varepsilon\ X\} \subset X,$

(2.12) $\quad \{V_i^+\ V_j^\mu\ V_i^-\} \subset V_j^{-\mu}, \qquad \{V_i^+\ X\ V_i^-\} \subset X,$

(2.13) $\quad P(X)X \subset X, \qquad P(X)V_i^\varepsilon \subset V_i^{-\varepsilon},$

(2.14) $\quad \{X\ X\ V_i^\varepsilon\} \subset V_i^\varepsilon,$

(2.15) $\quad$ all other types of products between V_i^ε and X vanish.

Following the procedure of §1 the first step towards a coordinatization theorem is to prove the existence of symmetries. Corresponding to the fact that $\mathcal{Q}_e(I)$ is parametrized by I and $\varepsilon = \pm$, there are two different kinds of symmetries: one, which operates on the indices i and fixes ε, and one, which switches ε and does not change the indices i. In the examples $[K,^-,2I]$ and $[K,^-,X \oplus 2I]$ the first type is given by ($\sigma \in \Sigma_I$, $a \in K$, $i \in I$, $\varepsilon = \pm$)

(2.16) $\quad T_\sigma(ae_i^\varepsilon) = ae_{\sigma(i)}^\varepsilon, T_\sigma|X = Id_X.$

These formulas motivate

<u>Theorem 2.1.</u> *Assume V satisfies (2.7). Then there exists a monomorphism*

$$\Sigma_I \to \text{Aut } V\colon \sigma \to T_\sigma,$$

where the automorphisms T_σ are defined by

$$T_\sigma(x_i^\varepsilon) = x_i^\varepsilon \text{ for } \sigma(i) = i,\ T_\sigma(x) = x \text{ for } x \in X \text{ and}$$

(2.17) $\quad T_\sigma(x_i^\varepsilon) = \{e_{\sigma(i)}^\varepsilon\ e_i^\varepsilon\ x_i^\varepsilon\} = -\{e_i^{-\varepsilon}\ e_{\sigma(i)}^{-\varepsilon}\ x_i^\varepsilon\} \in V_{\sigma(i)}^\varepsilon$ *for* $\sigma(i) \neq i$.

In particular $T_\sigma e_i^\varepsilon = e_{\sigma i}^\varepsilon$ *and*

(2.18) $\quad T_\sigma|V_i^\varepsilon = T_\tau|V_i^\varepsilon$ *if* $\sigma i = \tau i$.

For a transposition $\sigma = (ij)$ we have the formula

(2.19) $\quad T_{(ij)} = -B(e_i^\varepsilon - e_j^\varepsilon, e_i^\varepsilon - e_j^\varepsilon)$, *independent of* ε.

As the symmetries for the grids in §1 the map $T_{(ij)}$ here is also a product of a Peirce reflection and an exchange automorphism namely

$$T_{(ij)} = S_{(e_i^+, e_i^-, e_j^+, e_j^-)} \, T_{e_i^\varepsilon, e_j^\varepsilon}$$

Proof. First note, that the last equation in the definition (2.17) holds, since $L(c_1,c_2) = L(c_4,c_3)$ for every quadrangle (c_1,c_2,c_3,c_4) by (I.2.3). Moreover, (2.18) is an immediate consequence of (2.17).

Next we show $T_\tau T_\sigma = T_{\tau\sigma}$ for $\tau,\sigma \in \Sigma_I$. Clearly, this holds on X and is easy to check on x_i^ε in case $\sigma i = i$ or $\tau\sigma i = \sigma i$. So assume $\sigma i \neq i$ and $\tau\sigma i \neq \sigma i$. Then $T_\tau T_\sigma(x_i^\varepsilon) = \{e_{\tau\sigma i}^\varepsilon \; e_{\sigma i}^\varepsilon \{e_{\sigma i}^\varepsilon \; e_i^\varepsilon \; x_i^\varepsilon\}\}$ = (applying (I.1.14) and $\{e_{\tau\sigma i}^\varepsilon \; e_{\sigma i}^\varepsilon \; x_i^\varepsilon\} = 0$ by (2.15)) $\{e_{\tau\sigma i}^\varepsilon \; e_i^\varepsilon \; x_i^\varepsilon\} - \{e_{\sigma i}^\varepsilon \{e_{\sigma i}^\varepsilon \; e_{\tau\sigma i}^\varepsilon \; e_i^\varepsilon\} x_i^\varepsilon\}$. In case $\tau\sigma i = i$ this expression equals $x_i^\varepsilon = T_{\tau\sigma} x_i^\varepsilon$ and in case $\tau\sigma i \neq i$ we obtain $\{e_{\tau\sigma i}^\varepsilon \; e_i^\varepsilon \; x_i^\varepsilon\} = T_{\tau\sigma} \, x_i^\varepsilon$.

Before we prove, that the maps T_σ are automorphisms, we show (2.19). To this end, we recall that $(e_i^\varepsilon, -e_j^\varepsilon)$ for $i \neq j$ is a collinear pair and $B(e_i^\varepsilon - e_j^\varepsilon, e_i^\varepsilon - e_j^\varepsilon)$ is the exchange automorphism between e_i^ε and $-e_j^\varepsilon$. Thus $B: = -B(e_i^\varepsilon - e_j^\varepsilon, e_i^\varepsilon - e_j^\varepsilon)$ is an automorphism too and its operation can be read off from Theorem I.1.13. We get $B = Id$ on $V_k^\mu \oplus X$ for $i,j,k \neq$, because on this space $P(e_i^\varepsilon, e_j^\varepsilon) = 0$ by (2.15). Moreover $Bx_i^\varepsilon = \{e_j^\varepsilon \; e_i^\varepsilon \; x_i^\varepsilon\}$ and $Bx_i^{-\varepsilon} = -\{e_i^\varepsilon \; e_j^\varepsilon \; x_i^{-\varepsilon}\}$. Comparing these formulas with (2.17) and using symmetry in i and j, we obtain (2.19).

It remains to show that T_σ in general is an automorphism. Since T_σ maps the Peirce spaces onto each other, this follows as soon as T_σ behaves well with respect to the products (2.10)-(2.14). First we look at products within the subsystem $U = V_i^+ \oplus V_i^- \oplus X$. If $T_\sigma|U = Id$ nothing is to show. Otherwise $T_\sigma|U = T_{(i\sigma(i))}|U$. But we noted before that $T_{(i\sigma(i))}$ is an automorphism, hence $T_\sigma: U \to V$ is a homomorphism. It remains now to look at products in the subsystem $W = V_i^+ \oplus V_i^- \oplus V_j^+ \oplus V_j^-$. Again, we are done, as soon as $T_\sigma|W$ is a product of automorphisms $T_{(k\ell)}$, which indeed is true: If $\sigma i = i$ we have $T_\sigma|W = T_{(j\sigma(j))}|W$ or Id and similarly for $\sigma j = j$. For $i \neq \sigma i$, $j \neq \sigma j$ and $i \neq \sigma j$ resp. $j \neq \sigma i$ we get $T_\sigma|W = T_{(i\sigma i)} T_{(j\sigma j)}|W$ resp. $T_{(j\sigma j)} T_{(i\sigma i)}|W$ and finally $T_\sigma|W = T_{(ij)}|W$ for $\sigma i = j$, $\sigma j = i$. ∎

In the example $[K, \bar{\ }, X \oplus 2I]$ the second type of symmetries is given by $(\Delta \in \{\pm 1\}^I,\ a \in K,\ i \in I,\ \varepsilon = \pm)$

(2.20) $\qquad T_\Delta(ae_i^\varepsilon) = ae_i^{\Delta(i)\varepsilon}, \quad T_\Delta|X = Id,$

where clearly $\Delta(i)\varepsilon = \varepsilon$ if $\Delta(i) = 1$ and $= -\varepsilon$ if $\Delta(i) = -\varepsilon$.

Theorem 2.2. *Assume that V satisfies (2.7) and moreover*

(2.21) $\#I > 2$, *and if* $\#I = 2$ *there exists* $e_0 \in X$ *such that* $\{e_0\} \cup \mathcal{Q}_e(I)$ *is an odd-dimensional quadratic form grid.*

Then there exists a group monomorphism

$$\{\pm 1\}^I \to \operatorname{Aut} V\colon \Delta \to T_\Delta$$

where the automorphisms T_Δ *are defined by*

$$T_\Delta(x_i^\varepsilon) = x_i^\varepsilon \text{ if } \Delta(i) = 1, \quad T_\Delta(x) = x \text{ for } x \in X$$

(2.22) $$T_\Delta(x_i^\varepsilon) = -P(e_j^+, e_j^-)P(e_j^\varepsilon)x_i^\varepsilon = P(e_0)P(e_i^\varepsilon)x_i^\varepsilon \text{ if } \Delta(i) = -1,$$

independent of $j \neq i$.

In particular $T_\Delta e_i^\varepsilon = e_i^{\Delta(i)\varepsilon}$ *and*

(2.23) $$T_\Delta | V_i^\varepsilon = T_\Phi | V_i^\varepsilon \text{ for } \Delta(i) = \Phi(i).$$

Let Σ_I *operate on* $\{\pm 1\}^I$ *by* $(\sigma\cdot\Delta)(i) = \Delta(\sigma^{-1}(i))$. *Then*

(2.24) $$T_\sigma\, T_\Delta\, T_\sigma^{-1} = T_{\sigma\cdot\Delta}$$

Proof. As an abbreviation we put $e_i = e_i^+ + e_i^-$ for $i \in I$ and note $V = V_2(e_i)$, hence $P(e_i)$ and $P(e_0)$ are automorphisms of order 2 by (I.1.23). For $i,j,k \neq$ we let $y \in V_k^\varepsilon \oplus X$ and consider $P(e_i)P(e_j)y = P(e_i)\{e_j^+\ y\ e_j^-\}$ = (by (I.1.12)) $-\{y\ e_j^+\ P(e_i)e_j^-\} + \{e_i\ e_j^-\{y\ e_j^+\ e_i\}\}$ = (by (2.2) and (2.15)) $\{y\ e_j^+\ e_j^+\} = y$, thus

(2.25) $$P(e_i)y = P(e_j)y \text{ for } y \in V_k^\varepsilon \oplus X,\ i,j,k \neq,$$

and consequently T_Δ is well-defined. If X contains e_0, one proves similarly

(2.26) $$P(e_0)y = -P(e_j)y \text{ for } y \in V_k^\varepsilon,\ j \neq k$$

which gives the second equation for $T_\Delta(x_i^\varepsilon)$, $\Delta(i) = -1$. By definition, (2.23) holds and $T_\Delta e_i^\varepsilon = e_i^{\Delta(i)\varepsilon}$ follows from (2.2) resp. (2.5).

To show that $\Delta \to T_\Delta$ is a homomorphism of $\{\pm 1\}^I$ into $GL(V)$, we first prove the formula

(2.27) $$P(e_i)P(e_j) = P(e_j)P(e_i) \text{ for } i,j \in I.$$

Indeed, we may assume $i \neq j$, then $P(e_i)e_j^\varepsilon = -e_j^{-\varepsilon}$ and, using the fact that $P(e_i)$ is an automorphism, we get $P(e_i)P(e_j) = P(P(e_i)e_j)P(e_i) = P(e_j)P(e_i)$. Now $T_\Delta T_\Phi = T_{\Delta\cdot\Phi}$ can be proven in the following way: Since all three maps involved leave invariant $V_i^+ \oplus V_i^-$, it is enough to consider the restrictions to that space. The case $(\Delta(i), \Phi(i)) = (1,1)$, $(-1,1)$ or $(1,-1)$ trivially follow from (2.22), and for $\Delta(i) = -1 = \Phi(i)$ we have $T_\Delta T_\Phi(x_i^\varepsilon) = P(e_j)P(e_i)P(e_j)P(e_i)x_i^\varepsilon$ = (by (2.27)) $x_i^\varepsilon = T_{\Delta\cdot\Phi}(x_i^\varepsilon)$ because $\Delta(i)\cdot\Phi(i) = 1$.

The next aim is to show $T_\Delta \in \operatorname{Aut} V$. Since T_Δ mixes the Peirce spaces V_i^ε and X, this will follow, as soon as T_Δ behaves well with respect to the products (2.10)-(2.14). Let $U = V_i^+ \oplus V_i^- \oplus X$. Then the formulas (2.10)-(2.14) immediately reveal that the map $S \in \mathrm{End}\, U$, defined by $S|V_i^+ \oplus V_i^- = \mathrm{Id}$ and $S|X = -\mathrm{Id}$, is an automorphism of the subsystem U of V. We have either $T_\Delta|U = \mathrm{Id}$ or $T_\Delta|U = -P(e_j)P(e_i)S|U$ by (2.22) and (2.25). Consequently, $T_\Delta|U \in \mathrm{Aut}\, U$. It remains to show $T_\Delta|W \in \mathrm{Aut}\, W$ for the subsystem $W = V_i^+ \oplus V_i^- \oplus V_j^+ \oplus V_j^-$, $i \neq j$. Similar as for U, the map $R \in \mathrm{End}\, W$, given by $R|V_i^+ \oplus V_i^- = \mathrm{Id}$ and $R|V_j^+ \oplus V_j^- = -\mathrm{Id}$, is an automorphism of W, which we can use to write $T_\Delta|W$ as a product of obvious automorphisms of W in case $\Delta(i) = -1$, $\Delta(j) = 1$. In that case (2.25) implies $T_\Delta|W = -P(e_k)P(e_i)R|W$ for $i,j,k \neq$ or $T_\Delta|W = P(e_0)P(e_i)R|W$, so always $T_\Delta|W \in \mathrm{Aut}\, W$. The case $\Delta(i) = 1$, $\Delta(j) = -1$ follows by symmetry, and for $\Delta(i) = 1 = \Delta(j)$ we have $T_\Delta|W = \mathrm{Id}$. This leaves us with $\Delta(i) = -1 = \Delta(j)$. But here $T_\Delta|W = -P(e_i)P(e_j)|W$ by (2.27). Altogether we have shown $T_\Delta \in \mathrm{Aut}\, V$.

Finally we prove (2.24): It needs only to be checked on V_i^ε. If $\Delta(\sigma^{-1}i) = 1$ both sides are $\mathrm{Id}|V_i^\varepsilon$. If $\Delta(\sigma^{-1}i) = -1$ put $k = \sigma^{-1}i$ and take $j \neq k$. Then $T_\sigma T_\Delta T_\sigma^{-1} x_i^\varepsilon = -T_\sigma P(e_j)P(e_k^\varepsilon)T_\sigma^{-1}x_i^\varepsilon = -P(T_\sigma e_j)P(T_\sigma e_k^\varepsilon)x_i^\varepsilon = -P(e_{\sigma j})P(e_i^\varepsilon)x_i^\varepsilon = T_{\sigma\cdot\Delta}x_i^\varepsilon$ because $\sigma j \neq i$. ⊞

Remark 2.3. a) The only place, where we used our assumption (2.21), was to show that T_Δ for $\Delta(i) = -1$, $\Delta(j) = 1$ is an automorphism. Indeed, if $\#I = 2$ and $X = 0$, T_Δ is not always an automorphism: Observe that V is covered by a rectangular grid $\mathcal{R}(I,I)$ (take $e_{11} = e_1^+$, $e_{12} = e_2^+$, $e_{21} = -e_2^-$, $e_{22} = e_1^+$), hence $V \cong \mathrm{Mat}(2,2;D)$ by the Rectangular Coordinatization Theorem and T_Δ becomes $x \to x^t$, which is an automorphism iff D is commutative.

b) If $\#I = 1$ and $X = 0$ nothing more can be said but that V is a direct sum of two ideals both containing a tripotent. If $\#I = 1$ and $\mathcal{Q}_e(I) \subset \mathcal{Q}_0(I)$, then we are in the situation of a triangle which is treated in [43].

2.2. We now know enough symmetries to aim at coordinatization. We will define our coordinate algebra as in the fundamental 1 x 2 Coordinatization Theorem 1.3. So we get a unital alternative algebra over the base ring k with involution. However, in our present situation we will show that this algebra actually is an extension of k. Associativity will follow from

Lemma 2.4. (= [42] 2.6.(ii)). *If e_1, e_2, e_3 are tripotents in* V *where* $e_2 \top e_1 \top e_3 \perp e_2$, *then the product on* $V_2(e_1) \cap V_1(e_2) \cap V_1(e_3)$ *defined by* e_3 *is the opposite of that defined by* e_2, *and both products are associative.*

That our coordinate algebra is an extension is proven in

Lemma 2.5. *Assume* V *satisfies* (2.7) *and* (2.21). *Rename two distinct elements in* I *as* 1 *and* 2, *and define an algebra* K *on the* k-*module* V_1^+ *by*

(2.28) $\qquad a\cdot b = \{\{a\ e_1^+\ e_2^+\}e_2^-\ b\} \qquad (a,b \in V_1^+)$

Then K *is a commutative and associative algebra* (*i.e. an extension of* k) *with unit* e_1^+ *and involution* $a \to \bar{a} = P(e_1^+)a$.

Proof. Because $e_2^+ \top e_1^+ \top e_2^- \perp e_2^+$ we can apply Lemma 2.4 and obtain that K is associative. So all what is left to show, is commutativity. For this we prove some auxiliary identities:

(α) $\{u\ e_2^+\{e_2^-\ e_1^+\ v\}\} = \{e_2^-\ e_1^+\{\{v\ e_2^+\ e_1^+\}e_1^+\ u\}\}$ for $u,v \in V_2^+$.

By (I.1.14) we have

(*) $\qquad L(a,b)\{xyz\} = \{\{abx\}yz\} - \{x\{bay\}z\} + \{xy\{abz\}\}.$

We apply this formula to compute $L(e_2^-,e_1^+)\{\{v\ e_2^+\ e_1^+\}e_1^+\ u\}$. Using $L(e_i^\varepsilon,V_i^{-\varepsilon}) = 0$ because $e_i^+ \perp e_i^-$, the result is $\{\{L(e_2^-,e_1^+)v,e_2^+,e_1^+\}e_1^+\ u\} = L(u,e_1^+)\{e_1^+e_2^+\{e_2^-\ e_1^+\ v\}\}$. We apply (*) once more for $L(u,e_1^+) = L(a,b)$ and obtain the left side of (α).

(β) $L(e_2^-,e_1^+)L(e_2^+,e_1^+)c = P(e_2)P(e_1)c$ for $c \in V_1^+$, $e_i = e_i^+ + e_i^-$.

By (I.1.9) the left side of (β) equals $P(e_2^-,e_2^+)P(e_1^+)c + \{e_2^-\ P(e_1^+)e_2^+\ c\}$ $= P(e_2)P(e_1)c$ since $P(e_1^+)e_2^+ = 0$ and $P(e_1^+)c = P(e_1)c \in V_1(e_2^+) \cap V_1(e_2^-)$.

(γ) $\{\{e_2^+\ e_1^+\ b\}\ e_2^+\ P(e_2)P(e_1)a\} = P(e_2)P(e_1)(a\cdot b)$ for $a,b \in V_1^+$.

Because $P(e_2)P(e_1)a = \{e_2^-\ e_1^+\{e_2^+\ e_1^+\ a\}\}$ by (b) we can compute the left side of (γ) using (α) for $u = \{e_2^+\ e_1^+\ b\}$, $v = \{e_2^+\ e_1^+\ a\}$. Since $L(e_1^+,e_2^+)L(e_2^+,e_1^+)a = a$ the result is $L(e_2^-,e_1^+)\{a\ e_1^+\{e_2^+\ e_1^+\ b\}\}$. But $L(e_2^+,e_1^+)|V_1^+ = T_{(12)}|V_1^+$ by (2.17), hence $\{a\ e_1^+\ T_{(12)}b\} = T_{(12)}\ \{T_{(12)}a,e_2^+,b\} = T_{(12)}(a\cdot b)$, and thus $L(e_2^-,e_1^+)\{a\ e_1^+\{e_2^+e_1^+\ b\}\} = L(e_2^-,e_1^+)L(e_2^+,e_1^+)a\cdot b = P(e_2)P(e_1)a.b.$ by (β).

Next, we choose $f = e_i = e_i^+ + e_i^-$ for $1,2,i \neq$ (case $\#I > 3$) or $f = e_0$ (case $\#I = 2$). Then $P(e_2)|V_1^+ = \delta P(f)|V_1^+$ and $P(f)P(e_1^+)|V_2^+ = \delta\mathrm{Id}$ for $\delta = \pm 1$ by (2.25), (2.26). Applying the automorphism $P(f)P(e_1)$ to (γ)

yields

(δ) $\{\{e_2^+ \, e_1^+ \, b\} \, e_2^+ \, (P(f)P(e_1))^2 a\} = (P(f)P(e_1))^2(a \cdot b)$

For $f = e_i$ we have $(P(f)P(e_1))^2 = \mathrm{Id}$ by (2.27) and for $f = e_0$ we also have $(P(f)P(e_1))^2 = P(P(e_0)e_1)P(e_1) = P(e_1)^2 = \mathrm{Id}$. Hence (δ) becomes $b \cdot a = a \cdot b$. ⊞

We now can coordinatize the part of V which is covered by $\mathcal{Q}_e(I)$:

Theorem 2.6. Assume V satisfies (2.7) and (2.21). Define an extension K of k as in Lemma 2.5. Then

$$C_V(\mathcal{Q}_e(I)) = \bigoplus_{i \in I} (V_i^+ \oplus V_i^-) \cong [K, {}^-, 2I]$$

Proof. As in Lemma 2.5 we rename two distinct elements of I as 1 and 2 and define the extension K of k on V_1^+.

For easier notation we likewise write $T(\Delta)$ for T_Δ. Let $\Delta_i \in \{\pm 1\}^I$ be defined by $\Delta_i(i) = -1$ and $\Delta_i(j) = 1$ for $j \neq i$ and put

$$\text{(2.29)} \qquad \begin{array}{ll} \phi_1^+ = \mathrm{Id} \quad , & \phi_1^- = T(\Delta_1), \\ \phi_j^+ = T_{(1j)}, & \phi_j^- = T(\Delta_1)\phi_j^+ T(\Delta_1) = T(\Delta_j \Delta_1)\phi_j^+, \quad j \neq 1. \end{array}$$

Note that all ϕ_i^ε are automorphisms of order 2 mapping V_i^ε onto V_1^+ and that the second equation for ϕ_j^- follows from (2.24). Now we obtain a module isomorphism

$$\text{(2.30)} \qquad F: C_V(\mathcal{Q}_e(I)) \to [K, {}^-, 2I]: x_i^\varepsilon \to \phi_i^\varepsilon(x_i^\varepsilon)e_i^\varepsilon,$$

where $\{e_i^\varepsilon\}$ also denotes the natural base for $[K, {}^-, 2I]$. The proof that F actually is a triple isomorphism will be facilitated by the following formulas for $\sigma \in \Sigma_I$, $\Delta \in \{\pm 1\}^I$

$$\text{I)} \qquad T_\sigma F = F T_\sigma \quad , \qquad \text{i.e.} \ (\phi_i^\varepsilon x_i^\varepsilon) e_{\sigma i}^\varepsilon = (\phi_{\sigma i}^\varepsilon T_\sigma x_i^\varepsilon) e_{\sigma i}^\varepsilon,$$

$$\text{II)} \qquad T(\Delta) F = F T(\Delta) \ , \quad \text{i.e.} \ (\phi_i^\varepsilon x_i^\varepsilon) e_i^{\Delta(i)\varepsilon} = \phi_i^{\Delta(i)\varepsilon} \, T_\Delta(x_i^\varepsilon) e_i^{\Delta(i)\varepsilon}.$$

Note that the operations of the automorphisms T_σ resp. $T(\Delta)$ of $[K, {}^-, 2I]$ are given by (2.16) resp. (2.20). To prove I) and II) let $e = 1$ for $\varepsilon = -$ and $e = 0$ for $\varepsilon = +$, then ϕ_i^+ and ϕ_i^- can be simultaneously expressed as $\phi_1^\varepsilon = T(\Delta_1)^e$, $\phi_j^\varepsilon = T(\Delta_j\Delta_1)^e T_{(1j)}$, $j \neq 1$.

Re I): The case $\sigma i = i$ being clear by (2.17), we may assume $\sigma i \neq i$. For $i = 1$ we have $\phi_{\sigma 1}^\varepsilon T_\sigma x_1^\varepsilon = T(\Delta_1 \Delta_{\sigma 1})^e T_{(1\sigma 1)} \, T_\sigma x_1^\varepsilon =$ (by (2.18)) $T(\Delta_1\Delta_{\sigma 1})^e x_1^\varepsilon =$ (by (2.23)) $T(\Delta_1)^e x_1^\varepsilon = \phi_1^\varepsilon x_1^\varepsilon$. Similarly we get for $i \neq 1$, but $\sigma i = 1$ $\phi_{\sigma i}^\varepsilon T_\sigma x_i^\varepsilon = T(\Delta_1)^e T_{(i1)} x_i^\varepsilon = \phi_i^\varepsilon x_i^\varepsilon$, and for $i, \sigma i, 1 \neq$ we have $\phi_{\sigma i}^\varepsilon T_\sigma x_i^\varepsilon = T(\Delta_1\Delta_{\sigma i})^e T_{(1\sigma i)} T_{(i\sigma i)} x_i^\varepsilon = T(\Delta_1\Delta_{\sigma i})^e T_{(i1)} \, x_i^\varepsilon = \phi_i^\varepsilon x_i^\varepsilon$.

Re II): The case $\Delta(i) = 1$ is obvious, so we assume $\Delta(i) = -1$. For $i = 1$ we have $\phi_1^{-\varepsilon} T(\Delta) x_1^\varepsilon = T(\Delta_1)^{1-e} T(\Delta) x_1^\varepsilon = T(\Delta_1)^e x_1^\varepsilon = \phi_1^\varepsilon x_1^\varepsilon$ and for $i \neq 1$

we obtain $\phi_i^{-\varepsilon}T(\Delta)x_i^\varepsilon = T(\Delta_1\Delta_i)^{1-e}T_{(1i)}T(\Delta_i)x_i^\varepsilon = T(\Delta_1\Delta_i)^{1-e}T(\Delta_1)T_{(1i)}x_i^\varepsilon =$ $T(\Delta_1)^e T(\Delta_i)T_{(1i)}x_i^\varepsilon = \phi_i^\varepsilon x_i^\varepsilon$.

Now we are in the position to show that F is an isomorphism of triple systems. Comparing the multiplication rules (2.10)-(2.15) with the ones for $[K,\bar{\ },2I]$ given in Example I.1.6.a, we only have to prove for all $i,j \in I$, $i \neq j$:

(a) $$F(P(x_i^\varepsilon)y_i^\varepsilon) = (F(x_i^\varepsilon)\cdot\overline{F(y_i^\varepsilon)}\cdot F(x_i^\varepsilon))e_i^\varepsilon$$

(b) $$F\{x_i^\varepsilon\ y_i^\varepsilon\ z_j^\mu\} = (F(x_i^\varepsilon)\cdot\overline{F(y_i^\varepsilon)}\cdot F(z_j^\mu))e_j^\mu$$

(c) $$F\{x_i^+\ y_j^\mu\ z_i^-\} = -(F(x_i^+)\cdot\overline{F(y_j^\mu)}\cdot F(z_i^-))e_j^{-\mu}$$

Since the ϕ_i^ε are automorphisms, (a) reduces to $P(a)b = a\cdot\bar{b}\cdot a$ for $a,b \in V_1^+$, a formula which holds by (1.20). For (b) let $\sigma \in \Sigma_I$ such that $\sigma(i) = 1$, $\sigma(j) = 2$ and let $\Delta \in \{\pm 1\}^I$ satisfy $\Delta(1) = \varepsilon$, $\Delta(2) = \mu$. Applying $T_\Delta T_\sigma$ to both sides of (b) and using I) and II), reduces (b) to $T_{(12)}\{a\ b\ T_{(12)}c\} = a\cdot\bar{b}\cdot T_{(12)}c$ for $a,b,c \in V_1^+ = K$, which holds by (1.21). Finally, for (c) we take $\sigma \in \Sigma_I$ satisfying $\sigma i = 2$, $\sigma j = 1$ and $\Delta \in \{\pm 1\}^I$ with $\Delta(1) = \mu$, $\Delta(j) = 1$ for $j \neq 1$. Again we apply $T(\Delta)T_\sigma$ to (c) and use I) and II). We are left with the equation

(c') $$T(\Delta_1)\{\phi_2^+(a)\ b\ \phi_2^-(c)\} = -a\cdot\bar{b}\cdot c \quad \text{for } a,b,c \in K.$$

Now $\{\phi_2^+(a)\ b\ \phi_2^-(c)\} = \{\{e_2^+\ e_1^+\ a\}\ b\ \phi_2^- c\} =$ (by (I.1.14)) $\{e_2^+\ e_1^+\{a\ b\ \phi_2^-(c)\}\} + \{a\{e_1^+\ e_2^+\ b\}\phi_2^- c\} - \{a\ b\{e_2^+\ e_1^+\ \phi_2^- c\}\} =$ (by (2.12) and (2.15)) $\{e_2^+\ e_1^+\{a\ b\ \phi_2^-(c)\}\} =$ (by (2.17)) $- T_{(12)}\{a\ b\ \phi_2^-(c)\} =$ (by (b)) $- T_{(12)}\phi_2^-(a\cdot\bar{b}\cdot c) = -T(\Delta_1)(a\cdot\bar{b}\cdot c)$. ∎

Corollary 2.7. (even-dimensional quadratic form coordinatization) *For #I > 3 a Jordan triple system V is covered by an even-dimensional quadratic form grid $\mathcal{Q}_e(I)$ if and only if V is isomorphic to an even-dimensional quadratic form triple $[K,\bar{\ },2I]$.*

Finally, we coordinatize all of V - not only in the case of an odd-dimensional quadratic form triple, but also in the even-dimensional case.

Theorem 2.8. *Assume V satisfies (2.7) and (2.21). Define an extension K of k as in Lemma 2.5. Then*

$$V \cong [K,\bar{\ },X \oplus 2I].$$

Proof. Again we rename two distinct elements of I as 1 and 2 and define the extension K of k on V_1^+ by (2.28). We will use the automorphisms ϕ_i^ε defined in (2.29). First, we define an operation of K on X by

$$a \cdot x = \{a\ e_1^+\ x\} \qquad (a \in V_1^+,\ x \in X).$$

This turns X into a K-module: For $a,b \in V_1^+$ we have $(a\cdot b)\cdot x = L(x,e_1^+)\{\{a\ e_1^+\ e_2^+\}e_2^+\ b\}$ = (by (I.1.14)) $\{\{a\ e_1^+\ e_2^+\}e_2^+\{x\ e_1^+\ b\}\} = \{\phi_2^+(a),\phi_2^+(e_1^+),\phi_2^+(b\cdot x)\}$ (by (2.17)) $= \phi_2^+\{a\ e_1^+\ (b\cdot x)\} = a\cdot(b\cdot x)$.

Next we will make X into a quadratic form triple. We define a map

$$q\colon X \to K\colon x \to P(x)e_1^-$$

and an endomorphism

$$S\colon X \to X\colon x \to P(e_1)x = P(e_1^+,e_1^-)x$$

where as before $e_i = e_i^+ + e_i^-$. We claim that q is a quadratic form, i.e. $q(a\cdot x) = a^2\cdot q(x)$ and $q(x,y) = P(x,y)e_1^-$ is K-bilinear. The first equation is easily seen: A linearization of the identity (I.1.3) $P(P(x)y) = P(x)P(y)P(x)$ gives $P(\{xyz\}) + P(P(x)y,P(z)y) = P(x)P(y)P(z) + P(z)P(y)P(x) + P(x,z)P(y)P(x,z)$ (see [31] JP 20), which for $y = e_1^+$, $z = a \in K$, $x \in X$ applied to e_1^- yields $q(a\cdot x) = P(\{a\ e_1^+\ x\})e_1^- = P(a)P(e_1^+)P(x)e_1^- = P(a)\ \overline{q(x)}$ = (by the 1×2 coordinatization) $a^2\cdot q(x)$. To see that q(x,y) is K-bilinear, we need

(α) $\qquad \{x\ y\ e_i^\varepsilon\} = \{x\ e_i^{-\varepsilon}\ Sy\}$ for $x,y \in X$,

which is proven as follows: $\{x\ y\ e_i^+\} + \{x\ y\ e_i^-\} = \{x\ y\ e_i\} = \{x\ e_i\ P(e_i)y\}$ (by (I.1.25)) $= \{x\ e_i^+\ P(e_i)y\} + \{x\ e_i^-\ P(e_i)y\}$, and a comparison of components in V_i^ε gives $\{x\ y\ e_i^\varepsilon\} = \{x\ e_i^\varepsilon\ P(e_i)y\} = \{x\ e_i^\varepsilon\ Sy\}$ by (2.25). We use (α) to show

(β) $\qquad \{x\ y\ a\} = q(x,Sy)\cdot a$ for $x,y \in X$, $a \in K$.

The right side of (β) equals $\{\{\{x\ e_1^-\ Sy\}e_1^+\ e_2^+\}e_2^+\ a\} = \{\phi_2^+\{x\ e_1^-\ Sy\},e_2^+,a\} = \{\{x\ e_2^-\ Sy\}e_2^+\ a\} = \{\{x\ y\ e_2^+\}e_2^+\ a\}$ = (by (I.1.14)) $\{\{a\ e_2^+\ x\}y\ e_2^+\} - \{x\{e_2^+\ a\ y\}e_2^+\} + \{x\ y\{a\ e_2^+\ e_2^+\}\} = \{x\ y\ a\}$ by (2.15).

Now $q(a\cdot x,y) = \{\{a\ e_1^+\ x\}e_1^-\ y\}$ = (by (I.1.14)) $\{\{y\ e_1^-\ a\}e_1^+\ x\} - \{a\{e_1^-\ y\ e_1^+\}x\} + \{a\ e_1^+\{y\ e_1^-\ x\}\} = -\{a\ Sy\ x\} + \{a\ e_1^+\ q(y,x)\}$ = (by (β) and the 1×2 coordinatization 1.3) $-q(x,y)\cdot a + 2a\cdot q(x,y) = a\cdot q(x,y)$, hence q(x,y) is K-bilinear and q is a quadratic form.

The map S has the properties $S^2 = \mathrm{Id}$, $S(a\cdot x) = P(e_1)\{a\ e_1^+\ x\} = \bar{a}\cdot Sx$ and $\overline{q(x)} = P(e_1)P(x)e_1^- = P(P(e_1)x)P(e_1)e_1^- = q(Sx)$. Therefore X becomes a quadratic form triple as soon as we proved for $x,y \in X$

(γ) $\qquad P(x)y = q(x,Sy)\cdot x - q(x)\cdot Sy$

This will be done now: $P(x)y = P(x)P(e_1^-,e_1^+)Sy$ = (by (I.1.13)) $L(x,e_1^-)L(x,e_1^+)Sy - L(P(x)e_1^-,e_1^+)Sy$. The last term obviously is $q(x)\cdot Sy$ and the first equals $\phi_1^-\{P(x,Sy)e_1^+,e_1^-,x\}$ = (using $\phi_1^- \in \mathrm{Aut}\ V$)

$\{P(x,Sy)e_1^-, e_1^+, x\} = q(x,Sy)\cdot x$. Hence ($\gamma$).

Let $F\colon V \to [K,^-,X \oplus 2I]$ be the map, which on X is the identity and on $C_V(\mathcal{Q}_e(I))$ is the isomorphism (2.30). The multiplication rules for $[K,^-,X \oplus 2I]$, given in Example I.1.6.b, show that F is an isomorphism if the following 4 equations hold ($x,y \in X$, $x_i^\varepsilon \in V_i^\varepsilon$):

(a) $\phi_i^{-\varepsilon} P(x)x_i^\varepsilon = q(x)\cdot\overline{\phi_i^\varepsilon(x_i^\varepsilon)}$

(b) $\phi_i^\varepsilon\{x\ y\ x_i^\varepsilon\} = q(x,Sy)\cdot\phi_i^\varepsilon(x_i^\varepsilon)$

(c) $\{x_i^\varepsilon\ y_i^\varepsilon\ x\} = \phi_i^\varepsilon(x_i^\varepsilon)\cdot\overline{(\phi_i^\varepsilon\ (y^\varepsilon))}\ \cdot x$

(d) $\{x_i^\varepsilon\ x\ x_i^{-\varepsilon}\} = \phi_i^\varepsilon(x_i^\varepsilon)\cdot\phi_i^{-\varepsilon}(x_i^\varepsilon)\cdot Sx$

In the following let $x_i^\varepsilon = \phi_i^\varepsilon a$, then $\phi_i^{-\varepsilon}P(x)x_i^\varepsilon = P(x)\phi_i^{-\varepsilon}\phi_i^\varepsilon a = P(x)\phi_1^- a$ by (2.29),(2.24),(2.23). Thus (a) is reduced to

(a') $P(x)\phi_1^- a = q(x)\cdot\bar{a}$ for $x \in X$, $a \in K$.

We have $P(x)\phi_1^- a = P(x)P(e_2^-,e_2^+)P(e_2)\phi_1^- a =$ (by (I.1.13)) $= -\{P(x)e_2^-, e_2^+, P(e_2)\phi_1^- a\}$, $P(e_2)\phi_1^- a = -P(e_2)^2 P(e_1)a = -\bar{a}$ and $P(x)e_2^- = P(x)P(e_1^-,e_1^+)P(e_1)e_2^- =$ (by (I.1.13)) $-\{P(x)e_1^-, e_1^+, P(e_1)e_2^-\} = \{q(x)\ e_1^+\ e_2^+\}$, whence (a').

Since $\phi_i^\varepsilon\{x\ y\ x_i^\varepsilon\} = \{x\ y\ \phi_i^\varepsilon x_i^\varepsilon\} = \{x\ y\ a\} =$ (by (β)) $q(x,Sy)\cdot a$, we have (b). Next, $\{x_i^\varepsilon\ y_i^\varepsilon\ x\} = \{\phi_i^\varepsilon a,\ \phi_i^\varepsilon b,\ \phi_i^\varepsilon x\} = \{a\ b\ x\} =$ (by (I.1.9)) $L(a,e_1^+)L(\bar{b},e_1^+)x = a\cdot\bar{b}\cdot x$, whence (c).

Finally (d): $\{x_i^\varepsilon\ x\ x_i^{-\varepsilon}\} = \{\phi_i^\varepsilon a,\ x,\ \phi_i^\varepsilon b_1^-\} = \phi_i^\varepsilon\{a\ x\ b_1^-\} = \{a\ x\ b_1^-\} = \{b_1^-\{e_1^+\ e_1^+\ x\}a\} = -\{\{e_1^+\ e_1^+\ b_1^-\}x\ a\} =$ (by (I.1.14)) $-[L(e_1^+,e_1^+), L(b_1^-,x)]a = [L(b_1^-,x), L(e_1^+,e_1^+)]a = L(\{b_1^-\ x\ e_1^+\},e_1^+)a = a\cdot\{b_1^-\ x\ e_1^+\}$ and $\{b_1^-\ x\ e_1^+\} = \{e_1^+\{e_1^-\ e_1^-\ x\}b_1^-\} - \{\{e_1^-\ e_1^-\ e_1^+\}x\ b_1^-\} = [L(e_1^+,x), L(e_1^-,e_1^-)]b_1^- = L(\{e_1^+\ x\ e_1^-\},e_1^-)b_1^- = \{b_1^-\ e_1^-\ Sy\} = \phi_1^-\{b_1^-\ e_1^-\ Sy\} = (\phi_1^- b_1^-)\cdot Sy$, whence (d). ∎

<u>Corollary 2.9.</u> <u>(odd-dimensional quadratic form coordinatization)</u> <u>For #I > 2 a Jordan triple system V is covered by an odd-dimensional quadratic form grid</u> $\mathcal{Q}_o(I) = \{e_o\} \cup \mathcal{Q}_e(I)$ <u>if and only if V is isomorphic to an odd-dimensional quadratic form triple</u> $[K,^-,X \oplus 2I]$ <u>with</u> $e_o \in X$.

§3 Coordinatization Theorems for the exceptional types

In this section we prove coordinatization theorems for Jordan triple systems covered by a Bi-Cayley or an Albert grid. Unlike in the previous sections we do not prove symmetries theorems, instead we obtain the 2 coordinatization theorems using the previously proven coordinatization theorems.

3.1. The coordinatizations we will deal with are defined in terms of split octonion algebras. We recall their definition for an arbitrary base ring k.

We start out with A_0, the algebra direct sum of two copies of k, i.e. $A_0 = \{(\alpha,\beta);\ \alpha,\beta \in k\}$ with the multiplication

$$(\alpha_1,\alpha_2)(\beta_1,\beta_2) = (\alpha_1\beta_1,\alpha_2\beta_2).$$

Note that A_0 has an involution ' and a quadratic from q given by

$$(\alpha,\beta) \to (\alpha,\beta)' = (\beta,\alpha), \quad q(\alpha,\beta) = \alpha\beta.$$

The next step is to perform a doubling process which is the "split" version of the general Cayley-Dickson-construction: We define a new algebra A_1 on $A_0 \oplus A_0$ by

(3.1) $$(a_1,a_2)(b_1,b_2) = (a_1b_1 + b_2'a_2, b_2a_1 + a_2b_1')$$

This algebra A_1, called the <u>split quaternion algebra (over k)</u>, has the same unit as A_0. The involution ' and the quadratic form q are extended in the following way

(3.2) $$(a_1,a_2) \to (a_1,a_2)' = (a_1', -a_2)$$
$$q(a_1,a_2) = q(a_1) - q(a_2)$$

Finally, the same doubling process is performed once more. We obtain A_2, the <u>split octonion algebra (over k)</u>, as the algebra on $A_1 \oplus A_1$ whose product is given by (3.1). The involution ' and the quadratic form q are extended as in (3.2). Note that (A_2,q) is a hyperbolic space which justifies the attribute "split". It is easily seen that $x \in A_2$ satisfies

(3.3) $$xx' = q(x)1 = x'x, \quad 1 = \text{unit element}$$

We also remark that A_1 is still associative but no longer commutative and that A_2 is merely alternative. For more details about these algebras (which however will not be needed in the following) the reader is referred to [4]VI, [21]2.3 or [53]III.

For later use we exhibit a special base for A_2:

(3.4)
$$c_1^+ = (1,0), \qquad c_1^- = (0,1) \quad \in \quad A_0 \subset A_2,$$
$$c_2^+ = (0,c_1^-), \qquad c_2^- = (0,-c_1^+) \quad \in \quad A_1 \subset A_2,$$

$$c_3^+ = (0, c_1^-), \qquad c_3^- = (0, -c_1^+) \quad \in \quad A_2 = A_1 \oplus A_1,$$
$$c_4^+ = (0, c_2^-), \qquad c_4^- = (0, -c_2^+) \quad \in \quad A_2 = A_1 \oplus A_1.$$

Obviously, the involution ' acts on this base as

$$(c_1^\varepsilon)' = c_1^{-\varepsilon}, \; (c_j^\varepsilon)' = -c_j^\varepsilon \text{ for } j > 1.$$

By (3.1) we have the following multiplication rules ($2 \leq j \leq 4$):

(3.5)
$$\begin{aligned}
&c_1^\varepsilon c_1^\varepsilon = c_1^\varepsilon, && c_1^\varepsilon c_1^{-\varepsilon} = 0,\\
&c_1^\varepsilon c_j^{-\varepsilon} = c_j^{-\varepsilon}, && c_j^\varepsilon c_1^\varepsilon = c_j^\varepsilon,\\
&c_1^\varepsilon c_j^\varepsilon = 0, && c_j^\varepsilon c_1^{-\varepsilon} = 0,\\
&c_j^\varepsilon c_j^\varepsilon = 0, && c_j^\varepsilon c_j^{-\varepsilon} = -c_1^{-\varepsilon},\\
&c_j^\varepsilon c_k^{-\varepsilon} = 0, && c_j^\varepsilon c_k^\varepsilon = -\operatorname{sgn}\begin{pmatrix}2&3&4\\ j&k&\ell\end{pmatrix} c_\ell^{-\varepsilon}
\end{aligned}$$

where $\operatorname{sgn}\begin{pmatrix}2&3&4\\ j&k&\ell\end{pmatrix}$ is the signature of the permutation $\begin{pmatrix}2&3&4\\ j&k&\ell\end{pmatrix}$.

This means that (c_1^+, c_1^-) are orthogonal idempotents with $c_1^+ + c_1^- = 1$ and c_j^+ resp. c_j^- lie in the Peirce spaces $(A_2)_{21}$ resp. $(A_2)_{12}$ (Actually the computation of (3.5) is facilitated by using Peirce identities derived in [53] Proposition 3.4). The complete multiplication table is

(3.6)

	c_1^+	c_1^-	c_2^+	c_2^-	c_3^+	c_3^-	c_4^+	c_4^-
c_1^+	c_1^+	0	0	c_2^-	0	c_3^-	0	c_4^-
c_1^-	0	c_1^-	c_2^+	0	c_3^+	0	c_4^+	0
c_2^+	c_2^+	0	0	$-c_1^-$	$-c_4^-$	0	c_3^-	0
c_2^-	0	c_2^-	$-c_1^+$	0	0	$-c_4^+$	0	c_3^+
c_3^+	c_3^+	0	c_4^-	0	0	$-c_1^-$	$-c_2^-$	0
c_3^-	0	c_3^-	0	c_4^+	$-c_1^+$	0	0	$-c_2^+$
c_4^+	c_4^+	0	$-c_3^-$	0	c_2^-	0	0	$-c_1^-$
c_4^-	0	c_4^-	0	$-c_3^+$	0	c_2^+	$-c_1^+$	0

In the following we denote by $\mathbb{O}$ the split octonion algebra over K where $K \supset k$ is an extension of k. We further assume that K has a k-linear automorphism $\kappa: K \to K: a \to \bar{a}$ which is of period 2 ($\kappa = \mathrm{Id}$ allowed). It then follows from (3.5) that

(3.7) $S: \mathbb{O} \to \mathbb{O} : ac_i^\varepsilon \to \bar{a}c_i^{-\varepsilon}$ is a κ-linear automorphism of $\mathbb{O}$ of order 2.

Since S commutes with the involution ' we get

(3.8) $*: \mathbb{O} \to \mathbb{O} : u \to u^* = Su'$ is a κ-linear involution of $\mathbb{O}$ commuting with '.

There is the following close connection between $\mathbb{O}$ and quadratic form triples:

Lemma 3.1. Let $(e_i^\varepsilon;\ 1 \leq i \leq 4,\ \varepsilon = \pm)$ be the natural grid in the quadratic form triple $[K,\bar{\ },8]$ exhibited in Example I.1.6.a. We define the structure of a split octonion algebra over K on the module $[K,\bar{\ },8]$ by multiplication table (3.6) identifying the e_i^ε with the c_i^ε. Then the triple product of $[K,\bar{\ },8]$ is given by

(3.9) $$P(x)y = q(x,Sy)x - q(x)Sy = xy^*x$$

where q, S and * are defined in (3.2), (3.7), (3.8) respectively. Conversely, $P(x)y = xy^*x$ defines on $\mathbb{O}$ a Jordan triple system which has $(c_i^\varepsilon;\ 1 \leq i \leq 4,\ \varepsilon = \pm)$ as a covering 8-dimensional quadratic form grid.

Proof. By definition $P(x)y = \tilde{q}(x,\tilde{S}y)x - \tilde{q}(x)\tilde{S}y$ where the quadratic form $\tilde{q}$ and the map $\tilde{S}$ are given by the formulas in Example I.1.6.a. A comparison of these formulas with (3.2), (3.7) shows $\tilde{q} = -q$ and $\tilde{S} = -S$. This proves the first equation of (3.9). Linearizing (3.3) gives $xy' + yx' = q(x,y)1$, whence $xy'x = q(x,y)x - q(x)y$ because $yx'\cdot x = y\cdot x'x = q(x)y$ holds in $\mathbb{O}$. Now $q(x,Sy)x - q(x)Sy = xy^*x$ follows.

Conversely, $\mathbb{O}$ together with $P(x)y = xy^*x$ is a subsystem of the Jordan triple system $Mat(1,2;\mathbb{O},*)$ defined in Example I.1.3.a, hence a Jordan triple system. That indeed $(c_i^\varepsilon;\ 1 \leq i \leq 4,\ \varepsilon = \pm)$ is a covering 8- dimensional quadratic form grid is straightforward to check using (3.5) and Theorem II.2.12. ∎

3.2. As just mentioned $Mat(1,2;\mathbb{O},*)$ is a Jordan triple system (over k). Its underlying module is $Mat(1,2;\mathbb{O})$ and its product is $P(u)v = u(v^{*t}u)$. Using the usual abbreviation $a.bc = a(bc)$ for $a,b,c \in \mathbb{O}$ and writing $u,v \in Mat(1,2;\mathbb{O})$ in the form $u = (u_1,u_2)$, $v = (v_1,v_2)$ we get

(3.9) $$P(u)v = (u_1v_1^*u_1 + u_2\cdot v_2^*u_1,\ u_2v_2^*u_2 + u_1\cdot v_1^*u_2)$$

and in linearized form

(3.10) $$\{uvw\} = (u_1\cdot v_1^*w_1 + w_1\cdot v_1^*u_1 + u_2\cdot v_2^*w_1 + w_2\cdot v_2^*u_1,\\ u_2\cdot v_2^*w_2 + w_2\cdot v_2^*u_2 + u_1\cdot v_1^*w_2 + w_1\cdot v_1^*u_2)$$

The following results suggest to call $Mat(1,2;\mathbb{O},*)$ for $\mathbb{O}$ the split Cayley algebra over $K \supset k$ and * as defined in (3.8) a Bi-Cayley triple $B(K,\kappa)$.

Lemma 3.2. Using the notations defined above we put for $1 \leq i \leq 4$

(3.11) $$e_i^\varepsilon = (c_i^\varepsilon,0),\ e_{i+4}^\varepsilon = (0,c_i^\varepsilon).$$

Then $\mathcal{B} = (e_i^\varepsilon,\ 1 \leq i \leq 8,\ \varepsilon = \pm)$ is a covering Bi-Cayley grid of the Bi-Cayley triple $B(K,\kappa)$.

Proof. Throughout let $2 \leq j \leq 4$. We note

(3.12) $\qquad c_1^{\varepsilon *} = c_1^\varepsilon, \quad c_j^{\varepsilon *} = -c_j^{-\varepsilon},$

whence by (3.4) $c_1^\varepsilon c_1^{\varepsilon *} c_1^\varepsilon = c_1^\varepsilon$ and $c_j^\varepsilon c_j^{\varepsilon *} c_j^\varepsilon = c_j^\varepsilon c_1^\varepsilon = c_j^\varepsilon$. Therefore (3.9) shows that $\mathcal{B}$ is a family of non-zero tripotents in $V = B(K,\kappa)$.

Next we determine the Peirce decomposition for e_i^ε, $1 \leq i \leq 4$: Since $P(e_i^\varepsilon)v = (c_i^\varepsilon v_1^* c_i^\varepsilon, 0)$ we get $V_2(e_j^\varepsilon) = \operatorname{im} P(e_i^\varepsilon) = Ke_i^\varepsilon$. By (3.10) $L(e_1^\varepsilon, e_1^\varepsilon)w = (e_1^\varepsilon w_1 + w_1 e_1^\varepsilon, e_1^\varepsilon w_2)$ which implies $Ke_1^{-\varepsilon}, Ke_5^{-\varepsilon}, Ke_{j+4}^\varepsilon \subset V_0(e_1^\varepsilon)$ and $Ke_j^\mu, Ke_{j+4}^{-\varepsilon} \subset V_1(e_1^\varepsilon)$. Similarly $L(e_j^\varepsilon, e_j^\varepsilon)(w_1,w_2) = (-c_j^\varepsilon \cdot c_j^\varepsilon w_1 + w_1 c_1^\varepsilon, -c_j^\varepsilon \cdot c_j^{-\varepsilon} w_2)$ gives for $i \neq j$ $Ke_i^\mu, Ke_{i+4}^{-\varepsilon}, Ke_{j+4}^\varepsilon \subset V_1(e_j^\varepsilon)$ and $Ke_j^{-\varepsilon}, Ke_{j+4}^{-\varepsilon}, Ke_{i+4}^\varepsilon \subset V_0(e_j^\varepsilon)$.

We obtain the Peirce decomposition for e_{i+4}^ε by applying the automorphism $(u_1,u_2) \to (u_2,u_1)$. As a result we see that $\mathcal{B}$ satisfies all the relations (II.3.7) and that the spaces Ke_i^ε are Peirce spaces relative to $\mathcal{B}$. Consequently each collinear pair in $\mathcal{B}$ is rigid-imbedded in V. Lemma I.3.7 now implies that $\mathcal{B}$ is pure. That it is indeed a Bi-Cayley grid follows from Theorem II.3.1, because we have the following multiplication rules:

$$e_j^- = -\{e_1^+\ e_j^+\ e_1^-\},\quad e_{j+4}^+ = -\{e_5^+\ e_{j+4}^-\ e_j^-\}$$

(which both follow from $c_j^- = -c_1^+ \cdot c_j^+ * c_1^- + c_1^- \cdot c_j^+ * c_1^+$) and

$$e_5^+ = \{e_3^-\ e_2^+\ e_8^-\},\quad e_{j+4}^- = -\{e_1^+\ e_j^+\ e_5^-\}$$

(which are equivalent to $c_1^+ = c_3^- \cdot c_2^+ * c_4^-$, $c_j^- = -c_1^+ \cdot c_j^+ * c_1^-$). ∎

We now come to the converse of Lemma 3.2: the Bi-Cayley coordinatization. A simple application of the 1×2 coordinatization shows that a Jordan triple covered by a Bi-Cayley grid is isomorphic to $\mathrm{Mat}(1,2;D,\iota)$ where D is an alternative algebra over k with involution ι. The main work now is to prove that D actually is a split octonion algebra over an extension of k. To do this we will make use of the quadratic form coordinatization.

Bi-Cayley Coordinatization Theorem 3.3. A Jordan triple system is covered by a Bi-Cayley grid if and only if it is isomorphic to a Bi-Cayley triple.

More precisely, let V be covered by $\mathcal{B} = (e_i^\varepsilon;\ 1 \leq i \leq 8,\ \varepsilon = \pm)$. Put $e_i = e_i^+ + e_i^-$ and define an algebra D on $V_2(e_1)$ with involution ι by

(3.13) $\quad u \cdot v = \{\{u e_1 e_5\} e_5 v\}, \quad u^{\iota} = P(e_1)u$

Then there exists an extension K _of_ k _with a_ k-_linear automorphism_ κ _of period_ 2 _and an algebra isomorphism_ $\phi: D \to \mathbb{O}$ _satisfying_

(3.14) $\quad \phi(u^{\iota}) = \phi(u)^*$

($\mathbb{O}$ = _split octonion algebra over_ K, * _as in_ (3.8)). _The map_

$$\phi: V \to B(K,\kappa) = Mat(1,2;\mathbb{O},*): x = x_2 + x_1 \to (\phi x_2, \phi\{e_1\ e_5\ x_1\})$$

is a Jordan triple isomorphism mapping $\mathcal{B}$ _onto the Bi-Cayley grid_ (3.11).

Proof. Since V is covered by $\mathcal{B}$ we have

$$V = \oplus_{i=1}^{8} (V_i^+ \oplus V_i^-),$$

the V_i^{ε} being the Peirce spaces relative to $\mathcal{B}$, namely $V_i^{\varepsilon} = V_2(e_i^{\varepsilon})$. It is easy to see that (e_1, e_5) is a collinear pair satisfying $V = V_{(21)} \oplus V_{(12)}$. Indeed,

$$V_{(21)} = \oplus_{i=1}^{4} (V_i^+ \oplus V_i^-), \quad V_{(12)} = \oplus_{i=5}^{8} (V_i^+ \oplus V_i^-)$$

By the 1 × 2 Coordinatization Theorem 1.3.

$$V \to Mat(1,2;D,\iota): x_{21} \oplus x_{12} \to (x_{21}, \{e_1\ e_5\ x_{12}\})$$

is an isomorphism where D and ι are given by (3.13).
Note $\{e_1\ e_5\ e_{j+4}^{\varepsilon}\} = \{e_1^{\varepsilon}\ e_5^{\varepsilon}\ e_{j+4}^{\varepsilon}\} = e_j^{\varepsilon}$ by (II.3.5) and (II.3.6). Now we use the fact that $(e_j^{\varepsilon};\ 1 \le j \le 4,\ \varepsilon = \pm)$ is an 8-dimensional quadratic form grid covering $V_{(21)}$. Hence, by Theorem 2.6, there exists a Jordan triple isomorphism

$$\phi: V_{(21)} \to [K,^-,8]: x_i^{\varepsilon} \to \phi_i^{\varepsilon}(x_i^{\varepsilon})e_i^{\varepsilon}$$

where $x_i^{\varepsilon} \in V_i^{\varepsilon}$ and the ϕ_i^{ε} are defined in (2.29). Moreover, the extension K of k is defined on V_1^+ by

$$ab = \{\{a\ e_1^+\ e_2^+\}\ e_2^+\ b\}$$

It has an automorphism $a \to \bar{a} = P(e_1^+)a$. By Lemma 3.1 we know that $[K,^-,8]$ carries the structure of a split octonion algebra $\mathbb{O}$ over K such that the product of $[K,^-,8]$ is given by $P(x)y = xy^*x$. Since ϕ is an isomorphism of triple systems we have $\phi(uv^{\iota}u) = \phi(u)\phi(v)^*\phi(u)$ and since $\phi(1_D) = 1_{\mathbb{O}}$ we get (3.14):

$$\phi(v^{\iota}) = \phi(v)^* \text{ for all } v \in D$$

Therefore (3.9) shows that the map:

$$Mat(1,2;D,\iota) \to Mat(1,2;\mathbb{O},*): (x_1,x_2) \to (\phi x_1, \phi x_2)$$

is a triple isomorphism iff $\phi: D \to \mathbb{O}$ is an isomorphism of algebras, i.e. if

(*) $\quad \phi(a_i^{\varepsilon} \cdot b_{\ell}^{\delta}) = \phi(a_i^{\varepsilon})\phi(b_{\ell}^{\delta})$ for $a_i^{\varepsilon} \in V_i^{\varepsilon}$, $b_{\ell}^{\delta} \in V_{\ell}^{\delta}$.

The proof of (*) is divided into several cases:

(i) $\quad \phi(a_1^{\varepsilon} \cdot b_1^{\varepsilon}) = \phi(a_1^{\varepsilon})\phi(b_1^{\varepsilon}), \quad \phi(a_1^{\varepsilon} \cdot b_1^{-\varepsilon}) = \phi(a_1^{\varepsilon})\phi(b_1^{-\varepsilon})$

We have $a_1^\varepsilon \cdot b_1^\varepsilon$ = (since $e_1^\varepsilon \perp e_5^{-\varepsilon}$) $\{\{a_1^\varepsilon\ e_1^\varepsilon\ e_5^\varepsilon\}e_5^\varepsilon\ b_1^\varepsilon\}$ = (by Lemma 2.4) $\{\{a_1^\varepsilon\ e_1^\varepsilon\ e_2^\varepsilon\}e_2^\varepsilon\ b_1^\varepsilon\}$. Since ϕ is a Jordan triple isomorphism we get $\phi(a_1^\varepsilon \cdot b_1^\varepsilon) = (\phi_1^\varepsilon(a_1^\varepsilon)\cdot\phi_1^\varepsilon(b_1^\varepsilon))e_1^\varepsilon = \phi(a_1^\varepsilon)\phi(b_1^\varepsilon)$ by (3.3). Further $a_1^\varepsilon \cdot b_1^{-\varepsilon} = 0$ in D, but also $e_1^\varepsilon e_1^{-\varepsilon} = 0$ in $\mathbb{O}$.

In the following let $j > 2$:

(ii) $\phi(a_1^\varepsilon \cdot b_j^{-\varepsilon}) = \phi(a_1^\varepsilon)\phi(b_j^{-\varepsilon})$, $\phi(a_j^\varepsilon \cdot b_1^\varepsilon) = \phi(a_j^\varepsilon)\phi(b_1^\varepsilon)$

We have $a_1^\varepsilon \cdot b_j^{-\varepsilon} = \{\{a_1^\varepsilon\ e_1^\varepsilon\ e_5^\varepsilon\}e_5^\varepsilon\ b_j^{-\varepsilon}\}$ = (by (I.1.14)) $\{a_1^\varepsilon\ e_1^\varepsilon\{e_5^\varepsilon\ e_5^\varepsilon\ b_j^{-\varepsilon}\}\}$ - $\{a_1^\varepsilon\{e_1^\varepsilon\ b_j^{-\varepsilon}\ e_5^\varepsilon\}e_5^\varepsilon\} + \{\{a_1^\varepsilon\ e_1^\varepsilon\ b_j^{-\varepsilon}\}e_1^\varepsilon\ e_5^\varepsilon\} = \{a_1^\varepsilon\ e_1^\varepsilon\ b_j^{-\varepsilon}\}$ because $\{e_1^\varepsilon\ b_j^{-\varepsilon}\ e_5^\varepsilon\} = 0 = \{c_j^{-\varepsilon}\ e_1^\varepsilon\ e_5^\varepsilon\}$ in V. Thus $\phi(a_1^\varepsilon \cdot b_j^{-\varepsilon}) = \{\phi a_1^\varepsilon, e_1^\varepsilon, \phi b_j^{-\varepsilon}\}$ (in [K, ¯,8]) $= \phi_1^\varepsilon(a_1^\varepsilon)\cdot\phi_1^{-\varepsilon}(b_j^{-\varepsilon})e_j^{-\varepsilon} = \phi(a_1^\varepsilon)\phi(b_j^{-\varepsilon})$ by (3.3). The second equation follows by applying (3.14): $\phi(a_j^\varepsilon \cdot b_1^\varepsilon)^* = \phi(P(e_1)b_1^\varepsilon \cdot P(e_1)a_j^\varepsilon)$ = (by the first equation of (ii)) $\phi(P(e_1)b_1^\varepsilon)\phi(P(e_1)a_j^\varepsilon) = \phi(b_1^\varepsilon)^*\phi(a_j^\varepsilon)^* = [\phi(a_j^\varepsilon)\phi(b_1^\varepsilon)]^*$.

(iii) $\phi(a_1^\varepsilon \cdot b_j^\varepsilon) = \phi(a_1^\varepsilon)\phi(b_j^\varepsilon)$, $\phi(a_j^\varepsilon \cdot b_1^{-\varepsilon}) = \phi(a_j^\varepsilon)\phi(b_1^{-\varepsilon})$

Indeed, $a_1^\varepsilon \cdot b_j^\varepsilon = \{\{a_1^\varepsilon\ e_1^\varepsilon\ e_5^\varepsilon\}e_5^\varepsilon\ b_j^\varepsilon\} = 0$ since $e_5^\varepsilon \perp e_j^\varepsilon$, and also $\phi(a_1^\varepsilon)\phi(b_j^\varepsilon) = 0$ because $e_1^\varepsilon e_j^\varepsilon = 0$ in $\mathbb{O}$. The second equation follows in the same way.

(iv) $\phi(a_j^\varepsilon \cdot b_j^\varepsilon) = \phi(a_j^\varepsilon)\phi(b_j^\varepsilon)$

For both sides are zero.

(v) $\phi(a_j^\varepsilon \cdot b_j^{-\varepsilon}) = \phi(a_j^\varepsilon)\phi(b_j^{-\varepsilon})$

We have $a_j^\varepsilon \cdot b_j^{-\varepsilon} = \{\{a_j^\varepsilon\ e_1^\varepsilon\ e_5^\varepsilon\}e_5^\varepsilon\ b_j^{-\varepsilon}\}$ = (by (I.1.14)) $\{a_j^\varepsilon\ e_1^\varepsilon\{e_5^\varepsilon\ e_5^\varepsilon\ b_j^{-\varepsilon}\}\}$ - $\{a_j^\varepsilon\{e_1^\varepsilon\ b_j^{-\varepsilon}\ e_5^\varepsilon\}e_5^\varepsilon\} + \{\{a_j^\varepsilon\ e_5^\varepsilon\ b_j^{-\varepsilon}\}e_1^\varepsilon\ e_5^{-\varepsilon}\} = \{a_j^\varepsilon\ e_1^\varepsilon\ b_j^{-\varepsilon}\}$, whence $\phi(a_j^\varepsilon \cdot b_j^{-\varepsilon}) = \{\phi a_j^\varepsilon\ e_1^\varepsilon\ \phi b_j^{-\varepsilon}\} = -(\phi_j^\varepsilon(a_j^\varepsilon)\cdot\phi_j^{-\varepsilon}(b_j^{-\varepsilon}))e_1^{-\varepsilon} = \phi(a_j^\varepsilon)\phi(b_j^{-\varepsilon})$.

In the following let $j \neq k > 2$. Then

(vi) $\phi(a_j^\varepsilon \cdot b_k^{-\varepsilon}) = \phi(a_j^\varepsilon)\phi(b_k^{-\varepsilon})$

because both sides vanish. By (i) - (vi) ϕ will be an automorphism as soon as

(vii) $\phi(a_j^\varepsilon \cdot b_k^\varepsilon) = \phi(a_j^\varepsilon)\phi(b_k^\varepsilon)$

It is enough to show (vii) for $\varepsilon = -$, since this implies the case $\varepsilon = +$ by applying the involutions in D resp. $\mathbb{O}$. We have $a_j^- \cdot b_k^-$ = $\{\{a_j^-\ e_1^-\ e_5^-\}e_5^+\ b_k^-\} \in V_\ell^+$ where $\{j,k,\ell\} = \{2,3,4\}$. Therefore $\phi(a_j^- \cdot b_k^-) = \phi_\ell^+(a_j^- \cdot b_k^-)e_\ell^+$, where $\phi_\ell^+(a_j^- \cdot b_k^-) = L(e_1^+, e_\ell^+)(a_j^- \cdot b_k^-)$ = (by (I.1.14))

$\{\{e_1^+\ e_5^+\{a_j^-\ e_1^-\ e_5^-\}\}e_5^+\ b_k^-\} - \{\{a_j^-\ e_1^-\ e_5^-\}\{e_\ell^+\ e_1^+\ e_5^+\}b_k^-\} +$ $\{\{a_j^-\ e_1^-\ e_5^-\}e_5^+\{e_1^+\ e_\ell^+ b_k^-\}\}$. Since $\{a_j^-\ e_1^-\ e_5^-\} \in V_{j+4}^- \subset V_1(e_1^+)$ the first summand vanishes, also $\{e_1^+\ e_\ell^+\ b_k^-\} = 0$. Because $\{e_\ell^+\ e_1^+\ e_5^+\} = e_{\ell+4}^+$ we are left with $-\{\{a_j^-\ e_1^-\ e_5^-\}e_{\ell+4}^+\ b_k^-\}$ = (by (I.1.14)) $- \{a_j^-\ e_1^-\{e_5^-\ e_{\ell+4}^+\ b_k^-\}\}$ $- \{e_5^-\{e_1^-\ a_j^-\ e_{\ell+4}^+\}b_k^-\} + \{e_5^-\ e_{\ell+4}^+\{a_j^-\ e_1^-\ b_k^-\}\} = - \{a_j^-\ e_1^-\{e_5^-\ e_{\ell+4}^+\ b_k^-\}\}$. By (II.3.3) $(e_j^-, e_k^+, e_5^-, \delta e_{\ell+4}^+)$, $\delta = \operatorname{sgn}\begin{pmatrix}1 & 2 & 3 & 4\\ j & k & 1 & \ell\end{pmatrix} = \operatorname{sgn}\begin{pmatrix}2 & 3 & 4\\ j & k & \ell\end{pmatrix}$ is a quadrangle. Therefore by (I.2.3) $\{e_5^-\ e_{\ell+4}^+\ b_k^-\} = \delta\{e_k^+\ e_j^-\ b_k^-\}$, hence $\phi(a_j^-\cdot b_k^-) = -\delta(\{a_j^-\ e_1^-\{e_k^+\ e_j^-\ b_k^-\}\})e_\ell^+$. On the other hand $\phi(a_j^-)\phi(b_k^-) =$ $(\phi_j^-(a_j^-)\cdot\phi_k^-(b_k^-))e_j^- e_k^- = -\delta(\phi_j^-(a_j^-)\cdot\phi_k^-(b_k^-))e_\ell^+$, hence (vii) follows from

(viii) $\qquad \phi_j^-(a_j^-)\cdot\phi_k^-(b_k^-) = \{a_j^-\ e_1^-\{e_k^+\ e_j^-\ b_k^-\}\}$

Let $b = \phi_k^- b_k^- \in V_1^+$. Then $\phi_j^-(a_j^-)\cdot b = \{\{\phi_j^-(a_j^-)\ e_1^+\ e_5^+\}e_5^+\ b\}$ = (Lemma 2.4) $\{\{\phi_j^-(a_j^-)\ e_1^+\ e_2^+\}e_2^+\ b\} = \{\phi_2^+\phi_j^-(a_j^-)\ e_2^+\ b\}$ by (2.29), where $\phi_2^+\phi_j^-(a_j^-) =$ $T_{(12)}T(\Delta_1)T_{(1j)}a_j^-$ = (by (2.24)) $T_{(12)}T_{(1j)}T(\Delta_j)a_j^- = T_{(1j2)}T(\Delta_j)\ a_j^-$ = (by (2.18)) $T_{(2j)}T(\Delta_j)\ a_j^-$. Hence $\phi_j^-(a_j^-)\cdot\phi_k^-(b_k^-) = \{T_{(2j)}T(\Delta_j)a_j^-, e_2^+, b\} =$ $T_{(2j)}\{T(\Delta_j)a_j^-, e_j^+, b\}$ = (since $\{T(\Delta_j)a_j^-, e_j^+, b\} \in V_1^+$) $\{T(\Delta_j)a_j^-, e_j^+, b\}$ = (by (2.22)) $T(\Delta_j)\{a_j^-\ e_j^-\ b\} = \{a_j^-\ e_j^-\ b\}$. We further compute $b = \phi_k^-(b_k^-) = T(\Delta_1)T_{(1k)}b_k^- = T(\Delta_1)\{e_1^- e_k^-\ b_k^-\} = \{e_1^+\ e_k^-\ b_k^-\}$ because $T(\Delta_1)|V_k^- = \mathrm{Id}$. Therefore $\phi_j^-(a_j^-)\cdot\phi_k^-(b_k^-) = \{a_j^-\ e_j^-\ b\} = \{\{e_1^+\ e_k^-\ b_k^-\}e_j^-\ a_j^-\}$ = (by (I.2.3) for $(e_1^+, e_k^+, e_1^-, -e_k^-)$) $- L(\{e_k^+\ e_1^-\ b_k^-\}, e_j^-)a_j^-$ = (by (I.1.14)) $- [L(e_k^+, e_1^-), L(b_k^-, e_j^-)]a_j^- = [L(b_k^-, e_j^-), L(e_k^+, e_1^-)]a_j^-$ = (by (I.1.14)) $\{\{b_k^-\ e_j^-\ e_k^+\}e_1^-\ a_j^-\}$. ∎

3.3. We are now in the position to tackle the problem of the Albert coordinatization. But, first of all, we need to give a realization of an Albert grid - a point which we had postponed in II§3.2.

As before we start out with a split octonion algebra $\mathbb{O}$ over an extension K of k, which has an automorphism κ: K → K: a → $\bar{a}$ of period 2. Clearly K1, $1 = c_1^+ + c_1^-$, lies in the nucleus of $\mathbb{O}$ (using (3.6) it is straightforward to check that the nucleus actually equals K1). Hence (3.3) says that the standard involution ' is nuclear and that K1 is an ample subspace of $\mathbb{O}$. Since * - defined in (3.8) - is a second involution of $\mathbb{O}$ commuting with the standard involution, we are in the setting of Example I.1.5 and may consider the hermitian Jordan triple system $H_3(\mathbb{O}, K, ', *) =: A(K,\kappa)$ which we will call an Albert triple. We recall that the underlying module of this triple is spanned by elements

$$a[ii] = aE_{ii},\ a \in K, \text{ and}$$
(3.15)
$$u[ij] = uE_{ij} + u'E_{ji},\ u \in \mathbb{O},\ 1 \leq i < j \leq 3.$$

Lemma 3.4. The family $A = (1[ii]; 1 \leq i \leq 3) \cup (c_r^\varepsilon[ij]; 1 \leq r \leq 4,\ \varepsilon = \pm,\ (ij) = (12),(13),(32))$, where the c_r^ε are defined in (3.4), is an Albert grid, which covers the Albert triple $A(K,\kappa)$.

Proof. We put $V = A(K,\kappa)$ and note

$$c_1^\varepsilon[ij] = c_1^{-\varepsilon}[ji],\ c_r^\varepsilon[ij] = -c_r^\varepsilon[ji] \text{ for } r \geq 2,$$

i.e. (II.3.10) is fulfilled. The multiplication rules given in Example I.1.5 show the following:

(i) $\mathbb{O}[ij] \oplus \mathbb{O}[ik]$ for $i,j,k \neq$ is a subsystem of V isomorphic to $\mathrm{Mat}(1,2; \mathbb{O}, *),\ = B(K,\kappa)$,

(ii) all elements $e \in A$ are tripotents with $V_2(e) = Ke$,

(iii) $\mathbb{O}[ij] \subset V_1(1[ii]) \cap V_0(1[kk])$, $K[ii] \subset V_1(c_r^\varepsilon[ij]) \cap V_0(1[jj])$ for $i,j,k \neq$.

By (i)-(iii) A is an ortho-collinear family and the spaces Ke are Peirce spaces relative to A, in particular A covers V. Since all collinear pairs in A are rigid-imbedded A is pure by Lemma I.3.7. That A is an Albert grid follows from Theorem II.3.4: Choosing $c = 1[11]$ and using Lemma 3.2 the only assumptions which remain to be checked are the formulas for $1[22]$, $1[33]$ and $c_r^\varepsilon[32]$. But this is easily done. ∎

We finally come to the Albert coordinatization which we will obtain as an application of both the hermitian as well as the Bi-Cayley coordinatization.

Albert Coordinatization Theorem 3.5. A Jordan triple system V is covered by an Albert grid A iff V is isomorphic to an Albert triple. There exists an isomorphism mapping A onto the canonical Albert grid defined in Lemma 3.4.

Proof. Let V be covered by an Albert grid $A = ([i]; 1 \leq i \leq 3) \cup ({}_r^\varepsilon[ij];\ \varepsilon = \pm,\ 1 \leq r \leq 4,\ 1 \leq i < j \leq 3)$. Put

$$h_{ii} = [i]\ (1 \leq i \leq 3),\ h_{ij} = {}_1^+[ij] + {}_1^-[ij] = h_{ji}\ (1 \leq i < j \leq 3).$$

Then $h_{11} \dashv h_{12} \top h_{23} \perp h_{11}$ and $\{h_{11}\, h_{12}\, h_{23}\} = h_{13}$ by the properties of an Albert grid, whence $(h_{11}, h_{12}, h_{23}, h_{13})$ is a diamond by the Diamond Criterion I.2.7. Because $h_{ii} = P(h_{1i})h_{11}$ for $i = 2,3$ Theorem I.2.11

implies that $(h_{ij};\ 1\leq i\leq j\leq 3)$ is a family of 3×3 hermitian matrix units. We denote by V_{ij} its Peirce spaces (they are the same as the usual Peirce spaces relative to the orthogonal system (h_{11},h_{22},h_{33})). To apply the Hermitian Coordinatization Theorem we need to show for $1\leq i,j\leq 3,\ i\neq j$

$$L(h_{ij},h_{ii}):\ V_{ii}\to V_{ij}\quad \text{is injective.}$$

Indeed, $L(h_{ij},h_{ii})x_{ii} = L(\overset{+}{_1}[ij],\ [i])x_{ii} \oplus L(\overset{-}{_1}[ij],\ [i])x_{ii}$ with $L(\overset{\varepsilon}{_1}[ij],\ [i])x_{ii} = Bx_{ii} \in V_2(\overset{\varepsilon}{_1}[ij])$ where B is the exchange automorphism between the collinear tripotents $[i]$ and $\overset{\varepsilon}{_1}[ij]$. By the Hermitian Coordinatization Theorem 1.9 there exists an isomorphism ψ between V and a hermitian triple system $H_3(D,D_0,\pi,^-)$ mapping (h_{ij}) onto the canonical hermitian matrix units (H_{ij}). We may define the data $(D,\pi,^-)$ on V_{12} by

$$u\bullet v = \{\{u\ h_{12}\ h_{13}\}h_{13}\ v\},\quad \bar{u} = P(h_{12})u,\quad u^{\pi} = P(h_{11},h_{22}^-)\bar{u}$$

Since $(\overset{\varepsilon}{_r}[12],\overset{\varepsilon}{_s}[13];\ \varepsilon=\pm,\ 1\leq r,s\leq 4)$ is a covering Bi-Cayley grid for $V_{12}\oplus V_{13}$ we can also apply the Bi-Cayley Coordinatization Theorem 3.3 to see that there exists an algebra isomorphism ϕ: $D\to \mathbb{O}$, $\mathbb{O}$ split, such that

(*) $$\phi(\bar{u}) = \phi(u)^{*}\quad\text{and}\quad \phi(u^{\pi}) = \phi(u)'$$

Whereas the first equation follows from (3.13) and (3.14), the second is proven by showing $\phi(\bar{u}^{\pi}) = \phi(\bar{u})'$, i.e. $\phi\{h_{11}\ u\ h_{22}\} = S\phi(u)$ for S defined in (3.7). Recall that ϕ is the isomorphism ϕ: $V_{12}\to[K,^-,8]$ constructed in (2.29). Because $(\overset{\varepsilon}{_r}[12])\cup([1],-[2])$ is a 10-dimensional quadratic form grid covering $V_{12}\oplus V_{11}\oplus V_{22}$ we can extend ϕ to the isomorphism $\tilde{\phi}$: $V_{12}\oplus V_{11}\oplus V_{22}\to[K,^-,10]$ with $\tilde{\phi}(h_{11}) = e_5^+$, $\tilde{\phi}(h_{22}) = -e_5^-$, then $\phi\{h_{11}u\ h_{22}\} = \tilde{\phi}\{h_{11}\ u\ h_{22}\} = -\{e_5^+,\tilde{\phi}u,e_5^-\} = -\{e_5^+,\phi u,e_5^-\}$. We compute this triple product in $[K,^-,10]$ as defined in Example I.1.6 using $\mathfrak{S}$ for the S there. Then $-\{e_5^+\ \phi u\ e_5^-\} = q(e_5^+,e_5^-)\mathfrak{S}\phi u = -\mathfrak{S}\phi u$. Comparing the definitions of $\mathfrak{S}$ and S shows $-\mathfrak{S}\phi(u) = S\phi(u)$ proving the second equation in (*).

The ample subspace $D_0 = L(h_{12},h_{11})V_{11}$ is mapped by ϕ onto $K1\subset\mathbb{O}$. Therefore (*) shows that applying ϕ to all entries of a matrix in $H_3(D,D_0,\pi,^-)$ produces an isomorphism Φ: $H_3(D,D_0,\pi,^-)\to A(K,\kappa)$. It is easy to see that $\Phi\psi$ maps A onto the canonical Albert grid in $A(K,\kappa)$. ∎

CHAPTER IV. CLASSIFICATIONS

In this final chapter we apply the results of the previous chapters to obtain classifications of the following classes of Jordan triple systems:

§1: Simple Jordan triple systems over a ring which are covered by a grid.

§2: Jordan triple systems on real or complex Hilbert spaces whose scalar product is associative with respect to the continuous triple product (called Hilbert triples).

§3: Atomic JBW*-triples, a certain class of Jordan triple systems on dual complex Banach spaces.

§1 Simple Jordan triple systems

In this section we classify simple Jordan triple systems which are covered by a grid.

We start out with

Lemma 1.1. *Let V be a simple Jordan triple system which is covered by a grid E. Then E is connected.*

Proof. Otherwise there exists a non-trivial decomposition $E = E_1 \cup E_2$ such that $e_1 \perp e_2$ for $e_i \in E_i$. Since V is covered by E we get a corresponding decomposition $V = V_1 \oplus V_2$ where V_i is the sum over all Peirce spaces $V_I(E)$ which contain an element of E_i. Obviously $V_1 \subset V_0(e_2)$ for any $e_2 \in E_2$. We even have

$$V_1 = \bigcap_{e_2 \in E_2} V_0(e_2),$$

since the right side is a sum of Peirce spaces relative to E. But then V_1 is a subsystem of V and every product involving factors of V_1 and V_2 vanishes. Therefore V_1 is a non-trivial ideal of V. ∎

We can now apply the results of chapter II: Since associated cogs have the same cover, V is covered by one of the 7 standard grids (see the Grid Classification Theorem at the end of II§3). Then the coordinatization theorems proven in chapter III reveal the structure of V - except in the 2 cases where $E = \{e\}$ or E is a triangle. Whereas the structure of V in the first case is unknown (V need not be a division triple - see Example I.2.4.a), the structure of a simple V covered by a triangle was determined in [43]. To explain the result we need to generalize the quadratic form example:

<u>Example 1.2</u> (quadratic forms, revisited). Let $V = [K,^{-},X\oplus 2I]$, $I = \{1\}$ be a quadratic form example as defined in Example I.1.6.b. Hence V contains 2 orthogonal tripotents (e_1^+,e_1^-) with Peirce decomposition

$$\begin{aligned} V &= V_2(e_1^+) \oplus V_1(e_1^+) \cap V_1(e_1^-) \oplus V_2(e_1^-) \\ &= Ke_1^+ \quad \oplus \quad X \quad \oplus Ke_1^-, \end{aligned}$$

and $q = q_X$ and S satisfy

$$q(ae_1^+ \oplus x \oplus be_1^-) = -ab + q(x),$$

$$S(ae_1^+ \oplus x \oplus be_1^-) = -\overline{b}e_1^+ \oplus Sx \oplus -\overline{a}e_1^-$$

The generalization of this construction arises in the following way: Let $K_0 \subset K$ be a k- submodule of K. Then

$$\tilde{V} = K_0e_1^+ \oplus X \oplus K_0e_1^-$$

is a subsystem of V (and hence a Jordan triple system, denoted by $[K,K_0,^{-},X]$) iff for $a_0,b_0 \in K_0$, $x \in X$

(1.1) $\quad 1 \in K_0$, $a_0^2\overline{b}_0 \in K_0$ and $a_0q(x) \in K_0$

If (1.1) is satisfied one calls K_0 an <u>ample subspace</u>. In case (e_1^+,e_1^-) imbeds into a triangle $(e_0;e_1^+,e_1^-)$ we get $q(e_0) = 1$ and (1.1) is equivalent to

(1.1') $\quad 1 \in K_0$ and $\overline{a}_0b^2 \in K_0$ for $a_0 \in K_0$, $b \in K$.

We note that ample subspaces are a typical "characteristic - 2 - phaenomena": If K has no 2-torsion, then (1.1') implies that K is the only ample subspace of K. We further note that both, the even- and the odd-dimensional quadratic form triples can be viewed as a triple $[K,K_0,^{-},X]$. Before we can state the result from [43] we recall the following notations: If $(D,\pi,^{-})$ is an algebra with two involutions $d \to d^\pi$ and $d \to \overline{d}$, one calls B an <u>ideal</u> of $(D,\pi,^{-})$ if B is an ideal of D satisfying $B = \overline{B} = B^\pi$. One says $(D,\pi,^{-})$ is <u>simple</u> if the multiplication of D is non-zero and $(D,\pi,^{-})$ has only the trivial ideals $\{0\}$ and D.

<u>Triangle Coordinatization Theorem 1.3.</u>([43]) <u>V is a simple Jordan triple system covered by a triangle iff V is isomorphic to</u>

(i) <u>a hermitian matrix system</u> $H_2(D,D_0,\pi,^{-})$ <u>as defined in Example</u> I.1.5, <u>where</u> $(D,\pi,^{-})$ <u>is simple associative and</u> D_0 <u>generates</u> D <u>as an algebra</u>,

<u>or</u> (ii) <u>a quadratic form triple</u> $[K,K_0,^{-},X]$ <u>where</u> $q_X:X \to K$ <u>is non-degenerate</u>, $(K,^{-})$ <u>is simple and</u> X <u>contains an element</u> e_0 <u>satisfying</u> $q(e_0) = 1$.

We remark that one can coordinatize in such a way that also in the quadratic form case the ample subspace K_0 generates K.

We come back to the general object of this section, namely the classification of simple Jordan triple systems covered by a grid E with $\# E > 2$. By the theorem above we can assume that E is not a triangle. Then the coordinatization theorems of Chapter III give the general structure of V: It is isomorphic to one of the seven standard examples. Since these triple systems are in general not simple we investigate their ideals in the following lemmata.

Lemma 1.4. *The ideals of a rectangular matrix system* $\mathrm{Mat}(I,J;D,\bar{\ })$, $\#I + \#J > 2$, *are exactly the subspaces* $\mathrm{Mat}(I,J;C,\bar{\ })$ *where* C *is an ideal of* $(D,\bar{\ })$.

Proof. It is clear from the definition of the triple product (see Example I.1.5.) that any ideal C of $(D,\bar{\ })$ gives rise to the ideal $\mathrm{Mat}(I,J;C,\bar{\ })$ of $V = \mathrm{Mat}(I,J;D,\bar{\ })$. Conversely, let U be an ideal of V. Then for any $u = \sum_{ij} u_{ij}E_{ij} \in U$ we have $P(E_{ij})^2 u = u_{ij}E_{ij} \in U$, so $U = \oplus U_{ij}$, where $U_{ij} = U \cap V_2(E_{ij})$. Moreover, for $i \neq k$ we get $\{E_{kj},E_{ij},u_{ij}E_{ij}\} = u_{ij}E_{kj} \in U_{kj}$ and, by symmetry, $u_{ij}E_{i\ell} \in U_{i\ell}$ for any $\ell \in J$. It follows $U = \mathrm{Mat}(I,J;C)$ for a subspace C of D. Let $c \in C$, $d \in D$ and $\ell, j \in J$ be distinct. Then $\{cE_{ij}\ E_{ij}\ d\bar{E}_{il}\} = cdE_{i\ell}$ shows $cd \in C$. By symmetry $dc \in C$ and finally $P(E_{ij})cE_{ij} = \bar{c}E_{ij}$, so C an ideal of $(D,\bar{\ })$. ∎

The corresponding results for the remaining types of triple systems are proven in the same way. We therefore omit the proofs.

Lemma 1.5. *The ideals of a symplectic matrix system* $A(I;K,\bar{\ })$ *with* $\#I > 4$ *are exactly the subspaces* $A(I;B,\bar{\ })$ *where* B *is an ideal of* $(K,\bar{\ })$.

Lemma 1.6. *The ideals of a hermitian matrix system* $H_I(D,D_0,\pi,\bar{\ })$ *with* $\#I > 3$ *are exactly the subspaces* $H_I(B,B_0,\pi,\bar{\ })$ *where* B *is an ideal of* $(D,\pi,\bar{\ })$ *and the subspace* B_0 *satisfies*

$$B_0 = \bar{B}_0 \subset D_0 \cap B, \qquad aB_0a^\pi \subset B_0 \text{ for } a \in D,$$

$$bD_0b^\pi \subset B_0 \text{ and } b + b^\pi \in B_0 \text{ for } b \in B.$$

In particular, $H_I(D,D_0,\pi,\bar{\ })$ *is simple iff* $(D,\pi,\bar{\ })$ *is simple.*

Lemma 1.7. *The ideals of a quadratic form triple* $[K,\bar{\ },X\oplus 2I]$ *with*

$X = 0$ *allowed and* $\#I > 2$ *are exactly the subspaces* $U_0 \oplus [B,^-,2I]$ *where* B *is an ideal of* $(K,^-)$ *and* $U_0 \subset X$ *satisfies*

$$q(U_0) \subset B,\ q(U_0,X) \subset B,\ KU_0 \subset U_0,\ BX \subset U_0 \text{ and } SU_0 = U_0.$$

In particular, for $X \neq 0$ $[K,^-,X+2I]$ *is simple iff* $(K,^-)$ *is simple and* q_X *is non-degenerate, and* $[K,^-,2I]$ *is simple iff* $(K,^-)$ *is simple.*

Lemma 1.8. *Let* $V = B(K,\kappa)$ *be a Bi-Cayley triple and* $(e_i^{\varepsilon};\ \varepsilon = \pm,\ 1 \leq i \leq 8)$ *be the natural grid, so that* $V = \oplus_{i=1}^{8}(Ke_i^+ \oplus Ke_i^-)$. *For any ideal* B *of* $(K,^-)$ *the subspace* $\oplus_{i=1}^{8}(Be_i^+ \oplus Be_i^-)$ *is an ideal of* V. *Conversely, any ideal of* V *arises in this way.*

Lemma 1.9. *Let* $V = A(K,\kappa)$ *be an Albert triple and* $(e_i;\ i = 1,\dots,27)$ *be the canonical Albert grid. Then for any ideal* B *of* $(K,^-)$ *the space* $Be_1 \oplus \dots \oplus Be_{27}$ *is an ideal of* V, *and every ideal of* V *arises in this way.*

It is natural to call $(D,^-)$ in the rectangular case, $(D,\pi,^-)$ in the hermitian matrix case and $(K,^-)$ in the remaining cases the *coordinate algebra*.

We recall that two algebras with involutions are called *isomorphic* if there exists an algebra isomorphism commuting with the involutions involved. Using this notion the structure of a simple coordinate algebra is described in the following lemma whose first part is well-known and whose second part is proven in the same way as the first.

Lemma 1.10. a) *Let* (D,π) *be an algebra with involution. Then* (D,π) *is simple iff either* D *is simple or* (D,π) *is isomorphic to* $(C \oplus C^{op}$, exchange) *where* C *is simple.*

b) *Let* $(D,\pi,^-)$ *be an algebra with two commuting involutions. Then* $(D,\pi,^-)$ *is simple iff either* (D,π) *is simple or* $(D,\pi,^-)$ *is isomorphic to* $(C \oplus C^{op}, \pi$, exchange) *where* $C = C^{\pi}$ *and* (C,π) *is simple.*

We recall that the coordinate algebra can be alternative, namely in the rectangular matrix case for $\#I + \#J = 3$ and in the hermitian matrix case for $\#I < 3$. The structure of simple properly alternative algebras was determined by Kleinfeld:

Theorem 1.11.([27]) *An algebra* D *over a ring* k *is simple and properly alternative iff* D *is a Cayley algebra over an extension field* of k.

In the remaining examples the coordinate algebra is always associative, and sometimes even commutative. Trivially, a simple commutative associative algebra is a field.

For the classification of hermitian matrix systems we need

Lemma 1.12. *Let* (D,π) *be a simple alternative algebra with involution and* $D_0 \subset \mathrm{Fix}\,\pi$ *an ample subspace (see* Example I.1.5.). *Then there are only the following possibilities:*

(i) $D = C \oplus C^{op}$ *for* C *a simple associative algebra,* $(a \oplus b)^\pi = b \oplus a$ *and* $D_0 = \{d \in D;\ d^\pi = d\} = H(D,\pi)$,

(ii) D *is simple associative,*

(iii) D *is a Cayley algebra over an extension field* K *of* k, π *is the standard involution and* $D_0 = K$.

Proof. Since (D,π) is simple either D is simple or $D = C \oplus C^{op}$ for C a simple alternative algebra and π the exchange involution (Lemma 1.10.a). In the last case $H(D,\pi) = \{a + a^\pi;\ a \in C\} = \{d + d^\pi;\ d \in D\} \subset D_0$ because D_0 contains all traces $d + d^\pi$. On the other hand, $D_0 \subset H(D,\pi)$, so $D_0 = H(D,\pi)$. The nucleus of D splits corresponding to $D = C \oplus C^{op}$. Since it contains D_0 it also contains C and C^{op}, thus D is associative.

For the remainder of the proof we may assume that D is simple and nonassociative. Then, by Theorem 1.11, D is a Cayley algebra over an extension field K of k. Such an algebra does not contain absolute zero divisors $\neq 0$. Moreover, for any $a \in D$ we have $aa^\pi \in D_0 \subset K$ since K is the nucleus of D. Thus (D,π) is a composition algebra in the sense of [21]6.2.2. Their structure has been determined in [21]6.2.3. The result shows that π is the standard involution and $D_0 = K$. ∎

We remark that the lemma above is a weak form of the Herstein-Kleinfeld-Osborn-McCrimmon-Theorem ([21]6.2.1.). That theorem assumes that every non-zero element of D_0 is invertible and obtains a more precise structure in case (ii).

The phaenomena in Lemma 1.10 that the coordinatizing algebra is a direct sum of two ideals exchanged by the involution $^-$ appears exactly when the triple system is polarized, i.e. the Jordan triple system of a Jordan pair:

Lemma 1.13. *A standard example (i.e. a rectangular matrix system* resp. ... resp. an Albert triple) is polarized iff its coordinate algebra is a direct sum of two ideals which are exchanged by the involution $^-$ and left invariant by π in the hermitian matrix case.

Proof. For $V \cong [K,K_0,{}^-,X]$ the lemma can be found in [43]. Thus, in what follows $V \not\cong [K,K_0,{}^-,X]$. It is a routine calculation to show that V is polarized, if the conditions of the lemma are met.

Conversely, if V is a polarized Jordan triple system, then every Peirce space $V_2(e)$ is polarized too. In each of the six types of examples there always exists a tripotent e so that $V_2(e)$ is isomorphic to $(D,{}^-)$ resp. $(K,{}^-)$ viewed as a Jordan triple system. Therefore the problem is to show the following:

Let D be an alternative algebra over k with k-linear involution $^-$ and suppose $D = D^+ \oplus D^-$ is a polarization of D with respect to $P(x)y = x\bar{y}x$. Then the D^ε are ideals of D exchanged by $^-$ and left invariant by π in the hermitian matrix case.

Indeed, $P(1)d^\varepsilon = \overline{d^\varepsilon} \in D^{-\varepsilon}$ shows $\overline{D^\varepsilon} = D^{-\varepsilon}$. For $1 = 1^+ \oplus 1^-$ we have $1 = \bar{1}$ whence $\overline{1^\varepsilon} = 1^{-\varepsilon}$, also $1^+1^- + 1^-1^+ = 1^+1^-1 + 1\,1^-1^+ = \{1^+1^+1\} = 0$ by the polarization property. Therefore $1 = 1^2 = 1^{+2} + 1^{-2}$. Because $1^{\varepsilon 2} = P(1^\varepsilon)1 \in D^\varepsilon$ we can conclude $1^\varepsilon = 1^{\varepsilon 2}$, i.e. $(1^+,1^-)$ are orthogonal idempotents of D. Since $\overline{d^\varepsilon} = P(1)d^\varepsilon = P(1^{-\varepsilon})d^\varepsilon = 1^{-\varepsilon}\,\overline{d^\varepsilon}\,1^{-\varepsilon}$ and therefore $d^\varepsilon = 1^\varepsilon d^\varepsilon 1^\varepsilon$, the Peirce multiplication rules in alternative algebras show that D^+ and D^- are ideals of D. In the hermitian case $D_0[11]$ is polarized, so $1^\varepsilon \in D_0 \cap D^\varepsilon$. Since $1^\pi = 1$ we have $1^{-\varepsilon}[11] = P(1[12])1^\varepsilon[22] = P(1^{-\varepsilon}[12])1^\varepsilon[22] = 1^{-\varepsilon}(1^{-\varepsilon}1^{-\varepsilon\pi})[11]$, thus $1^{-\varepsilon} = 1^{-\varepsilon\pi}$ and therefore also $D^{\varepsilon\pi} = D^\varepsilon$. ∎

We have now gained enough information to obtain a classification of simple Jordan triple systems which are covered by a grid E with $\# E > 2$.

We first treat the polarized case, i.e. $V = V^+ \oplus V^-$ is the Jordan triple of the Jordan pair (V^+,V^-). We will phrase our classification in Jordan pair notation, i.e. specify (V^+,V^-) and its quadratic representation $Q(x)y = (Q(x^+)y^-,\ Q(x^-)y^+)$. Surprisingly, it turns out that with one exception only always $V^+ = V^- = \mathcal{V}$ is a nonpolarized simple Jordan triple system, say with quadratic representation $\mathcal{P}$, and

$$Q(x)y = (Q(x^+)y^-,\ Q(x^-)y^+) = (\mathcal{P}(x^+)y^-,\ \mathcal{P}(x^-)y^+)$$

One says that the Jordan pair $(\mathcal{V},\mathcal{V})$ is associated to the Jordan triple system $\mathcal{V}$.

Jordan Pair Classification Theorem 1.14. A Jordan triple system over

k *is simple, polarized and covered by a grid* E *with* $\# E > 2$ *iff it is isomorphic to the Jordan triple system of one of the following Jordan pairs:*

(I) $(\mathrm{Mat}(I,J;D), \mathrm{Mat}(I,J;D^{op}))$, $\#I + \#J > 3$, *with* D *a simple associative algebra over* k *and* $Q(x)y = (x^+(y^-)^t x^+, x^-(y^+)^t x^-)$

(II) $(A_I(K), A_I(K))$, $\#I > 5$, K *an extension field of* k *and* $Q(x) = -(x^+y^-x^+, x^-y^+x^-)$

(III) $(H_I(D,D_0,^-), H_I(D,D_0,^-))$, $\#I > 2$, D *a simple associative algebra with involution* $^-$ *and ample subspace* D_0, $Q(x)y = (x^+y^-x^+, x^-y^+x^-)$

(IV) (V,V), *the Jordan pair associated to the quadratic form triple* V *with product* $P(x)y = q(x,y)x - q(x)y$, *where the quadratic form* q *on* V *has values in an extension field* K *of* k *and* (V,q) *are*

(IVa) $V = K^{2I}$, $\#I > 3$, $q = q_{2I}$ (see Example I.1.6)

(IVb) $V = K^{2I} \oplus X$, $\#I > 2$, X *a vector space over* K *with a nondegenerate quadratic form* $q_X : X \to K$ *and* $e_0 \in X$ *with*

$$q_X(e_0) = 1, \quad q = q_{2I} \oplus q_X$$

(IVc) $V = K_0 \oplus X \oplus K_0$, K_0 *an ample subspace of* K, X, q_X, e_0 *as in* (IVb) *and* $q(a_0 + x + b_0) = -a_0 b_0 + q_X(x)$.

(V) $(\mathrm{Mat}(1,2;C,^-), \mathrm{Mat}(1,2;C,^-))$, *the Jordan pair associated to the rectangular matrix system* $\mathrm{Mat}(1,2;C,^-)$ *where* C *is a Cayley algebra over an extension field* K of k *and* $^-$ *is the standard involution of* C .

(VI) $(H_3(C,K,^-), H_3(C,K,^-))$, *the Jordan pair associated to the exceptional Jordan algebra* $H_3(C,K,^-)$ *with* $(C,K,^-)$ *as in* (V).

In proving this classification theorem we will find out the isomorphism type of the cover $C_V(E)$ of a connected standard grid E in case $C_V(E)$ is simple. The result is given in the following table:

E	$R(1,2)$	$R(I,J)$, $\#I+\#J > 4$	$S(I)$, $\#I > 5$
$C_V(E)$	(I)or(V)	(I)	(II)

E	$H(2)$	$H(3)$	$H(I)$, $\#I > 4$
$C_V(E)$	(I),(III) or (IVc)	(I),(III) or(VI)	(I) or (III)

E	$\mathcal{Q}_e(I), \#I \geq 3$	$\mathcal{Q}_o(I)$, $\#I \geq 2$	$\mathcal{B}$ or $\mathcal{A}$
$C_V(E)$	(IVa)	(IVb)	(V) or (IV) for C split

Proof. Since E is connected by Lemma 1.1., the association type of E has been determined in Ch.II, thus the general structure of V follows from the results of Ch.III and Theorem 1.3: V is isomorphic to one of the standard examples. Therefore we only have to find out when are standard examples simple and polarized. The latter condition is described in Lemma 1.13 : The coordinate algebra is always a sum of 2 ideals exchanged by the involution $^-$. We now consider the standard examples individually:

If $V \cong Mat(I,J;D,^-)$ Lemmata 1.4 and 1.10.a imply $(D,^-) \cong (C \oplus C^{op}$, exchange) and consequently $Mat(I,J;D,^-) \cong Mat(I,J;C \oplus C^{op}$, exchange), but the latter is the triple system of the Jordan pair $(Mat(I,J;C), Mat(I,J;C^{op}))$ with quadratic representation as described in (I). We know that D, and thus also C is alternative for $\#I + \#J = 3$ and associative otherwise. Since C is simple, Theorem 1.11 shows that we have type (I) or C is a Cayley algebra, in which case we get type (V) because of the following observation:

(*) If C is an algebra with involution ι, then $(C \oplus C^{op}$, exchange) $\cong$ $(C \oplus C, (a,b) \to (b^\iota, a^\iota))$ via $(a,b) \to (a,b^\iota)$.

If $V \cong A_I(K,^-)$ then $(K,^-) \cong (L \oplus L$, exchange) where L is an extension field of k. Applying (*) with $\iota = Id$ shows that V is of type (II).

If $V \cong H_I(D,D_0,\pi,^-)$ then again $(D,^-) \cong (C \oplus C^{op}$, exchange) and $C^\pi = C$ by Lemma 1.13. We claim that the ample subspace D_0 of D splits correspondingly: Let $1 = e_1 \oplus e_2$ with $e_i \in C$, $e_2 = \overline{e}_1$. Then $e_i^\pi = e_i$ whence $e_i d_0 e_i = e_i d_0 e_i^\pi \in D_0$ shows $D_0 = C_0 \oplus C_0^{op}$ for $C_0 = D_0 \cap C$. Clearly, C_0 is an ample subspace of (C,π). Using (*) we now see that (V^+,V^-) is isomorphic to the Jordan pair associated to the hermitian matrix triple $H_I(C,C_0,\pi)$. Since (C,π) is simple we can apply Lemma 1.12: If C is nonassociative we get type (VI), otherwise there are 2 possibilities: Either C is simple and we get type (III), or $C = B \oplus B^{op}$ for a simple associative B and π is the exchange involution. Then every $x \in H_I(C,C_0,\pi)$ is of the form $x = a \oplus a^t$ with $a \in Mat(I,I;B)$ and $H_I(B \oplus B^{op}$; exchange) $\cong Mat(I,I;B)^J$ as triple systems, where $Mat(I,I;B)^J$ has the product $P(x)y=xyx$ (so $Mat(I,I;B)^J$ is the Jordan triple associated to the associative algebra $Mat(I,I;B)$). On the Jordan pair level we get a pair of type (I): $(Mat(I,I;B)^J, Mat(I,I;B)^J) \cong (Mat(I,I;B), Mat(I,I;B^{op}))$ via $(x^+,x^-) \to (x^+,x^{-t})$ using $(ab)^t = a^t {}_{op}\, b^t$

for $a,b \in Mat(I,I;B)$ where $a^t op\, b^t$ denotes the matrix product in $Mat(I,I;B^{op})$.

If V is isomorphic to a quadratic form triple with coordinate algebra K, we have $K = L \oplus L$ for an extension field L of k by Lemma 1.7 and the Triangle Coordinatization Theorem 1.3. Moreover, $\kappa: K \to K$ is the exchange involution, whence in the even-dimensional case $S : K^{2I} \to K^{2I}$ exchanges the two summands L^{2I}, so V is of type (IVa). If $V \cong K^{2I} \oplus X$ or $V \cong K_0 \oplus X \oplus K_0$ with $X \neq 0$ the splitting of $1 \in K = L \oplus L$ induces one of X: $X = X_+ \oplus X_-$ with $q(X_+,X_-) = 0$, $SX_\varepsilon = X_{-\varepsilon}$, and also a splitting of L_0. It then follows that V is of type (IVb) or (IVc).

Finally, if $V \cong B(K,\kappa)$ or $V \cong A(K,\kappa)$, then again $K = L \oplus L$ where L is an extension field of k and $\kappa : K \to K$ is the exchange involution (Lemmata 1.8 and 1.9). Correspondingly the data $\mathbb{O}$, *, ' needed to form $B(K,\kappa)$ resp. $A(K,\kappa)$ also split: $\mathbb{O} = \mathbb{O}_L \oplus \mathbb{O}_L$ where $\mathbb{O}_L$ is the split octonion algebra over L, $(a,b)^* = (b^*,a^*)$ and $(a,b)' = (a',b')$. The map $(\mathbb{O},*) \to (\mathbb{O}_L \oplus \mathbb{O}_L, (a,b) \to (b',a')) : (a,b) \to (a,b'^*) = (a,Sb)$ is an isomorphism. Therefore $B(K,\kappa) = Mat(1,2;\, \mathbb{O},*) \cong Mat(1,2;\, \mathbb{O}_L \oplus \mathbb{O}_L, (a,b) \to (b',a'))$, i.e. $B(K,\kappa)$ is of type (V). Similarly, $A(K,\kappa)$ is of type (VI). ∎

<u>Remark 1.15.</u> Since a polarized Jordan triple system is simple iff it is the Jordan triple system of a simple Jordan pair, the theorem above gives a classification of simple Jordan pairs which are covered by a grid. By Corollary I.6.7 this class contains all the simple and semisimple Jordan pairs with dcc on all inner ideals.

Another class of Jordan pairs for which a classification is known are simple and semisimple Jordan pairs with dcc and acc on principal inner ideals (see [31]12.12). These two classes are distinct: A Jordan pair of type $A(M,R,\phi)^J$ (notation of [31]) with R a simple Artinian k-algebra and ϕ non-degenerate is simple and has dcc and acc on principal inner ideals. However, if M is not a finitely generated R-module $A(M,R,\phi)^J$ need not be covered by a grid as Example I.6.5 shows. We note that $A(M,R,\phi)^J$ is the only type in Loos' list [31]12.12 which is not always covered by a grid. On the other hand, the types (I), (II) and (III) of Theorem 1.14 with infinite I and J have infinite rank and hence do not satisfy acc on principal inner ideals.

<u>Theorem 1.16.</u> <u>A Jordan triple system over k is simple, non-polarized and covered by a grid</u> E <u>with</u> $\#E > 2$ <u>iff it is isomorphic to one of the following triple systems:</u>

(I) <u>rectangular matrix system</u> $Mat(I,J;D,\bar{\ })$, $\#I + \#J > 3$, D <u>a simple associative algebra over</u> k,

(II) symplectic matrix system $A_I(K,^-)$, $\#I \geq 5$, K an extension field of k

(III) hermitian matrix system, more precisely

(IIIa) $H_I(D,D_0,\pi,^-)$, $\#I \geq 2$ and D a simple associative algebra over k,

(IIIb) $Mat(I,I;D)$, $\#I \geq 2$, D as in (IIIa) and product $P(x)y = xy^{\alpha}x$ for α a k-linear automorphism of D with $\alpha^2 = Id$,

(IV) quadratic form triple, more precisely, for K an extension field of k there are the following possibilities:

(IVa) $[K,^-,2I]$ (see Example I.1.6.a) with $\#I \geq 3$,

(IVb) $[K,^-,X \oplus 2I]$ (Example I.1.6.b) for $\#I \geq 2$, X a vector space over K, $q: X \to K$ a nondegenerate quadratic form and X contains e_0 satisfying $q(e_0) = 1$, $Se_0 = e_0$,

(IVc) $[K,K_0,^-,X]$ (see Example 1.2) where X is as in (IVb),

(V) rectangular matrix system $Mat(1,2;C,^-)$, C a Cayley algebra over an extension field K of k and $^-$ an involution of C,

(VI) hermitian matrix system $H_3(C;K,\pi,^-)$ where $(C,^-)$ is as in (V) and π is the standard involution.

We remark that for the isomorphism type of $C_V(E)$ the same table as in the polarized case holds.

Proof. As in the proof of Theorem 1.14 we know the general structure of V by the results of Chap.II and III and Theorem 1.3. Moreover, by Lemma 1.13 "non-polarized" is equivalent to the non-splitting of the coordinate algebra. This together with the lemmata on simplicity of the standard examples immediately gives the precise structure of V in all cases except when V is isomorphic to a hermitian matrix system $H_I(D,D_0,\pi,^-)$. In this case (D,π) is simple by Lemma 1.10.b which allows us to apply Lemma 1.12: the cases (iii) resp. (ii) there give type (VI) resp. (IIIa). In case (i), i.e. $D = C \oplus C^{op}$ for a simple associative C and π = exchange, every $x \in H_I(D,D_0,\text{exchange},^-)$ is of the form $x = a \oplus a^t$ for some $a \in Mat(I,I;C)$ and we get 2 possibilities for the product $P(x)y = x\bar{y}^t x = (a \oplus a^t)\overline{(b \oplus b^t)}(a \oplus a^t) = (a \oplus a^t)\overline{(b^t \oplus b)}(a \oplus a^t)$: Since $\bar{C}$ is an ideal of $D = C \oplus C^{op}$ we either have $\bar{C} = C$, in which case $V \cong Mat(I,I;C,^-)$, i.e. type (I), or we have $\bar{C}$

$= C^{op}$ which gives type (IIIb). ∎

Remark 1.17. The two theorems above give a classification of simple Jordan triple systems which are covered by a grid. A classification of simple Jordan triple systems over fields of characteristic $\neq 2$ and $\neq 3$ was recently obtained by E.I. Zel'manov in a series of articles ([56]). Obviously, these two classes of simple Jordan triple systems are distinct. Moreover the methods are quite distinct: Zel'manov uses for example Jordan algebra theory and the "classical" finite-dimensional Jordan triple theory.

Remark 1.18. For the standard examples in our Classification Theorems 1.14 and 1.16 the "natural" grid need not be a grid with minimal tripotents. Conditions for that have been derived in Lemma I.7.8 and Corollary I.7.9. For example: The Albert grid in the Albert triple $H(3;\mathbb{O},K,*)$ for K a field consists of minimal tripotents. Or, $(1,0) \in Mat(1,2;C,^{-})$ for an alternative algebra C over k is minimal if $H(C,^{-}) = \{c \in C;\ \bar{c} = c\}$ is a domain.

§2 Hilbert triples

In this section we completely determine the structure of Hilbert triples, a category of Jordan triple systems which includes hermitian Hilbert triples studied by Kaup in [25]. The main results are: spectral decomposition for elements in a Hilbert triple, decomposition of Hilbert triples in an orthogonal sum of a negative, trivial and positive ideal, decomposition of positive Hilbert triples in a Hilbert sum of simple positive Hilbert triples and classification of simple positive Hilbert triples.

In the following, whenever H is a Hilbert space over $\mathbb{K} = \mathbb{R}$ or $\mathbb{C}$, its scalar product and norm is denoted by $(.,.)$ and $\|.\|$ respectively. The algebra $B(H)$ of all bounded linear operators on H will always be equipped with the operator norm, also denoted by $\|.\|$, which makes $B(H)$ into a (real or complex) C*-algebra.

A <u>Hilbert triple</u> is a real Jordan triple system defined on a real Hilbert space H such that

(2.1) the left multiplication $L: H \times H \to B(H)$ is continuous, i.e.

$$\|L(x,y)\| \leq \lambda \|x\| . \|y\|$$

for all $x,y \in H$ and a universal non-negative constant λ, and

(2.2) the scalar product $(.,.)$ is associative, i.e.

$$(L(x,y)u,v) = (u,L(y,x)v)$$

for all $x,y,u,v \in H$.

A <u>hermitian Hilbert triple</u> is a real Jordan triple system defined on a complex Hilbert space H such that

(2.3) H is a Jordan - * - triple, i.e. the triple product $(x,y,z) \to \{xyz\}$ is $\mathbb{C}$-linear in x and z and conjugate-linear in y,

(2.4) (2.1) and (2.2) hold correspondingly, i.e. $L:H\times H \to B(H)$ is continuous and the hermitian scalar product $(.,.)$ is associative.

The connection between Hilbert triples and hermitian Hilbert triples is described in

<u>Lemma 2.1</u>. a) ("Reellification") <u>Every hermitian Hilbert triple is a Hilbert triple relative to the real scalar product</u> $\langle .,. \rangle = Re(.,.)$. <u>A Hilbert triple</u> H <u>is hermitian iff</u> H <u>has a complex structure</u> $i: H \to H$ <u>satisfying</u> for $x,y \in H$

$$(ix,iy) = (x,y) \quad \underline{\text{and}} \quad iL(x,y) = L(x,y)i$$

<u>In this case, the complex scalar product</u> σ <u>on the complex vector space</u> H

is given by $\sigma(x,y) = (x,y) - i(ix,y)$.

b) ("Hermitefication") *Every Hilbert triple* H *is a real form of a hermitian Hilbert triple, namely the one defined on* $\mathbb{C} \otimes H = H \oplus iH$ *with scalar product*

$$(x_1+ix_2, y_1+iy_2) = (x_1,y_1) + (x_2,y_2) + i[(x_2,y_1)-(x_1,y_2)]$$

for $x_i, y_i \in H$ *and Jordan triple product*

$$P(x_1+ix_2)(y_1+iy_2) = P(x_1)y_1 - P(x_2)y_1 + \{x_1y_2x_2\} \oplus i[-P(x_1)y_2 + P(x_2)y_2 + \{x_1y_1y_2\}]$$

Proof. All claims are easily checked. In b) note that the triple product actually is the κ-mutation of the natural complex Jordan triple system on $\mathbb{C} \otimes H$, where κ is the conjugation of $\mathbb{C} \otimes H$ in H: $\kappa(x_1+ix_2) = x_1 - ix_2$. ∎

Although by this lemma one could develop a theory for Hilbert triples by first studying hermitian Hilbert triples and then look at real forms, it is easier and more natural to study Hilbert triples right away and consider hermitian Hilbert triples as a special case of Hilbert triples. Before doing so we feel it is appropriate to make the following remark which shows the geometric significance of Hilbert triples and compares our notion to others already occuring in the literature.

Remark 2.2. The interest in Hilbert triples comes from the prominent role they play in the applications of Jordan triple systems in geometry: Hermitian Hilbert triples were introduced and classified by Kaup in [24] (under the name Hilbert triple), giving rise to a classification of symmetric hermitian Banach manifolds. Earlier the finite-dimensional case had been treated by Loos in [32] which is based on Koecher's lecture notes [28]: Finite-dimensional hermitian Hilbert triples with positive-definite trace form $\sigma(u,v) = \text{trace } L(u,v)$ (called positive hermitian Jordan triple systems in [32]) are in one-to-one correspondence with circled bounded symmetric domains. Arbitrary finite-dimensional Hilbert triples have been used by Backes and Reckziegel in [2] (under the name euclidean Jordan triple system): They are exactly the "initial data" of symmetric submanifolds in standard spaces of constant curvature. This algebraic characterization led to a classification of symmetric submanifolds, see also [1]. The work of Backes and Reckziegel was preceded by the work of Ferus on symmetric submanifolds in the euclidean space ([7]), which turned out to be standard imbedded symmetric R-spaces. The relevant Jordan triple structures here are finite-dimensional Hilbert triples with positive-definite trace form (called compact Jordan triple systems in [29], [33] and I §6) which are known to be in one-to-one correspondence with symmetric R-spaces. This fact was announded by Loos in [29] and proven in [33] where symmetric R-spaces are characterized among all compact symmetric spaces by a simple geometric property of their unit lattices.

The first aim in our investigation of Hilbert triples is to establish the existence of non-zero tripotents in Hilbert triples H satisfying $L(H,H) \neq 0$ (Theorem 2.3) - as Kaup did in [24]. After that, our approach is different from Kaup's: We will prove the main results (Decomposition Theorem 2.10, Classification Theorem 2.14 for positive Hilbert triples) in a different order and we will not use the theory and classification of Jordan pairs. Instead we will work with grids.

If in the following we attribute a result to Kaup ([23], [24]), this means that its proof is a straightforward generalization of his proof in the hermitian case. We include these proofs to be complete (and sometimes we will also give more details).

We start out with some immediate observations for Hilbert triples:

(2.5) $P(x)$ and $P(x,y)$ are selfadjoint operators, (indeed, $(\{xuy\},v) = (y,\{uxv\}) = (y,\{vxu\}) = (\{xvy\},u)$ by (2.2)),

(2.6) every closed subsystem is again a Hilbert triple,

(2.7) ([24](3.1)) A closed subspace $T \subset H$ is an ideal iff $L(H,H)T \subset T$. If $T \lhd H$ is an ideal, its orthogonal complement $T^{\perp}$ is an ideal too.

Lemma 2.3.([23] 1.10, 1.11) *Let H be a Hilbert triple and* $x,y \in H$. *Then*

(2.8) $P(x)y = \{xyy\} = \{yxx\} = 0 \leftrightarrow L(x,y) = 0 \leftrightarrow L(y,x) = 0$

For a single element we have

(2.9) $P(x)x = 0 \leftrightarrow L(x,x) = 0 \leftrightarrow P(x) = 0.$

Proof. Suppose $P(x)y = \{xyy\} = \{yxx\} = 0$. Then (I.1.12) becomes $L(x,y)P(x) + P(x)L(y,x) = P(x,\{xyx\}) = 0$. On the other hand $L(x,y)P(x) = P(x)L(y,x)$ is the defining Jordan triple identity (I.1.1), whence $L(x,y)P(x) = P(x)L(y,x) = 0$. Also, $L(x,y)^2 = 2P(x)P(y)$ by (I.1.13) and thus $L(x,y)^3 = 2L(x,y)P(x)P(y) = 0$. Because $[L(x,y),L(y,x)] =$ (by(I.1.14)) $L(\{xyy\},x) - L(y,\{yxx\}) = 0$ the operator $L(x,y)$ is normal, whence $L(x,y)^3 = 0$ implies $L(x,y) = 0$ and therefore also its adjoint $L(x,y)^* = L(y,x)$ vanishes. The remaining implications are obvious.

For a single element $L(x,x) = 0$ gives $P(x)^2 = 0$ by (I.1.9), whence $P(x) = 0$ since $P(x)$ is selfadjoint by (2.5). Clearly, $P(x) = 0 \rightarrow P(x)x = 0 \rightarrow L(x,x) = 0$ by (2.8). ∎

It is now necessary to generalize the notion of a tripotent and to deal (only for a short while) with generalized tripotents, i.e. elements $e \in H$ satisfying $P(e)e = \varepsilon e$ for $\varepsilon = \pm 1$. In this context, a tripotent in the usual sense is called a positive tripotent whereas e satisfying $P(e)e = -e$ is called a negative tripotent. If we denote by $-H$ the Hilbert triple arising from H by changing the sign of the quadratic representation P_H of H:

(2.10) $$-P_H = \text{quadratic representation of } -H,$$

then e is a negative tripotent in H iff $-e$ is a positive tripotent in $-H$. Using this remark it is straightforward to carry over properties of positive tripotents to the case of generalized tripotents. For example, we have a Peirce decomposition

(2.11) $$H = H_2(e) \oplus H_1(e) \oplus H_0(e)$$

where (since ½ is in the ground ring) for $e = \varepsilon P(e)e$

$$H_j(e) = \{x \in H;\ \varepsilon L(e,e)x = jx\}, \quad j = 0,1 \text{ or } 2,$$

are the eigenspaces of the selfadjoint operator $L(e,e)$ which are therefore orthogonal in the scalar product, as indicated in (2.11) by $\perp$.

Theorem 2.4.([24]3.3) Every Hilbert triple H with $L(H,H) \neq 0$ contains a non-zero generalized tripotent.

Proof. We may actually assume $y \in H$ with $P(y)y \neq 0$. Indeed, $P(x)x = 0$ for all $x \in H$ implies $P(x) = 0$ by (2.9), and thus $P(H,H) = 0$, i.e. $L(H,H) = 0$. For our particular y, the real span $\mathbb{R}\{y\}$ of the odd powers $y,\ y^3 = P(y)y\ ,\ldots,\ y^{2n+1} = P(y)y^{2n-1}$ of y is a nontrivial subsystem, which is flat: $\{uvw\} = \{vuw\}$ for all $u,v,w \in \mathbb{R}\{y\}$. Since the same properties hold for the closure of $\mathbb{R}\{y\}$, which is a Hilbert triple, it is more than enough to prove:

Every real flat Hilbert triple F with $L(F,F) \neq 0$ contains a
(*) generalized tripotent $e \neq 0$ such that

$$F = \mathbb{R}e \oplus F_0(e), \qquad F_2(e) = \mathbb{R}e.$$

To show (*), let $\hat{A} < B(F)$ = bounded operators on F be the closed subalgebra generated by $L(F,F)$. Note $a^* = a$ and $ab = ba$ for all $a,b \in \hat{A}$

by flatness of F. Then $A = \mathbb{R}\,\mathrm{Id} + \hat{A}$ is norm closed (being the sum of the closed $\hat{A}$ and a one-dimensional subspace) and obviously a subalgebra of symmetric operators, whence a real C*-algebra with identity involution ([10]8.2). We imbed A into a real commutative Banach algebra B by putting

$$B = A \oplus F, \quad \|a \oplus f\| = \|a\| + \|f\|,$$
$$(a_1 \oplus f_1)(a_2 \oplus f_2) = (a_1 a_2 + L(f_1,f_2)) \oplus a_1(f_2) + a_2(f_1)$$

(In order to have $\|b_1 b_2\| \leqslant \|b_1\| \cdot \|b_2\|$ for $b_i \in B$ we multiply the given scalar product on F by a positive constant so that $\lambda = 1$ holds in (2.1).)

We claim that there exists a complexified character of B, i.e. a unital $\mathbb{R}$-algebra homomorphism χ: $B \to \mathbb{C}$, such that $\chi|F \neq 0$. Assume otherwise: Since all χ's are continuous by [10]10.3, we get $\chi(\hat{A} \oplus F) = 0$, whence $\mathrm{Id} \notin \hat{A}$, $A = \mathbb{R}\,\mathrm{Id} \oplus \hat{A}$ and there is exactly one complexified character of B and exactly one maximal ideal of B (namely $\hat{A} \oplus F$): B is a local ring forcing A to be local too. Indeed, $\hat{A} \lhd A$ is a maximal ideal and every element $1 + \hat{a}$, $\hat{a} \in \hat{A}$, is invertible in B (by localness of B) and then, as is easily seen, also in A. This implies localness of A, hence A has exactly one character χ_A, and this χ_A satisfies $\ker \chi_A = \hat{A}$. However, by assumption there exists $f \in F$ with $0 \neq L(f,f) \in \hat{A} \subset A$. Therefore, by [10]11.1 and 10.4, there exists also a complexified character χ: $A \to \mathbb{C}$ such that $\chi(L(f,f)) \neq 0$, contradiction.

We may now assume that χ: $B \to \mathbb{C}$ is a complexified character such that $\chi(F) \neq 0$. Then, for every $x \in F$, we have $\chi(x)^2 = \chi(x^2) = \chi(L(x,x)) \in Sp_A^{\mathbb{C}}(L(x,x))$ by [10]10.4 since $\chi|A$ is a complexified character of A. But A is a real C*-algebra and hence for $L(x,x) = L(x,x)^*$ we have $Sp_A^{\mathbb{C}}(L(x,x)) \subset \mathbb{R}$ by [10]11.3. Altogether, we obtain $\chi(x) \in \mathbb{R}$ or $\chi(x) \in i\mathbb{R}$ for all $x \in F$. Since $\chi(F)$ is a real subspace of $\mathbb{C}$, which is non-zero by our choice of χ, we have $\chi(F) = \mathbb{R}$ or $\chi(F) = i\mathbb{R}$. Therefore, $T = \ker(\chi|F)$ is a closed ideal of F of codimension 1 and there exists $x \in F$ with $(x,T) = 0$ and $\chi(x) \neq 0$. By (2.9), $F = T \oplus \mathbb{R}x$ is a direct sum of ideals, in particular $L(x,x)T = 0$ and $0 \neq L(x,x)x \in \mathbb{R}x$. So $\mathbb{R}x$ contains a non-zero generalized tripotent e satisfying $L(e,e)T = 0$, which finally shows (*). ∎

Knowing now the existence of non-trivial generalized tripotents, we will improve on this by showing that there even exist minimal generalized tripotents, where naturally a non-zero generalized tripotent $e \in H$ is called <u>minimal</u> if every non-zero generalized tripotent $c \in H_2(e)$ is associated to e, i.e. has the same Peirce spaces. To this end we will, among other characterizations, show that e is minimal iff e is primitive in the sense of I§7: e cannot be written as the sum of two non-zero generalized orthogonal tripotents, where of course two generalized tripotents e,f are orthogonal if $e \in H_0(f)$, equivalently $f \in H_0(e)$. All these definitions sound very general, however we have

Lemma 2.5. Tripotents of opposite signs are always orthogonal (as tripotents, hence also in the scalar product). In particular, any generalized tripotent in $H_2(e)$ has the same sign as e.

Proof. Let e resp. f be a positive resp. negative tripotent and $e = e_2 + e_1 + e_0$ resp. $f = f_2 + f_1 + f_0$ the Peirce decomposition of e relative to f resp. of f relative to e. Then $0 \leq 2(f_2,f_2) + (f_1,f_1) = (\{eef\},f) = (e,\{eff\}) = -2(e_2,e_2) - (e_1,e_1) \leq 0$, so $f_2 = f_1 = 0$ and $e \perp f$. ∎

We will also characterize minimality of $e = \varepsilon P(e)e$ in terms of the Peirce-2-space using the splitting of $H_2(e)$ relative to the involution $P(e)$:

$$H_2(e) = H_2^+(e) \oplus H_2^-(e)$$

for $H_2^\delta(e) = \{x \in H_2(e);\ \varepsilon P(e)x = \delta x\}$, $\delta = \pm$.
Note that in the hermitian case

$$H_2^{-\delta}(e) = iH_2^\delta(e)$$

where $i = \sqrt{-1}$ is the imaginary unit. However, in general there is no connection between $H_2^+(e)$ and $H_2^-(e)$: $H_2^-(e)$ may vanish (as it does for idempotents in formally real Jordan algebras) or it may be of arbitrary dimension although $H_2^+(e) = \mathbb{R}e$ (as it does in Example I.2.4.a for V = H a real Hilbert space with scalar product q = (.,.) and $P(x)y = 2(x,y)x - (x,x)y$). This phenomenon is one of the major differences between the hermitian and the real case. Nevertheless we have the following list of equivalences for generalized minimal tripotents, the first two of which are already stated in [24]3.4:

Lemma 2.6. For a generalized tripotent $e \neq 0$ in a Hilbert triple H the following are equivalent:

i) e is minimal,

ii) e is primitive,

iii) $H_2^+(e) = \mathbb{R}e$,

iv) for $x,y \in H_2(e)$ we have

$$P(x)y = 2\sigma(x,y)x - \sigma(x,x)y$$

where σ and the scalar product $(.,.)|H_2(e)$ coincide up to a non-zero real constant, and

(2.12) $\sigma(\alpha e+x_-, \beta e+y_-) = \alpha\beta - \frac{1}{2}\{x_- e y_-\}$.

v) $H_2(e)$ is a division triple: every $x \neq 0$ is invertible.

Proof. The implication i) → ii) is a general fact (Lemma I.7.1). Assume ii), then e is $H_2^+(e)$-minimal (by Theorem I.7.5). For $x \in H_2^+(e)$ the subsystem $\mathbb{R}\{e,x\} \subset H_2^+(e)$ which is generated by e and x, is flat. Indeed, $\mathbb{R}\{e,x\}$ is just the span of all powers of x in the Jordan algebra $H_2^+(e)^{(e)}$, which is known to be flat. Hence $F = \overline{\mathbb{R}\{e,x\}} \subset H_2^+(e)$ satisfies the assumptions of (*) in the proof of Theorem 2.4 and thus contains a non-zero generalized tripotent c with $F = \mathbb{R}c \oplus F_0(c)$. Then $c \approx e$ by $H_2^+(e)$-minimality of e, so $H_2^+(e) \subset H_2(c)$ implies $F = \mathbb{R}c = \mathbb{R}e$ proving $H_2^+(e) = \mathbb{R}e$.

For iii) → iv) it is no harm to assume e positive. Then $P(x)y = 2\sigma(x,y)x - \sigma(x,x)y$ for σ as in (2.12) was shown in [50]2.1: It follows essentially from $2P(u)v = P(u)\{e\,\overline{v}\,e\} =$ (by(I.12)) $\{\{\overline{v}\,e\,u\}e\,u\} - \{\overline{v}\,e\,P(u)e\}$. Normalizing $(e,e) = 1$, iv) holds since $\mathbb{R}e \perp H_2^-(e)$ and $(\{x_- e y_-\},e) = (y_-,\{e x_- e\}) = -2(y_-,x_-)$.

The implications iv) → v) → i) are straightforward to prove. ∎

Corollary 2.7. ([24]3.5) In a Hilbert triple H every non-zero generalized tripotent can be written as a finite sum of orthogonal minimal generalized tripotents. If $L(H,H) \neq 0$ then H contains minimal tripotents.

Proof. For any tripotent $c \neq 0$ we have $2\|c\| = \|\{ccc\}\| \leq \lambda\|c\|^3$ by (2.1), hence $\|c\|^2 \geq 2/\lambda$. On the other hand, if $e = e_1 + e_2$ is a decomposition of orthogonal tripotents then $(e_1,e_2) = 0$, so $\|e\|^2 = \|e_1\|^2 + \|e_2\|^2$. Therefore, any e can be written as a finite sum of orthogonal primitive tripotents, implying the first claim in view of Lemma 2.6. The second follows from Theorem 2.4. ∎

In the following, if $(H_k)_{k\in K}$ is a family of Hilbert spaces H_k, their Hilbert sum is defined to be

$$\oplus^2_{k\in K} H_k = \{(x_k)_{k\in K};\ x_k \in H_k,\ \textstyle\sum_{k\in K} (x_k,x_k) < \infty\},$$

which is again a Hilbert space with the scalar product $((x_k),(y_k)) = \sum_{k\in K}(x_k,y_k)$.

Corollary 2.8. Every flat Hilbert triple F contains a (possibly empty) orthogonal system $(e_k)_{k\in K}$ of non-zero generalized tripotents such that

(2.13) $$F = (\oplus^2_{k\in K} \mathbb{R}e_k) \oplus^2 F_0$$

where F_0 has trivial multiplication: $L(x_0,y_0) = 0$ for $x_0,y_0 \in F_0$. *This decomposition is unique: If* $F = (\oplus^2_{\ell\in L} \mathbb{R}c_\ell) \oplus^2 F_0'$ *for an orthogonal system* $(c_\ell)_{\ell\in L}$ *and a trivial* F_0', *then* $(e_k)_{k\in K} = (c_\ell)_{\ell\in L}$ *up to sign and order and* $F_0 = F_0'$.

Proof. If F is trivial, we are done. Otherwise $L(F,F) \neq 0$, so F contains a non-zero minimal generalized tripotent by Theorem 2.4. Let $(e_k)_{k\in K} \subset F$ be a maximal orthogonal system of non-zero generalized tripotents. Then $F_0 = \bigcap_{k\in K} F_0(e_k)$ is trivial: Otherwise F_0 would contain a non-zero minimal tripotent contradicting maximality of $(e_k)_{k\in K}$. Since $(e_k,e_\ell) = 0$ for $k \neq \ell$, the result follows if $(x,e_k) = 0$ for all $k \in K$ implies $x \in F_0$. But this follows from the Peirce decomposition $F = \mathbb{R}e_k \oplus F_0(e_k)$ for every k: $F_1(e_k) = 0$, since $\varepsilon x = \{e_k e_k x\}$ for $e_k = \varepsilon P(e_k)e_k$ by flatness forces $\varepsilon x = \{e_k x e_k\} = 2P(e_k)x = 0$, and similarly $F_2(e_k) = F_2^+(e_k)$, so $F_2(e_k) = \mathbb{R}e_k$ by minimality of e_k. To show uniqueness, note that every c_ℓ can be written as $c_\ell = (\sum_k \alpha_{k\ell} e_k) + x_0$ and $P(c)c = \pm c$ implies $x_0 = 0$, $\alpha_{k\ell} = 0, \pm 1$. Now $F_2(c_\ell) = \mathbb{R}c_\ell$ shows $c_\ell = \pm e_k$ for exactly one k and the rest immediately follows. ∎

Corollary 2.9. Every element x *of a Hilbert triple has a spectral decomposition*

$$x = \sum_{n\in N} \lambda_n e_n + x_0 \tag{2.14}$$

where $(e_n)_{n\in N}$, $N \subset \mathbb{N}$ *is an orthogonal system of non-zero generalized tripotents in* $\overline{\mathbb{R}\{x\}}$, *the closure of the span of all powers of* x, *and the eigenvalues* λ_i *form a nonnegative decreasing sequence*

$$\lambda_1 \geq \lambda_2 \geq \dots > 0 \tag{2.15}$$

and x_0 *is trivial* ($P(x_0) = 0$).

Proof. We apply Corollary 2.8. to the flat Hilbert triple $\overline{\mathbb{R}\{x\}}$ and obtain a decomposition (2.14) with real λ's and some orthogonal system $(e_k)_{k\in K}$. Since the series for x is summable, K is countable and possibly switching from e_k to $-e_k$ allows us to assume $\lambda_i > 0$, so after a rearrangement we have (2.15). ∎

Note that one cannot strengthen (2.15) to $\lambda_1 > \lambda_2$ since the sum of two generalized tripotents of opposite sign is not a tripotent. Also note that because of (2.15) and the uniqueness statement in Corollary 2.8 the spectral decomposition (2.14) is unique if we require that in case $\lambda_n = \lambda_{n+1}$ e_n should be the positive and e_{n+1} the negative tripotent.

We will now show that every Hilbert triple can be decomposed into 3 parts: a positive, negative and zero part. One calls a Hilbert triple H positive, if H is nondegenerate ($P(x)H = 0 \to x = 0$) and all operators $L(x,x)$ have non-negative spectrum, i.e. $(L(x,x)y,y) \geq 0$ for all $x,y \in H$. Naturally, H is called negative if its negative -H is positive. Note, that in the finite-dimensional case a positive Hilbert triple is the same as a compact Jordan triple system.

For hermitian Hilbert triples the following result was proven in [24] (3.12) using classification:

Decomposition Theorem 2.10. Every Hilbert triple H has a unique decomposition as an orthogonal sum of 3 closed ideals

(2.16) $$H = H_+ \oplus H_0 \oplus H_-$$

where i) H_+ is positive and is the closure of the span of all positive minimal tripotents of H,

ii) H_- is negative and is the closure of the span of all negative minimal tripotents of H, and

iii) H_0 is the set of all absolute zero divisors. Every $x_0 \in H_0$ is extremely trivial:

$$L(x_0,H) = 0 = L(H,x_0)$$

Every closed ideal I of H splits corresponding to (2.16):

$$I = I \cap H_+ \oplus I \cap H_0 \oplus I \cap H_-$$

Proof. We define H_0 as the set of all absolute zero divisors (so H_0 is closed) and show that H_0 is an ideal. It is a subspace, since obviously $x \in H_0$, $\alpha \in \mathbb{R} \to \alpha x \in H_0$, and $x,y \in H_0$ implies $P(x+y)(x+y) =$ (by $P(x)=0=P(y)$) $\{xxy\} + \{xyy\} = 0$ by (2.9), hence (2.9) again says $x + y \in H_0$. Moreover, $P(H)H_0 \subset H_0$ by the Fundamental Formula (I.1.3), therefore H_0 is an ideal as soon as $\{H\,H\,H_0\} \subset H_0$ which will follow from $B(H,H)H_0 \subset H_0$ where B is the Bergman operator. However, this holds by (I.1.16). By (2.7) we now obtain an orthogonal decomposition

(*) $$H = H_0 \oplus H_0^\perp$$

where $H_0^\perp$ is also an ideal. Therefore $L(H_0,H) = L(H_0,H_0) = 0$ by Lemma 2.3.

Let H_+ (resp. H_-) be the closure of the span of all positive (resp. negative) minimal tripotents. Then

(**) $$L(H_+,H_-) = 0 = L(H_-,H_+)\ , \quad (H_+,H_-) = 0$$

by Lemma 2.5. Because of the decomposition (*) every generalized tripotent lies in $H_0^\perp$ whence $H_+ \oplus H_- \subset H_0^\perp$. On the other hand, every $x \in H_0^\perp$ has a spectral decomposition without trivial part (Corollary 2.9), and since every generalized tripotent is a finite sum of minimal ones, (2.14) shows $H_0^{\perp} \subset H_+ \oplus H_-$, so

(***) $\qquad H = H_0 \oplus H_+ \oplus H_-$

The spaces H_ε, $\varepsilon = \pm$, are ideals of H: By (2.7) it is enough to show $\{H\, H\, H_\varepsilon\} \subset H_\varepsilon$. We know already $\{H\, H\, H_\varepsilon\} = \{H_\varepsilon\, H_\varepsilon\, H_\varepsilon\} \subset H_0^{\perp}$ by (*), (**) and (***), and so $\{H_\varepsilon H_\varepsilon\, H_\varepsilon\} \subset H_\varepsilon$ because $(\{H_\varepsilon\, H_\varepsilon\, H_\varepsilon\}, H_{-\varepsilon}) = (H_\varepsilon, \{H_\varepsilon\, H_\varepsilon\, H_{-\varepsilon}\}) = 0$ by (**). Finally, positivity of H_+ (and then negativity of H_-) follows from the spectral decomposition of $x \in H_+$: $x = \sum_{n>0} \lambda_n e_n$ where $(e_n) \subset H_+$ is an orthogonal system of positive tripotents, so $(L(x,x)y,y) = \sum_{n>0} \lambda_n^2 (L(e_n,e_n)y,y) > 0$ because each summand $(L(e_n,e_n)y,y) > 0$.

Every closed ideal I is itself a Hilbert triple whence has a decompostion $I = I_+ \oplus I_0 \oplus I_-$ as in (2.16). Because minimal tripotents of $I_\pm$ stay minimal in H and absolute zero divisors of I are also absolute zero divisors in H by Lemma 2.3, we have $I_\mu \subset I \cap H_\mu$ for $\mu = \pm 1, 0$. Thus $I = I_+ \oplus I_0 \oplus I_- \subset (I \cap H_+) \oplus (I \cap H_0) \oplus (I \cap H_-) \subset I$ establishes the last claim. ∎

The Decomposition Theorem reduces the study of general Hilbert triples to studying positive Hilbert triples which we will do now. We will need the following

Lemma 2.11. ([24]3.2) *Let $\mathcal{O} = (e_i)_{i \in I}$ be an orthogonal system of generalized tripotents in a Hilbert triple H and denote by H_{ij} the Peirce spaces of $\mathcal{O}$ (see(II.4.4)). Then every closed ideal G of H satisfies*

$$G = \oplus^2_{i,j \in I \cup \{0\}} (G \cap H_{ij}).$$

In particular,

$$H = \oplus^2_{i,j \in I \cup \{0\}} H_{ij}$$

Proof. We have to show that $(x, G \cap H_{ij}) = 0$ for all $i,j \in I \cup \{0\}$ and $x \in G$ implies $x = 0$. To this end, fix $i \in I$ and decompose x relative to e_i: $x = x_2 + x_1 + x_0$ with $x_\mu \in H_\mu(e_i)$. Then each component x_μ satisfies $(x_\mu, G \cap H_{k\ell}) = 0$, whence $x_2 = P(e_i)^2 x \in G \cap H_{ii}$ vanishes. For $j \in I$, $j \neq i$, we have $H_1(e_i) = H_{ij} \oplus (H_1(e_i) \cap H_0(e_j))$, thus $x_1 = \{e_i e_i x\} \in G \cap H_1(e_i)$ splits: $x_1 = x_{11} + x_{10}$ with $x_{11} = \{e_j e_j x_1\} \in G \cap H_{ij}$, forcing $x_{11} = 0$. Therefore $x_1 \in G \cap H_{i0}$ vanishes too, and we get $x = x_0 \in G \cap H_{00}$, so $x = 0$. ∎

With this lemma we finish the brief interlude in the area of generalized tripotents. From now on, a tripotent will be a tripotent in the original sense: $P(e)e = e$.

<u>Structure Theorem 2.12 for positive Hilbert triples.</u> <u>Every positive Hilbert triple H has a unique representation as a Hilbert sum of topologically simple positive Hilbert triples</u> H_α

(2.17) $\qquad H = \oplus^2 H_\alpha, \qquad H_\alpha \lhd H.$

<u>Every ideal of H is a Hilbert sum of some of the</u> H_α<u>'s. Every</u> H_α <u>contains a Peirce-dense atomic standard grid</u> E_α <u>with minimal tripotents and a topologically dense cover</u>

(2.18) $\qquad H_\alpha = \oplus^2 (H_\alpha)_I(E_\alpha), \qquad (H_\alpha)_I(E_\alpha) \cap E_\alpha \neq \emptyset.$

$E = \bigcup_\alpha E_\alpha$ <u>is a Peirce-dense atomic grid in H with minimal tripotents and a topologically dense cover. If H is hermitian</u> E <u>is a complex Hilbert space basis.</u>

<u>Proof.</u> We want to apply the Main Construction Theorem II.4.7 and therefore check the conditions (II.4.1) - (II.4.3). The first two follow from Criterion I.1.12 and Lemma 2.6. For the third we note that every Peirce space $H_I(E)$ of some cog E is a closed subsystem and hence a Hilbert triple. It also inherits positivity from H. Therefore, every $H_I(E) \neq 0$ contains a non-zero tripotent, even a minimal one (by Corollary 2.7). In particular, for a maximal orthogonal system O of minimal tripotents we have $\bigcap_{e \in O} H_0(e) = 0$, establishing (II.4.3.a). The same argument also shows (II.4.3.b).

By the Decomposition Lemma II.4.2 the Peirce sum of O is a direct sum of ideals: $PS_H(O) = \oplus_\alpha V_\alpha$. Since $H = \oplus^2 H_{ij} = \overline{PS_H(O)}$ by Lemma 2.11, the closure $H_\alpha = \overline{V}_\alpha$ is an ideal of H. It is the Hilbert sum of all Peirce spaces H_{ij} which lie in V_α. This proves the orthogonal ideal decomposition (2.17). Every closed ideal G of H satisfies G = $\oplus^2(G \cap H_{ij})$. Since every non-zero $G \cap H_{ij}$ is a Hilbert triple and therefore contains a non-zero tripotent, the Decomposition Lemma II.4.2 implies that G is a Hilbert sum of some of the H_α's, in particular every H_α is topologically simple.

By the Main Construction Theorem II.4.7 every V_α contains an atomic standard grid E_α with H-minimal tripotents which covers V_α, in which case (2.18) obviously holds, or which is of rectangular resp. quadratic form type and has the density properties of Lemma II.4.5 resp. II.4.4. To prove (2.18) in this case we note that because of $H_\alpha = \overline{PS_F(O_\alpha)}$ for $F = V_\alpha$ and $O_\alpha = O \cap V_\alpha$ it is enough to show that $C_F(E_\alpha)$ is dense in $PS_F(O_\alpha)$: Every $x \in PS_F(O_\alpha)$ with $(x, C_F(E_\alpha)) = 0$ has to vanish. Suppose E_α is rectangular and decompose x relative to a fixed $e_i \in O_\alpha$, say $x = x_1 + x_0$ with $x_\mu \in H_\mu(e_i)$. Since also $(x_\mu, C_F(E_\alpha)) = 0$ we see $x_1 \in H_{i0}$ and x_1 is orthogonal to all $H_2(e_{ij})$ for $e_{ij} \in E_\alpha = R(I,J)$, $j \in J\setminus I$, so $x_1 \in \bigcap_{j \in J\setminus I} (H_{i0})_1(e_{ij}) = 0$ because e_{ij}, $j \in J\setminus I$, is a maximal collinear system in H_{i0}. This shows $x \in \bigcap_i H_0(e_i) = 0$. The same type of argument applies in the quadratic form case, proving (2.18) in all cases.

The last claim follows from the fact that in the hermitian case Peirce-2-spaces of minimal tripotents are 1-dimensional over $\mathbb{C}$. ∎

Examples 2.13 of topologically simple positive Hilbert triples: We first note that all examples will in an obvious way be the closure of the cover of a standard grid, whence they will be topologically simple by the Structure Theorem 2.12. We will encounter six different types with various subcases, the type distinction following essentially the distinction of grids (up to some exceptions for small grids). The numbering of the types is the same as in the Classification Theorems 1.14 and 1.16, and also the same as Cartan's notation for bounded symmetric domains, which in view of Remark 2.2 is no surprise. The examples of type X.a) will be the hermitian examples.

type I) (rectangular operators): Let $\mathbb{K} = \mathbb{R}$, $\mathbb{C}$ or $\mathbb{H}$ (= real division quaternions) and endow $\mathbb{K}$ with the natural real scalar product $(a,b) = \mathrm{Re}(a\bar{b})$, $\bar{\ }$ = standard involution of $\mathbb{K}$ (so $\bar{\ }$ = Id for $\mathbb{R}$ and $\bar{\ }$ = complex conjugation for $\mathbb{C}$). For arbitrary index sets I,J we form the real Hilbert spaces $E = \oplus^2_{j\in J}\mathbb{K}$ and $F = \oplus^2_{i\in I}\mathbb{K}$. Let $L_2(I,J;\mathbb{K})$ be the Hilbert space of all $\mathbb{K}$-linear Hilbert-Schmidt operators from E to F, i.e.

I.a) $L_2(I,J;\mathbb{C}) = I_{I,J}$

I.b) $L_2(I,J;\mathbb{R})$

I.c) $L_2(I,J;\mathbb{H})$

We regard $L_2(I,J;\mathbb{K})$ as the real Hilbert space of all square-summable $(I\times J)$ - matrices over $\mathbb{K}$ with scalar product $(x,y) = \sum_{i,j}(x_{ij},y_{ij})$ for $x = \sum_{i,j} x_{ij}E_{ij}$, $y = \sum_{i,j} y_{ij}E_{ij}$, and adjoint $x^* = \sum_{i,j}\bar{x}_{ij}E_{ji}$. The triple product $P(x)y = xy^*x$ makes $L_2(I,J;\mathbb{K})$ into a positive Hilbert triple (for example $\|L(x,y)\| \leq 2\|x\|\cdot\|y\|$ holds, and $L_2(I,J;\mathbb{K})$ is the closure of the span of the minimal tripotents aE_{ij}, $a\bar{a} = 1$).

type II) (alternating operators) For $x = \sum_{i,j} x_{ij}E_{ij} \in L_2(I,I;\mathbb{K})$ we put $x^t = \sum_{i,j} x_{ji}E_{ij}$ and consider the following subsystems of $L_2(I,I;\mathbb{K})$:

II.a) $A^2_I(\mathbb{C}) = \{x \in L_2(I,I;\mathbb{C});\ x^t = -x\} = II_I$

II.b) $A^2_I(\mathbb{R}) = \{x \in L_2(I,I;\mathbb{R});\ x^t = -x\}$.

(Note that $\{x \in L_2(I,I;\mathbb{H});\ x^t = -x\}$ is not closed under the triple product $P(x)y = xy^*x$). Being subsystems of $L_2(I,I;\mathbb{K})$, $A_I^2(K)$ are positive Hilbert triples.

type III) (hermitian operators) For any involution π of $\mathbb{K} = \mathbb{R},\mathbb{C}$ or $\mathbb{H}$ and $x = \sum_{i,j} x_{ij}E_{ij} \in L_2(I,I;\mathbb{K})$ we put $x^\pi = \sum_{i,j} x_{ij}^\pi E_{ij}$. Then

$$H_I^2(\mathbb{K},\pi) = \{x \in L_2(I,I;\mathbb{K});\ x^t = x^\pi\},$$

considered as subsystem of $L_2(I,I;\mathbb{K})$ (i.e. with triple product $P(x)y = x\bar{y}^t x = xy^*x$), is a topologically simple positive Hilbert triple. Up to isometric isomorphy we obtain by this general construction the following triple systems:

III.a) $H_I^2(\mathbb{C},Id) = III_I$

III.b) $H_I^2(\mathbb{R},Id) = H_I^2(\mathbb{R},^-)$

III.c) $H_I^2(\mathbb{C},^-)$ ($^-$ = complex conjugation)

III.d) $H_I^2(\mathbb{H},^-)$ ($^-$ = standard involution)

III.e) $H_I^2(\mathbb{H},\iota_{II})$ where ι_{II} is an involution of second kind, i.e. the standard involution of $\mathbb{H}$ followed by an orthogonal reflection in a complex subalgebra of $\mathbb{H}$. (Note that $H_I^2(\mathbb{H},\iota_{II})$ is isometrically isomorphic to $U_I^2(\mathbb{H}) = \{x \in L_2(I,I;\mathbb{H});\quad x^* = -x\}$.)

type IV) (spin factors) In the notation of Example I.1.6 these are the triple systems $[\mathbb{K},^-,E,q,S]$ for $\mathbb{K} = \mathbb{R}$ or $\mathbb{C}$ and E a $\mathbb{K}$-Hilbert space with scalar product $2\sigma(u,v) = q(u,Sv)$. Hence we get 2 different types:

IV.a) E a complex Hilbert space with hermitian scalar product σ and a conjugation of E, traditionally denoted by $^-$ (so $(\bar{x},\bar{y}) = \overline{(x,y)}$), such that the triple product is $P(x)y = 2\sigma(x,y)x - \sigma(x,\bar{x})\bar{y}$.

IV.b) E a real Hilbert space with scalar product σ and $S: E \to E$ a reflection (= involutorical isometry) such that the product of E is given by $P(x)y = 2\sigma(x,Sy)x - \sigma(x,x)Sy$.

type V) All 3 examples in this class are rectangular matrix systems $Mat(1,2;D,^-)$, namely:

V.a) The Bi-Cayley triple $B(\mathbb{C},\kappa)$ (see III§3.2) where κ is complex conjugation,

V.b) The Bi-Cayley triple $B(\mathbb{R},Id)$,

V.c) $Mat(1.2;\mathbb{C}a,^-)$ where $\mathbb{C}a$ is the real division octonion algebra (= Cayley algebra) and $^-$ is the standard involution of $\mathbb{C}a$. By Theorem 1.16 these triple systems are simple. They are also positive Hilbert triples (in fact compact Jordan triple systems), because every type V.x) is a subsystem of the positive Hilbert triple VI.x).

type VI) All 3 examples in this class are special cases of hermitian matrix triples $H_3(D,D_0,\pi,^-)$:

VI.a) the Albert triple $A(\mathbb{C},\kappa)$ with κ = complex conjugation (see III §3.3),

VI.b) the Albert triple $A(\mathbb{R},Id)$,

VI.c) $H_3(Ca,\mathbb{R},^-,^-)$ with $(Ca,^-)$ as in V.c), so VI.c) is the exceptional formally real Jordan algebra. As for type V) simplicity follows from Theorem 1.16. That VI.c) is a positive Hilbert triple is well-known: [4]XISatz 3.4. Because $A(\mathbb{C},\kappa)$ is the hermitification of $H_3(Ca,\mathbb{R},^-)$, it is also a positive Hilbert triple. But then the same holds for $A(\mathbb{R},Id)$ since it is another real form of $A(\mathbb{C},\kappa)$.

The Structure Theorem combined with the Coordinatization Theorems of Chapter III will allow us to derive a classification of positive Hilbert triples up to isometric isomorphy, where an isometric isomorphism between Hilbert triples is to be understood as an isomorphism of the triple systems which is also an isometry. A coarser equivalence relation is induced by the notion of Hilbert morphisms which are simply continuous homomorphisms between Hilbert triples. It seems to be remarkable that the isometric classification of Hilbert triples shows that Hilbert isomorphy already determines the scalar product of a simple Hilbert triple up to a positive factor. To express this the following concept is helpful:

Given a Hilbert triple H with triple product $\{...\}$ and scalar product $(.,.)$ we can define a new Hilbert triple H^s for $s > 0$ by taking $(H,\{...\})$ together with the new scalar product $(.,.)_s = s(.,.)$:

$$H^s = (H,\{...\}, s(.,.))$$

Then $Id: H \to H^s$ is a Hilbert isomorphism with norm $\sqrt{s}$, but H^s is isometric isomorphic to H^r iff $s = r$. (We remark that a similar concept was introduced in [23]: ${}^tH = (H,t\{...\},(.,.))$ for any real t. For $t > 0$ ${}^tH$ and H^s, $s^{-1} = \sqrt{t}$, are isometrically isomorphic via $\sqrt{t}Id: {}^tH \to H^s$.)

Classification Theorem 2.14. Up to isometric isomorphy the topologically simple positive Hilbert triples are exactly the triples G^s for $s > 0$ and G one of the Examples 2.13.

Proof. By the Structure Theorem 2.12 we know that every topologically simple positive Hilbert triple H contains a Peirce-dense atomic standard grid E with minimal tripotents such that $H = \oplus^2_I H_I(E)$. Let $O \subset E$ be a maximal orthogonal system of minimal tripotents. Then H fulfills (II.4.1)-(II.4.3') relative to $\mathcal{O}$, and we can apply Lemma II.4.2 to the Peirce sum V of $\mathcal{O}$: V is nondegenerate since $\overline{V} = H$ is and every $H_2(e) \subset V$ for $e \in \mathcal{O}$ is simple by Lemma 2.6, thus

(1) V is (algebraically) simple.

In applying the Coordinatization Theorems of Chap. III to the cover C of E, we will obtain coordinate algebras for which the following remark holds:

(2) If $H_2(e)$ for a minimal e carries the structure of an alternative algebra D with unit e and involution $^- = P(e)$ such that $P(x)y = x\bar{y}x$ for $x,y \in D$, then D is one of the real division algebras $\mathbb{R}$, $\mathbb{C}$, $\mathbb{H}$ or Ca, $^-$ is the standard involution and there exists a $s > 0$ such that $(x,y) = s\mathrm{Re}(x\bar{y})$.

Proof: First note that P(x) is invertible iff x is invertible in D, so D is a division algebra by Lemma 2.6. Since $x\bar{x} = \overline{x\bar{x}} \in \mathbb{R}e$ and since D does not have absolute zero divisors, we can apply [21] Theorem 6.2.3 saying that D is one of the composition division algebras $\mathbb{R}, \mathbb{C}, \mathbb{H}$ or Ca with norm $n(x,y) = \mathrm{Re}(x\bar{y})$. For $x = \alpha e + x_0$, $y = \beta e + y_0$, $x_0,y_0 \in H_2^-(e) = D_0$ we get $n(x,y) = \alpha\beta - n(x_0,y_0)$ with $\mathrm{Re}(x_0y_0) = \frac{1}{2}(x_0y_0 + y_0x_0) = \frac{1}{2}\{x_0ey_0\}$ whence $n = \sigma$ as defined in (2.12), and Lemma 2.6.iv) shows $(.,.) = sn$ for some $s > 0$.

Recall the general coordinatization procedure of the cover C of E: one coordinatizes one particular Peirce space $H_I(E)$ and constructs an isomorphism ϕ from C onto a standard example S essentially by mapping any other Peirce space $H_J(E)$ onto $H_I(E)$ (exceptions occur only in the case of a hermitian grid or an odd-dimensional quadratic form grid which will be discussed below). These mappings $H_J(E) \to H_I(E)$ are in fact isometries since they are products of generalized Peirce reflections and exchange automorphisms between collinear tripotents (if $e \top f$ then $T_{e,f} = B(e+f,e+f)$ is an isometry because clearly $T_{e,f}^* = T_{e,f}$ and $T_{e,f}^2 = \mathrm{Id}$). Adjusting therefore the scalar product on $H_I(E)$ using (2), we may assume that ϕ: C $\to$ S is an isometry, hence also ϕ: $\overline{C} = H \to \overline{S}$. Therefore all what remains to do is to identify $\overline{S}$. For this we have to go through the various possibilities of E.

$E = \{e\}$: $H = H_2(e)$ is of type IV.b) with S = Id by Lemma 2.6. If H is hermitian we have $H \cong \mathbb{C}$ of type IV.a).

$E = R(1,2)$: $H \cong \mathrm{Mat}(1,2;D)$ for $D = \mathbb{R}, \mathbb{C}, \mathbb{H}$ or Ca by (2), so H is of type I or V.c).

$E = \mathcal{R}(I,J)$ for $\#I + \#J > 4$: $C \cong \mathrm{Mat}(I,J;\mathbb{K})$ for $\mathbb{K} = \mathbb{R}, \mathbb{C}$ or $\mathbb{H}$, so $H \cong L_2(I,J;\mathbb{K})$ is of type I).

$E = \mathcal{S}(I)$, $\#I > 4$: $C \cong A_I(K)$ for $\mathbb{K} = \mathbb{R}$ or $\mathbb{C}$, $H \cong A_I^2(K)$.

$E = \mathcal{H}(I)$, $\#I > 3$: First we note that the Hermitian Coordinatization Theorem III.1.9 is applicable, because (for $r \neq s$)

$$(L(h_{rs},h_{ss})x_{ss}, L(h_{rs},h_{ss})x_{ss}) = (L(h_{ss},h_{rs})L(h_{rs},h_{ss})x_{ss},x_{ss})$$
$$(3) \qquad = \text{(by I.1.14)}\ ((2L(h_{ss},h_{ss}) - L(h_{rs},h_{rs}))x_{ss},x_{ss})$$
$$= 2(x_{ss},x_{ss})$$

shows that $L(h_{rs},h_{ss})|V_{ss}$ is injective. Hence $C \cong H_I(D,D_0,\pi,\bar{\ })$ where this isomorphism can be assumed to be an isometry by (3) and the remarks above. Since $\mathcal{H}(I)$ is atomic, we get $D = \mathbb{R}, \mathbb{C}, \mathbb{H}$ or $\mathbb{C}a$ (only for $\#I = 3$) and $\bar{\ }$ = standard involution, leaving us exactly with the possibilities of type III, as is well-known perhaps except for $D = \mathbb{H}$ where one can argue as follows: If $\pi \neq \bar{\ }$ is an involution of $\mathbb{H}$ commuting with $\bar{\ }$, then $\pi \circ \bar{\ }$ is an automorphism of $\mathbb{H}$ of order 2, hence a reflection in a complex subalgebra by [18].

$E = \mathcal{Q}_e(I)$, $\#I > 3$, or $E = \mathcal{Q}_o(I)$, $\#I > 2$: In this case $V = H$ and C is an even- or odd-dimensional quadratic form triple with coordinate algebra $(K,\bar{\ }) = (\mathbb{C},\bar{\ })$ or $(\mathbb{R},\mathrm{Id})$ by (2), a $\mathbb{K}$-hermitian form $\sigma(u,v) = \frac{1}{2}q(u,Sv)$ (indeed, $2\sigma(v,u) = q(Su,v) = \overline{q(u,Sv)} = 2\overline{\sigma(u,v)}$) and a conjugation resp. an involutorical isometry S: $S^2 = \mathrm{Id}$, S is κ-linear and $\sigma(Su,Sv) = \overline{\sigma(u,v)}$. Thus, in the even-dimensional case we get $H = \overline{C} = G^s$ for $s > 0$ and G an example of type IV. In the odd-dimensional case the scalar product $(.,.)$ of H restricted to the even-dimensional part C_I resp. the X-part is always a multiple of $\sigma|C_I$ resp $\sigma|X$. That both factors are the same follows from $\sigma(e_i^\varepsilon,e_i^\varepsilon) = \frac{1}{2}$, $\sigma(e_0,e_0) = 1$ and $(e_0,e_0) = (\{e_i^+e_0e_i^-\}, e_0) = (e_i^-,\{e_0e_i^+e_0\}) = 2(e_i^-,e_i^-)$. Therefore, $H = G^s$ for G of type IV holds also in the odd-dimensional case.

$E = \mathcal{B}$ or $E = \mathcal{A}$: Then $H = V = C$ is a Bi-Cayley or an Albert triple with coordinate algebra $(\mathbb{C},\bar{\ }) = (\mathbb{R},\bar{\ })$ or $(\mathbb{R},\mathrm{Id})$ by (2). So H is of type V or VI.

$E = \mathcal{H}(2)$: Here $H = V = C$ is algebraically simple by (1), so the structure of H follows from Theorem 1.3: H is a hermitian matrix system or a quadratic form triple yielding type III or IV. ∎

<u>Corollary 2.15</u>. (Kaup, [24]3.9) <u>Every hermitian Hilbert triple is isometrically isomorphic to a Hilbert sum</u> $\oplus^2_{\alpha\in A} H_\alpha$ <u>with</u> $0 \in A$, <u>where</u> H_0 <u>is trivial and every</u> H_α, $\alpha \neq 0$, <u>is a triple</u> $\pm G^s$, $s > 0$ <u>for</u> G <u>an example of type</u> X.a), i.e. I.a), ..., VI.a).

Proof. Decomposition Theorem 2.10, Structure Theorem 2.12 for positive Hilbert triples and Classification Theorem 2.14, picking out the topologically simple positive hermitian Hilbert triples. ∎

Remark 2.16. Restricted to the finite-dimensional case the Classification Theorem 2.14 gives a classification of compact Jordan triple systems. This classification has first been obtained by Loos in [32]XI using his classification of Jordan pairs (the result was announced in [30]p.65).

Remark 2.17. For a simple positive Hilbert triple H and a maximal orthogonal system $\mathcal{O} = (e_i)_{i\in I}$ of minimal tripotents in H we define the following cardinal numbers:

$r = \#\mathcal{O}$,

$a^{\varepsilon} = \dim_{\mathbb{R}} H_{ij}^{\varepsilon}$, where $i,j \in I$, $i \neq j$ and $H_{ij}^{\varepsilon} = \{x \in H;\ P(e_i,e_j)x = \varepsilon x\}$,
$\varepsilon = \pm$ or $= 0$ if $\#\mathcal{O} = 1$,

$b = \dim_{\mathbb{R}} H_{io}$,

$c = \dim_{\mathbb{R}} H_{ii}^{-}$ where $H_{ii}^{-} = \{x \in H;\ P(e_i)x = -x\}$,

$t = (e_i,e_i)$.

Because $\mathcal{O}$ can always be imbedded in a hermitian grid $\mathcal{H}(I)$ modulo association: $\mathcal{O} \approx \{h_{ii};\ i \in I\} \subset \mathcal{H}(I)$ it is easy to see using Theorem I.3.11 that $\{a^+,a^-\}$, b,c and t are well-defined. Moreover, we claim that the

$$\underline{\text{code}} \text{ of } H = (r,\{a^+,a^-\},b,c,t)$$

is actually an isometric invariant of H, i.e. independent of $\mathcal{O}$ modulo isometric isomorphisms. Indeed, for a finite-dimensional H this follows from the conjugacy theorem [32]11.8, and for an infinite-dimensional H by considering the Examples 2.13 (see [23]§3). For example, any $\mathcal{O}$ in $L_2(I,J;\mathbb{K})$ is of the form $(x_k \otimes y_k^{\star})_{k\in K}$ where $\{x_k; k\in K\} \subset F = \oplus_{i\in I}^2 \mathbb{K}$ and $\{y_k; k\in K\} \subset E = \oplus_{j\in J}^2 \mathbb{K}$ are orthonormal and at least one of these sets is a $\mathbb{K}$-Hilbert space basis.

It is a consequence of the following table and of the classification that the code is a complete isometric invariant system for simple positive Hilbert triples, i.e. the isometric isomorphism classes are determined by the code.

In this table always $d = \dim_{\mathbb{R}} \mathbb{K}$ where $\mathbb{K} = \mathbb{R}, \mathbb{C}$ or $\mathbb{H}$ is the coordinate algebra (depending on the type).

simple positive Hilbert triples	r	$\{a^+,a^-\}$	b	c
$L_2(1,J;\mathbb{K})$	1	$\{0,0\}$	$d(\#J-1)$	$d-1$
$L_2(I,J;\mathbb{K})$, $2 \leq \#I \leq \#J$	$\#I$	$\{d,d\}$	$d(\#J-\#I)$	$d-1$
$A^2_{2I+1}(\mathbb{K})$, $2 \leq \#I$,	$\#I$	$\{2d,2d\}$	$2d$	$d-1$
$A^2_{2I}(\mathbb{K})$, $3 \leq \#I$,	$\#I$	$\{2d,2d\}$	0	$d-1$
$H^2_I(\mathbb{C},Id)$, $2 \leq \#I$	$\#I$	$\{1,1\}$	0	1
$H^2_I(\mathbb{K},^-)$, $2 \leq \#I$	$\#I$	$\{d,0\}$	0	0
$H^2_I(\mathbb{H},\iota_{II})$, $2 \leq \#I$	$\#I$	$\{2,2\}$	0	2
IV.a), $5 \leq e: = \dim_{\mathbb{C}} E$	2	$\{e-2,e-2\}$	0	1
IV.b), $3 \leq e: = \dim_{\mathbb{K}} E \neq 4$, $S = Id$	1	$\{0,0\}$	0	$e-1$
IV.b), $5 \leq e: = \dim_{\mathbb{R}} E$, $\frac{1}{2}(e+1) \leq s = \dim_{\mathbb{R}} \mathrm{Fix}\, S < e$	2	$\{s-1,e-s-1\}$	0	0
V.a),b): $B(\mathbb{K},\kappa)$,	2	$\{3d,3d\}$	$4d$	$d-1$
V.c): $Mat(1,2,\mathbb{C}a,^-)$	1	$\{0,0\}$	8	7
VI.a),b): $A(\mathbb{K},\kappa)$,	3	$\{4d,4d\}$	0	$d-1$
VI.c): $H_3(\mathbb{C}a,^-)$	3	$\{8,0\}$	0	0

Remark 2.18. A Jordan H*-algebra is a complex Jordan algebra defined on a complex Hilbert space $(H,(\cdot,\cdot))$, such that the Jordan algebra product is continuous with respect to the Hilbert space topology and $(xy,z) = (x,\bar{y}z)$ for all $x,y,z \in H$ where $y \to \bar{y}$ is a $\mathbb{C}$-antilinear involution. Any Jordan H*-algebra is a hermitian Hilbert triple with respect to the triple product $\{xyz\} = x(\bar{y}z) - \bar{y}(xz) + (x\bar{y})z$, and their structure can easily be obtained from the results in this section. The structure of the more general noncommutative Jordan H*-algebras is determined in [58].

§3 JBW*-triples

In this section we study JBW*-triples - a category of Jordan triple systems investigated by several authors (see [8], [9], [15] or [25], or [51] for the special case of JBW*-algebras and their real forms, JBW-algebras). We prove that every JBW*-triple has a unique decomposition into an "atomic" and "non-atomic" part and completely determine the structure of the atomic part.

Several of our results are contained in or follow from [8] and [15], and a reference to these papers would have shortened this section. However, this would have spoiled our main goal, namely to provide the reader with an exposition which

- is essentially self-contained: We will use without proof only 3 results (A), B) and C) below), which the reader may easily take for granted and whose proof require more geometric or functional analytic methods than we wanted to develop here.
- uses less functional analytic methods (besides a basic knowledge about Banach spaces only the concept of weak-*-summability is needed) and more Jordan-theoretic methods, in particular makes consistent use of the theory of grids. As a result of this new approach several generalizations and simplifications of the theory are obtained (see e.g. Theorem 3.5 and its corollaries) and any reference to the theory of JB-algebras is avoided.

In this section all Banach spaces are complex. We therefore usually drop the affix "complex". The norm of a Banach space U will be denoted by $\|\cdot\|$. The algebra B(U) of bounded linear operators on U will always be equipped with the operator norm, also denoted by $\|\cdot\|$.

A JB*-triple is a real Jordan triple system on a complex Banach space U such that

(3.1) U is a Jordan-*-triple, i.e. the triple product $(x,y,z)\to\{xyz\}$ is $\mathbb{C}$-linear in x and z and $\mathbb{C}$-antilinear in y,

(3.2) the left multiplication $L\colon U\times U\to B(U)$ is continuous, i.e. $\|L(x,y)\| \leq \lambda\|x\|\cdot\|y\|$ for all $x,y\in U$ and a universal $\lambda > 0$,

(3.3) U is hermitian in the sense that all left multiplications $L(z,z)$, $z\in U$, are hermitian operators, i.e. $\exp itL(z,z)$, $t\in\mathbb{R}$, is an isometry. (Note that $iL(z,z) = \frac{1}{2}(L(iz,z)-L(z,iz))$ is a derivation of U, thus $\exp itL(z,z)$ is an isometric automorphism, called an inner automorphism.)

(3.4) U is positive, i.e. the spectrum of $L(z,z)$ (which is real by (3.3)) is non-negative,

(3.5) U fulfills the C*-condition:

$$\|P(z)z\| = \|z\|^3 \qquad \text{for all } z\in U.$$

The reason for calling (3.5) the C*-condition is

Example 3.1. Let A be a C*-algebra, i.e. an associative complex algebra which is defined on a Banach space and which has a $\mathbb{C}$-antilinear involution * such that $\|xy\| \leq \|x\|\|y\|$ and $\|xx^*\| = \|x\|^2$ hold for all $x,y \in A$ (see e.g. [10],[51] or [55]). Then A with the Jordan triple product $P(x)y = xy^*x$ is a JB*-triple, as follows immediately from well-known properties of C*-algebras: (3.1) and (3.2) are clear. For (3.3) and (3.4) let L resp. R denote the left resp. right multiplication in A. Then $L(z,z) = L(zz^*) + R(z^*z)$, $[L(zz^*), R(z^*z)] = 0$ and $L(zz^*)$ and $R(z^*z)$ are positive hermitian operators since zz^* and z^*z are positive. Now (3.3) immediately follows and (3.4) holds because the positive hermitian operator on A form a convex cone ([55]14.31). Finally, for (3.5) we have $\|P(z)z\|^2 = \|zz^*z\|^2 = \|zz^*zz^*zz^*\| = \|(zz^*)^3\| = \|zz^*\|^3 = \|z\|^6$.

JB*-triples were introduced in [25] whose fundamental result is the equivalence between the two categories of JB*-triples and of bounded symmetric domains with base points in Banach spaces. We will use without proof the following two results from the theory of JB*-triples:

Theorem A(= [25]5.3). *Let U be a Jordan triple system satisfying (3.1)-(3.3). Then (3.4) and (3.5) are equivalent to either one of the following two conditions:*

(3.6) *U is positive and $\|L(z,z)\| = 2\|z\|^2$ for all $z \in U$*

(3.7) *For every $z \in U$ the closed subtriple generated by z is isometrically isomorphic to the Jordan triple of a commutative C*-algebra (as in Example 3.1).*

We record an important consequence of (3.7):

(3.8) Every closed subtriple of a JB*-triple is again a JB*-triple with the induced norm.

In a JB*-triple U a tripotent e is called *maximal* if $U_0(e) = \{0\}$.

Theorem B ([25]5.4 and [26]3.5). *The maximal tripotents of a JB*-triple U are exactly the complex extremal points of the closed unit ball of U.*

In general, a JB*-triple need not contain non-zero tripotents. This follows from

Example 3.2. The space $C_0(X)$ of all complex-valued continuous functions on a locally compact Hausdorff space X, which vanish at infinity, is a C*- algebra in the sup-norm whence a JB*-triple by Example 3.1. In case X is connected and not compact, $C_0(X)$ does not contain non-zero tripotents.

Because of this deficiency of JB*-triples we will restrict our attention to a subcategory of JB*-triples, studied in [8], [9] and [15]:

A JB*-triple U is called a JBW*-triple if U is a dual Banach space, i.e. there exists a Banach space U_* such that $(U_*)^* = U$ where $(U_*)^*$ is the dual Banach space of U_*.

Looking at Example 3.1 we note that a C*-algebra is a JBW*-triple iff it is a W*-algebra, i.e. a C*-algebra with a predual.

Since a JBW*-triple U is a dual Banach space it carries an important second topology besides the norm topology: the weakest topology in which all linear forms $x \to x(\phi)$, $\phi \in U_*$, are continuous. This topology is called the weak-*-topology, for short: the w*-topology (see e.g. [11]§1). In the following topological notions with respect to the w*-topology will be preceded by "w*-", whereas without such a prefix they refer to the norm topology. The w*-topology makes U into a Hausdorff locally convex topological vector space. It is weaker than the norm topology. Since every w*-closed subspace of U is a dual Banach space (see e.g. [11]1.1.25) we have the following extension of (3.8):

(3.9) Every w*-closed subtriple of a JBW*-triple is again a JBW*-triple.

Because the closed unit ball of U is w*-compact (Alaoglu theorem), the Krein-Milman theorem in combination with Theorem B shows

(3.10) Every JBW*-triple contains a maximal tripotent.

Note that a maximal tripotent in a non-zero triple system is non-zero. The third and last result we shall use without proof is

Theorem C ([3],[5]). *The triple product of a JBW*-triple is separately w*-continuous: $(x,y,z) \to \{xyz\}$ is w*-continuous in each of the three variables if one fixes the remaining two.*

An immediate consequence of Theorem C is that the predual of a JBW*-triple is unique ([15]3.22) which justifies to speak of *the* w*-topology.

Theorem C also implies that Peirce spaces of a tripotent e, being defined as eigenspaces of L(e,e), are w*-closed. Hence also generalized Peirce spaces are w*-closed. Therefore (3.9) and (3.10) imply

(3.11) A JBW*-triple is tripotential in the sense of I§5.1.

For the application of the construction theorem for grids with minimal tripotents (II.4.7) we need a characterization of minimal tripotents.

Theorem 3.3. In a JBW*-triple U a tripotent e is minimal (in the sense of I.§5.1) iff $U_2(e) = \mathbb{C}\, e$.

Proof. Let $e \in U$ be a minimal tripotent. We will work with the mutation $U_2(e)^{(e)}$, i.e. the Jordan algebra defined on $U_2(e)$ by

$$xy = \tfrac{1}{2}\{xey\}, \quad e = \text{unit element.}$$

For any $z \in U_2^+(e) = \{x \in U_2(e);\ P(e)x = x\}$ the subtriple U(z,e), generated by z and e, coincides with the $\mathbb{C}$-span of all the powers $z^0 = e,\ z,\ z^2 = P(z)e,\ z^3 = P(z)z,\ \ldots$ of z in the Jordan algebra $U_2(e)^{(e)}$. It is therefore abelian: $[L(u,v),L(x,y)]|U(z,e) = 0$ for all $u,v,x,y \in U(z,e)$. By Theorem C and (3.9) the w*-closure $\overline{U(z,e)}^* = W \subset U_2(e)$ is an abelian JBW*-triple.

We claim that relative to the induced norm and the involution $w \to w^* = P(e)w$ the mutation $W^{(e)}$ is a commutative W*-algebra with unit e. Indeed, all conditions to be checked easily follow from the defining properties of JBW*-triples, except associativity and the two norm conditions $\|uv\| \leq \|u\|\|v\|$, $\|uu^*\| = \|u\|^2$. For associativity we use that W is abelian: $4(uv)w = L(w,e)L(u,e)v = L(u,e)L(w,e)v = 4u(vw)$. Since $W = W_2(e)$ the triple product is given by $\{uvw\} = 2[u(v^*w)-v^*(uw)+(uv^*)w]$, whence by associativity and commutativity $\{uvw\} = 2uv^*w$. By (3.6) we therefore have $\|u\|^2 = \|L(uu^*)\|$ where $L(\cdot)$ denotes the left multiplication in $W^{(e)}$. Consequently $\|uv\|^2 = \|L(uv(uv)^*)\| = \|L(uu^*)L(vv^*)\| \leq \|u\|^2\|v\|^2$, i.e. $\|uv\| \leq \|u\|\|v\|$. Since $\|e\| = 1$ this implies $\|L(u)\| = \|u\|$, whence $\|u\|^2 = \|L(uu^*)\| = \|uu^*\|$.

The proof will now be finished by showing that any commutative W*-algebra, whose unit is a minimal tripotent, is isomorphic to $\mathbb{C}$. Let A be such an algebra. Then A is isometrically isomorphic to C(K), the complex-valued continuous functions on K = spec A which is a compact Hausdorff space ([51]1.2.1). Mimimality of e forces K to be connected. By [51]1.7.5 spec A is stonean, i.e. the closure of every open set is open. By connectedness, every open set is dense. Hence, the Hausdorff-property shows K is a singleton, so $C(K) = \mathbb{C}$. ∎

Although by (3.10) every JBW*-triple contains tripotents in abundance, there are interesting examples which do not contain minimal one's:

Example 3.4. Let T be a locally compact space, μ a positive Radon measure on T and denote by $L^{\infty}(T,\mu)$ the commutative W*-algebra of all essentially bounded μ-measurable complex-valued functions (see e.g. [51]1.18 or [15]§6). Such a function f is a non-zero tripotent in the JBW*-triple $L^{\infty}(T,\mu)$ iff $\mu(\{t \in T;\ f(t) \neq 0,1\}) = 0$ and $T_f = \{t \in T;\ |f(t)| = 1\}$ has positive μ-measure, and f is minimal iff there is no subset $S \subset T_f$ satisfying $0 < \mu(S) < \mu(T_f)$. So in particular, $L^{\infty}(\mathbb{R},\lambda)$, λ = Lebesgue measure, does not contain minimal tripotents.

For a general JBW*-triple we will see later that one can always split off a part spanned by minimal tripotents. This will easily follow as soon as we know that for a JBW*-triple U L(U,U) operates completely reducibly on U. This result improves [15]5.2.

Complete-Reducibility-Theorem 3.5. *Let U be a JBW*-triple and W a w*-closed subspace such that*

$$L(x,x)W \subset W \quad \text{for all } x \in U$$

(equivalently, W is invariant under all inner automorphisms $\exp itL(x,x)$, $t \in \mathbb{R}$). *Then W is an ideal and there exists a unique w*-closed ideal* $W^{\perp}$, *namely*

(3.12) $$W^{\perp} = \cap \{U_0(e);\ e \text{ tripotent in } W\}$$

such that

(3.13) $$U = W \oplus^{\infty} W^{\perp}, \quad \|w+w^{\perp}\| = \max(\|w\|, \|w^{\perp}\|)$$

Here and in the following we use the symbol $\oplus^{\infty}$ to denote the "infinity-sum" of a family $(U_i)_{i \in I}$ of Banach spaces

(3.14) $$\oplus^{\infty}_{i \in I} = \{(u_i)_{i \in I};\ \sup \|u_i\| < \infty\}$$

which is a Banach space with respect to the norm

$$\|(u_i)_{i \in I}\|_{\infty} = \sup(\|u_i\|).$$

Proof. By the polarization formula

$$4L(x,y) = L(x+y,x+y)-L(x-y,x-y)+iL(x+iy,x+iy)-iL(x-iy,x-iy)$$

W is a L(U,U)-invariant subspace. Using (3.10) we can choose a maximal tripotent f of W. Then for $x \in U_j(f)$, j = 1 or 2, $jx = \{ffx\} = L(x,f)f \in W$, so

(*) $$W = U_2(f) \oplus U_1(f), \qquad U = W \oplus U_0(f)$$

and W is an ideal since $\{UWU\} \subset W$: indeed, by (*) and the Peirce multiplication rules

$$\{U\ U_2\ U\} = \{U_2\ U_2\ U_2\} + \{U_2\ U_2\ U_1\} + \{U_1\ U_2\ U_1\} \subset W$$

because $\{U_1\ U_2\ U_1\} \subset U_0 \cap W = \{0\}$, and

$$\{U\ U_1\ U\} = \{U_2\ U_1\ U_1\} + \{U_2\ U_1\ U_0\} + \{U_1\ U_1\ U_1\} + \{U_1\ U_1\ U_0\} \subset W$$

because $\{U_1\ U_1\ U_0\} \subset U_0 \cap W = \{0\}$.

Since $U_0(f) = \ker L(e,e)$ is a w*-closed subsystem, we can again apply (3.10) and choose a maximal tripotent g of $U_0(f)$, so in particular $f \perp g$. We will show that $U_0(f)$ is an ideal of U using the Peirce decomposition relative to the orthogonal pair (f,g):

$$U = U_{11} \oplus U_{10} \oplus U_{22} \oplus U_{20}, \quad W = U_{11} \oplus U_{10}, \quad U_0(f) = U_{22} \oplus U_{20}$$

Note $U_{12} = 0$, because otherwise there would exist a non-zero tripotent $h \in U_{12} \subset W$ which is compatible with g by the Compatibility Criterion I.1.8, whence g decomposes relative to h as $g = g_2 + g_1 + g_0$ where $g_i \in U_i(h) \cap U_2(g) = 0$ for $i = 1,2$, thus $g = g_0 \perp h$, contradiction. We claim $L(U_{10},U_{20}) = 0$. Indeed, $L(U_{10},U_{20})U = L(U_{10},U_{20})(U_{22} + U_{20}) = L(U_{10},U_{20})U_{20}$ since $\{U_{10}U_{20}U_{22}\} \subset U_{12} = 0$ and $L(U_{10},U_{20})U_{20} = \{\{U_{10}U_{20}g\}\ g\ U_{20}\}$ (by (I.1.27)) $= 0$ since $\{U_{10}U_{20}\ g\} \subset U_{12} = 0$. Also $L(U_{20},U_{10})U = L(U_{20},U_{10})U_{10} \subset U_{20} \cap W = 0$. From the Peirce multiplication rules and the fact that W is an ideal, it easily follows that $L(W,U_0(f)) = 0 = L(U_0(f),W)$, which implies that $U_0(f)$ is also an ideal of U_0.

The uniqueness of the decomposition (3.13) is now immediate: If $U = W \oplus W^\perp$ is a direct sum of two ideals, then necessarily $W^\perp \subset \cap\{U_0(e);$ e tripotent of $W\} =: Z$ and $Z \cap W = 0$ because W contains tripotents which are maximal in W. Therefore $Z = W^\perp$.

It remains to prove the assertion about the norm, for which we use some arguments from [8] Corollary 1.2, Lemma 1.3: Let $E_0\colon U \to U_0(f) = W^\perp$ be the Peirce projection. Then a straightforward computation shows $(Id + \exp(\frac{1}{2}\pi\, iL(e,e)))(Id + \exp(\pi i L(e,e)) = 4E_0$. By (3.3) $\exp(\pi i t L(e,e))$ is always an isometry, whence $\|E_0\| \leq 1$ and we get for $w \in W$, $z \in W^\perp$ $\|z\| \leq \|w+z\|$. By uniqueness $(W^\perp)^\perp = W$, so also $\|w\| \leq \|w+z\|$, thus $\max(\|w\|,\|z\|) \leq \|w+z\|$. Assuming $\max(\|w\|,\|z\|) = 1$ we obtain from (3.5) that

$$\|w+z\| = \|(w+z)^3\|^{3^{-1}} = \|w^3+z^3\|^{3^{-1}} = \ldots = \|w^{3^n}+z^{3^n}\|^{3^{-n}} \leq (2)^{3^{-n}} \to 1,$$

therefore $\|w+z\| = \max(\|w\|,\|z\|)$. ∎

<u>Corollary 3.6.</u> *Let (P) be an algebraic property for elements of a JBW*-triple U (more precisely, a property invariant under inner automorphisms* $\exp itL(x,x)$). *Then there exists a unique decomposition*

$$U = U_{(P)} \oplus^\infty U_{(P)}^\perp$$

of U into w-closed ideals* $U_{(P)}$ *and* $U_{(P)}^\perp$ *where* $U_{(P)}$ *is the w*-closure of the span of all elements of U having property* (P).

Proof. Obviously $V_{(P)}$, the span of all elements of U having property (P), is invariant under $\exp(itL(x,x))$, whence $L(x,x)\overline{V}_{(P)} \subset \overline{V}_{(P)}$ = norm-closure of $V_{(P)}$, and because of the w*-continuity of $L(x,x)$ also $L(x,x)U_{(P)} \subset U_{(P)}$. ∎

Remark 3.7. As a trivial application of this corollary we obtain the two known decomposition theorems for JBW*-triples: [15] 5.11 and [8] Theorem 2. The first states that $U = U_I \oplus^\infty U_I^\perp$ where U_I is the w*-closure of the span of abelian tripotents, i.e. tripotents e such that $[L(x,y),L(u,v)]|U_2(e) = 0$ for all x,y,u,v. To state the second decomposition theorem we call a JBW*-triple atomic iff it is the w*-closure of the span of its minimal tripotents. (We will see later that there is good reason for choosing this name: An atomic JBW*-triple is the w*-closure of the linear span of a special atomic grid.)

Corollary 3.8. ([8] Theorem 2) *Every JBW*-triple U has a unique decomposition*

$$U = U_a \oplus^\infty U_a^\perp \tag{3.15}$$

as a direct sum of two w-closed ideals where*

(i) *U_a is the w*-closure of the span of all minimal tripotents of U, in particular U_a is atomic,*

and (ii) *$U_a^\perp$ does not contain minimal tripotents.*

Every w-closed ideal splits corresponding to* (3.15).

Proof. By Corollary 3.6 for (P) = "e is a minimal tripotent" we have the decomposition (3.15) and (i) and (ii).

The assertion about w*-closed ideals I follows from the decomposition (3.15) for I and the observation that a tripotent $e \in I$ is minimal in I iff it is minimal in U. ∎

Our next aim is to derive a structure theorem for atomic JBW*-triples along the lines of the Structure Theorem 2.12 for positive Hilbert triples. To this end, we need several auxiliary results which are of independent interest. The first one generalizes [8] Prop. 6:

Lemma 3.9. *Let e be a tripotent in a JBW*-triple U and let $c \in U$ be a minimal tripotent with Peirce decomposition $c = c_2 + c_1 + c_0$ relative to e. Then c_2 and c_0 are scalar multiples of some minimal tripotent of U.*

Proof. To treat both cases simultaneously we put $\mu = 0$ or 2. We may assume $c_\mu \neq 0$. By minimality of c there exists $\lambda \in \mathbb{C}$ such that $P(c)c_\mu =$

λc. On the other hand, $P(c_2+c_1+c_0)c_\mu = P(c_\mu)c_\mu + \{c_1 c_\mu c_\mu\} + P(c_1)c_\mu$, whence by comparison of the Peirce components $P(c_\mu)c_\mu = \lambda c_\mu$, or $L(c_\mu,c_\mu)c_\mu = 2\lambda c_\mu$ forcing $\lambda \geq 0$ by (3.4). By (3.5) and $c_\mu \neq 0$ we have $\lambda > 0$ and $\sqrt{\lambda}c_\mu$ is a non-zero tripotent, which is also minimal because for every $x = x_2+x_1+x_0$ (decomposed relative to e) we have $P(c_\mu)x = P(c_\mu)x_\mu = (P(c)x_\mu)_{\mu\text{-comp}} = (\rho c)_{\mu\text{-comp}} = \rho c_\mu$ for some $\rho \in \mathbb{C}$ by minimality of c. Thus $P(c_\mu)U \subset \mathbb{C}\, c_\mu$, so $\sqrt{\lambda}c_\mu$ is minimal. ∎

We will have to consider arbitrary sums $\sum_\lambda x_\lambda$ of elements x_λ from a JBW*- triple U, which converge in the w*-topology. Because this topology does not satisfy the First Axiom of Countability we cannot use sequences but have to use nets to define convergence: For any set L the set L_{fin} of finite subsets of L is partially ordered and directed with respect to the inclusion relation. Therefore we can associate to every family $\{x_\lambda\}_{\lambda\in L} \subset U$ the net $(x_J = \sum_{\lambda\in J} x_\lambda;\ J \in L_{fin})$. If this net has a w*-limit, it is denoted by

$$\sum^*_{\lambda\in L} x_\lambda = \text{w*-lim}\,(\sum_{\lambda\in J} x_\lambda;\ J \in L_{fin})$$

Note that the limit is unique since the w*-topology is Hausdorff. To show w*-summability we will always use the following criterion, which is a refinement of [15](4.3) and (4.4). In fact, some of the arguments in the proof are taken from there. A similar criterion is known for W*-algebras ([51]1.7.4).

<u>w*-Summability-Criterion 3.10.</u> <u>Let U be a Banach space with pre-dual and assume $(P_\alpha;\ \alpha \in A) \subset B(U)$ is a family of w*-continuous projections which is</u>

i) <u>orthogonal</u>: $P_\alpha P_\beta = 0$ <u>for</u> $\alpha \neq \beta$,

ii) <u>faithful</u>: $P_\alpha x = 0$ <u>for all</u> $\alpha \ \Rightarrow\ x = 0$

<u>Assume</u> $(x_\alpha)_{\alpha\in A}$ <u>satisfies</u>

iii) $x_\alpha \in P_\alpha U$

<u>Then</u> $(x_\alpha)_{\alpha\in A}$ <u>is</u> w*-<u>summable iff</u>

iv) $\sup(\{\|\sum_{\alpha\in B} x_\alpha\|;\ B \in A_{fin}\}) =: M < \infty$.

<u>In this case</u>

(3.16) $\quad \|\sum^*_{\alpha\in A} x_\alpha\| \leq M.$

<u>If in addition to</u> i) - iv) <u>also</u>

v) $\|\sum_{\alpha\in B} x_\alpha\| = \max_{\alpha\in B} \|x_\alpha\|$ <u>for every</u> $B \in A_{fin}$,

vi) <u>every</u> P_α <u>is contractive</u>: $\|P_\alpha\| \leq 1$,

<u>then</u>

(3.17) $\quad \|\sum^*_{\alpha\in A} x_\alpha\| = \sup_{\alpha\in A} \|x_\alpha\|$

<u>Proof</u>. We denote by $U_*(\subset U^*)$ the pre-dual of U and define

$$V_\alpha = \{\phi \in U_*;\ \phi \circ P_\alpha = \phi\},\quad V = \sum_{\alpha\in A} V_\alpha$$

We claim: V separates points of U. Indeed, for $u \neq 0$ exists $\alpha \in A$ with $u_\alpha = P_\alpha u \neq 0$ by i) and consequently $\psi \in U_*$ with $\psi(u_\alpha) \neq 0$. Then $\psi \circ P_\alpha \in V_\alpha$ satisfies $(\psi \circ P_\alpha)(u) \neq 0$. Let $\bar{V}$ be the norm-closure of V in U_*. If $\bar{V} \neq U_*$ the Hahn-Banach-Theorem shows the existence of a non-trivial linear form $x \in U$ such that $\bar{V}(x) = 0$ contradicting the separation property of V. Thus $\bar{V} = U_*$ and (U,V) are in strict duality ([11]1.1.7). The weak topology $\sigma(U,V)$ on U, defined by the strict duality (U,V), is Hausdorff and weaker than the w*-topology, i.e. the weak topology $\sigma(U,U_*)$. Since $U^1 = \{x \in U;\ \|x\| \leq 1\}$ is w*-compact, the w*-topology on U^1 coincides with the topology induced by $\sigma(U,V)$ on U^1.

After these preparations assume now that i) - iii) hold. If $(x_\alpha)_{\alpha \in A}$ is w*-summable then $\{\phi(\sum_{\alpha \in B} x_\alpha);\ B \in A_{fin}\}$ is bounded for every $\phi \in U_*$ whence iv) follows from the uniform boundedness theorem. Conversely, assume iv) with $M = 1$, i.e. $\sum_{\alpha \in B} x_\alpha \in U^1$ for every $B \in A_{fin}$. Then the w*-limit $\sum^*_{\alpha \in A} x_\alpha$ exists iff the net $(\sum_{\alpha \in B} x_\alpha;\ B \in A_{fin})$ has a $\sigma(U,V)$-limit which is the case iff for all $\phi \in V$ the net $(\sum_{\alpha \in B} \phi(x_\alpha);\ B \in A_{fin})$ has a limit in $\mathbb{C}$. It is enough to show this for ϕ in some V_β where it is trivial: For every $B \in A_{fin}$ we have $\sum_{\alpha \in B \cup \{\beta\}} \phi(x_\alpha) = \sum_{\alpha \in B \cup \{\beta\}} (\phi \circ P_\beta)(x_\alpha) = \phi(x_\beta)$ by i) and iii).

For (3.16) and (3.17) first recall that

$$\|\sum^*_{\alpha \in A} x_\alpha\| = \sup_\phi \|\phi(\sum^*_{\alpha \in A} x_\alpha)\|$$

where the sup is taken over all $\phi \in U^1_* = \{\phi \in U_*;\ \|\phi\| \leq 1\}$. By continuity of ϕ and $\|\cdot\|$ we have

$$\|\phi(\sum^*_{\alpha \in A} x_\alpha)\| = \|\lim_B \sum_{\alpha \in B} \phi(x_\alpha)\| = \lim_B \|\sum_{\alpha \in B} \phi(x_\alpha)\|$$

where the limit is taken over all $B \in A_{fin}$. In general, $\|\sum_{\alpha \in B} \phi(x_\alpha)\| \leq \|\phi\| \cdot \|\sum_{\alpha \in B} x_\alpha\| \leq M\|\phi\|$, whence (3.16). To show (3.17) first note $M = \sup_{\alpha \in A} \|x_\alpha\|$ in the presence of v). By what we proved so far, (3.17) follows if there exists a sequence $(\phi_n) \subset U^1_*$ such that

(*) $$\lim_B \|\sum_{\alpha \in B} \phi_n(x_\alpha)\| \geq M - \frac{1}{n}$$

By definition of $\|x_\alpha\|$ there exists for every $\varepsilon > 0$ a $\phi \in U^1_*$ with $\|\phi(x_\alpha)\| \geq \|x_\alpha\| - \varepsilon$. Then $\psi_\alpha = \phi \circ P_\alpha \in V_\alpha$ satisfies $\|\psi_\alpha\| \leq \|\phi\|\|P_\alpha\| \leq 1$ by vi) and $\|\psi(x_\alpha)\| \geq \|x_\alpha\| - \varepsilon$. Because $M = \sup_\alpha \|x_\alpha\|$ there exists a sequence $(\alpha(n)) \subset A$ with $\|x_{\alpha(n)}\| \geq M - \frac{1}{2n}$. As shown before, for every $\alpha(n)$ there exists a $\phi_n \in U^1_* \cap V_{\alpha(n)}$ with $\|\phi_n(x_{\alpha(n)})\| \geq \|x_{\alpha(n)}\| - \frac{1}{2n} \geq M - \frac{1}{n}$. But for this ϕ_n we have $\lim_B \|\sum_{\alpha \in B} \phi_n(x_\alpha)\| = \|\phi_n(x_{\alpha(n)})\| \geq M - \frac{1}{n}$. By (*) we are done. ∎

The criterion above will be used frequently in the following, for example to give a different proof of [15]3.15:

Lemma 3.11. Let $\mathcal{O} = (e_i)_{i \in I}$ be a family of orthogonal tripotents in a JBW*-triple U. Then $\mathcal{O}$ is w*-summable and $\sum^*_{i \in I} e_i$ is a tripotent in U.

Proof. The main assertion is that $\mathcal{O}$ is w*-summable, because then $\sum^*_{i \in I} e_i$ is a tripotent because of the w*-continuity of the triple product.

To show w*-summability we note that the induced topology on a w*-closed subspace is the same as the w*-topology of that subspace, whence $\mathcal{O}$ is w*-summable iff $\mathcal{O}$ is $\sigma(W, W_*)$-summable for some w*-closed subspace W of U containing $\mathcal{O}$. We will construct a W in such a way which will allow us to apply the w*-Summability-Criterion.

To that end, assume $1, 0 \notin I$, denote the Peirce spaces relative to $\mathcal{O}$ by U_{ij} and let $e_1 \in U_{00}$ be a maximal tripotent, whose existence follows from (3.10). Then $\mathcal{O} \cup \{e_1\}$ is an orthogonal system whose Peirce spaces we denote by $\hat{U}_{ij}$. Note $\hat{U}_{ij} = U_{ij}$ for $i, j \in I$, $U_{00} = \hat{U}_{11} \oplus \hat{U}_{10}$ and $\hat{U}_{00} = 0$. There are natural projections onto some of the Peirce spaces $\hat{U}_{k\ell}$, namely for $k, \ell \in I \cup \{1\} =: \hat{I}$, $k \neq \ell$:

$$P_k = P(e_k)^2 , \quad P_{k\ell} = P(e_k, e_\ell)^2 .$$

All these projections have images in the w*-closed subspace

$$W = \cap \{\ker P(g);\ g \in \hat{U}_{k0} \text{ for some } k \in \hat{I}\}$$

because for such a g: $P(g)\hat{U}_{kk} \subset \hat{U}_{00} = 0 = P(g)\hat{U}_{\ell\ell} = P(g)\hat{U}_{k\ell}$, in particular $\mathcal{O} \cup \{e_1\} \subset W$. We therefore obtain a family of w*-continuous orthogonal projections which, as we claim, is also faithful: Suppose $P_k w = 0 = P_{k\ell} w$ for all $k, \ell \in \hat{I}$ and $w \in W$, hence $w \in U_1(e_k) \oplus U_0(e_k)$ for all $k \in \hat{I}$ and the Peirce decomposition of w with respect to (e_k, e_ℓ), $k \neq \ell$, is $w = w_{(10)} \oplus w_{(01)} \oplus w_{(00)}$, where $w_{(ij)} \in U_i(e_k) \cap U_j(e_\ell)$, since $P_{k\ell} w = w_{(11)} = 0$. Therefore $w_{(10)} \in U_0(e_m)$ for every $m \neq k$, i.e. $w_{(10)} \in \hat{U}_{k0}$. Then by definition of W, $0 = P(w_{(10)})w = P(w_{(10)})w_{(10)} \oplus P(w_{(10)})w_{(00)}$, consequently $P(w_{(10)})w_{(10)} = 0$ forcing $w_{(10)} = 0$ by (3.5). But then $w \in \bigcap_{k \in \hat{I}} U_0(e_k) = \hat{U}_{00} = \{0\}$, proving faithfulness of the family $(P_k|W, P_{k\ell}|W;\ k, \ell \in \hat{I})$. Finally, for every finite subset $B \subset I$ the element $e_B = \sum_{i \in B} e_i$ is a tripotent, whence $\|e_B\| = 1$ by (3.5). Therefore $\mathcal{O}$ is w*-summable by the w*-Summability-Criterion 3.10. ∎

As an obvious application we can prove that the Peirce sum of every orthogonal system is w*-dense ([15]3.17), more generally we have:

Corollary 3.12. Let $\mathcal{O} = (e_i)_{i \in I}$, $0 \notin I$, be a family of orthogonal tripotents in a JBW*-triple U with Peirce spaces U_{ij}. Then every w*-closed ideal W is the w*-closure of the span of its Peirce components

$$W = \overline{[\oplus_{i,j\in I\cup\{0\}} (W \cap U_{ij})]}^*,$$

in particular every $u \in U$ *is the* w*-*sum of its Peirce components relative to* $\mathcal{O}$:

$$u = u_{00} + \sum^*_{i\in I}[u_{ii} + u_{i0} + \sum^*_{j\in I,\ j\neq i} \tfrac{1}{2}u_{ij}]$$

Proof. ([15]3.17) By Lemma 3.11 $e = \sum^*_{i\in I} e_i$ is a tripotent. Since $L(e,e)W \subset W$ we have

$$W = W_2(e) \oplus W_1(e) \oplus W_0(e), \quad W_\rho(e) = W \cap U_\rho(e),$$

and it suffices to consider $w \in W_\rho(e)$.

For $\rho = 2$ we get $w = P(e)^2w =$ (by Theorem C) $\sum^*_{i\in I} [P(e_i)^2w + \frac{1}{2}\sum^*_{j\in I,\ i\neq j} P(e_i,e_j)^2w]$ where $w_{ii} = P(e_i)^2w \in W \cap U_{ii}$ and $w_{ij} = P(e_i,e_j)^2w \in W \cap U_{ij} = W \cap U_{ji}$ for $i \neq j$. For $\rho = 1$ we obtain $w = \{eew\} = \sum^*_{i\in I}\{e_ie_iw\}$ with $w_{i0} = \{e_ie_iw\} \in U_{i0} \cap W$ since $e_i \in U_2(e)$ and therefore $U_1(e) \subset U_1(e_i) \oplus U_0(e_i)$ and $U_1(e) \cap U_1(e_i) \subset U_0(e_j)$ for $i \neq j$. Finally, for $\rho = 0$ we get $w \in U_{00}$. ∎

For any family $(U_\alpha)_{\alpha\in A}$ of JB*-triples we may form the ∞-sum (see (3.14))

$$U^\infty = \oplus^\infty_{\alpha\in A} U_\alpha = \{u \in \Pi_{\alpha\in A} U_\alpha;\ \|u\|_\infty < \infty\}$$

where

$$\|u\|_\infty = \sup_{\alpha\in A} \|u_\alpha\|.$$

With respect to the norm $\|\cdot\|_\infty$ U^∞ is a JB*-triple ([15]2.20). Indeed, (3.1)-(3.3) are trivially verified and (3.4), (3.5) follow from (3.7): For every $u = (u(\alpha))_{\alpha\in A}$ the closed subtriple U_u generated by u is $\oplus^\infty_{\alpha\in A} (U_\alpha)_{u(\alpha)}$ which is isometrically isomorphic to the JB*-triple of a commutative C*-algebra. If every U_α has a pre-dual $U_{\alpha*}$ then $\{u \in \Pi_{\alpha\in A} U_{\alpha*};\ \sum_\alpha \|u_\alpha\| < \infty\}$ is a predual for U^∞, whence U^∞ is a JBW*-triple ([15]3.12).

Lemma 3.13. ([15]5.2) *Let* $(U_\alpha)_{\alpha\in A}$ *be a family of* w*-*closed ideals of a* JBW*-*triple* U *with* $U_\alpha \cap U_\beta = 0$ *for* $\alpha \neq \beta$. *Then* $\sum^*_{\alpha\in A} U_\alpha$, *the* w*-*closure of the algebraic sum* $\sum_{\alpha\in A} U_\alpha$, *is an ideal of* U *and isometrically isomorphic to* $\oplus^\infty_{\alpha\in A} U_\alpha$. *In particular, for every* $u \in \sum^*_{\alpha\in A} U_\alpha$

$$u = \sum^*_{\alpha\in A} u_\alpha, \quad u_\alpha \in U_\alpha$$

Proof. Since $L(x,x)$ leaves $\sum^*_{\alpha\in A} U_\alpha$ invariant, it is a w*-closed ideal of U by the Complete-Reducibility-Theorem 3.5. For the following we may assume $U = \sum^*_{\alpha\in A} U_\alpha$. For every $\alpha \in A$ define the projection P_α using the decomposition $U = U_\alpha \oplus U_\alpha^\perp$ from (3.13). Because $\|u\| \geq \|P_\alpha u\|$ we have $\|P_\alpha\| \leq 1$ and we get a well-defined linear map

$$\phi: \; U \to \oplus^\infty_{\alpha\in A} U_\alpha: \;\; u \to (P_\alpha u = u_\alpha)_{\alpha\in A}$$

We show that ϕ is injective: If $u_\alpha = 0$ for all $\alpha \in A$, then $u \in W = \bigcap_{\alpha\in A} U_\alpha^\perp$ which is a w*-closed ideal such that $\sum_{\alpha\in A} U_\alpha \subset U_0(g)$ for every tripotent $g \in W$. Since $U_0(g)$ is w*-closed we have $U_0(g) = U$ forcing $W = 0$.

Next we prove surjectivity of ϕ. Every family $(u_\alpha) \in \oplus^\infty_{\alpha\in A} U_\alpha$ is w*-summable by the w*-Summability-Criterion 3.10.: assumption i) follows from $U_\beta \subset U_\alpha^\perp$ for $\alpha \neq \beta$, ii) $\leftrightarrow$ injectivity of ϕ, and by repeated application of (3.13) we have $\|\sum_{\alpha\in B} u_\alpha\| = \max_{\alpha\in\beta} \|u_\alpha\| \leq \sup_{\alpha\in A} \|u_\alpha\| < \infty$, i.e. iv) and v) in the Criterion 3.10 hold. Since $P_\beta(\sum^*_{\alpha\in A} u_\alpha) = \sum^*_{\alpha\in A} P_\beta u_\alpha = u_\beta$ we see that ϕ is surjective, and at the same time that ϕ is an isometry: This is just (3.17).

Finally, ϕ is also a Jordan triple isomorphism: For $u, v \in U$ we have $P(u)v = P(u)(\sum^*_{\alpha\in A} v_\alpha) =$ (by (Theorem C) $\sum^*_{\alpha\in A} P(u)v_\alpha = \sum^*_{\alpha\in A} P(u_\alpha)v_\alpha$ whence $\phi(P(u)v) = (P(u_\alpha)v_\alpha)_{\alpha\in A} = P(\phi u)\phi v$. ∎

The existence of the decomposition (3.18) in the following Structure Theorem is also proven in [9] Prop. 2 using a different method.

Structure Theorem 3.14 for atomic JBW*-triples. (a) Every atomic JBW*-triple U has a unique decomposition

(3.18) $\qquad U = \oplus^\infty_\alpha U_\alpha$

where $\{U_\alpha;\ \alpha \in A\}$ is the set of all w*-closed and w*-simple ideals of U. Every w*-closed ideal of U is a sum of some of the U_α's.

(b) Every atomic JBW*-triple U contains a grid E with a w*-dense cover,

(3.19) $\qquad U = \overline{C_U(E)}^*$.

E is a union of standard grids E_α:

(3.20) $\qquad E = \bigcup_\alpha E_\alpha, \quad E_\alpha = E \cap U_\alpha.$

Moreover:

i) E is a Peirce-dense atomic grid with minimal tripotents.

ii) U is w*-simple iff E is a standard grid.

iii) $(P(e)^2;\ e \in E)$ is a faithful family of orthogonal contractive w*-continuous projections.

Proof. a) and b) are applications of the results in II § 4, so we verify the assumptions needed there: By (3.11) U is tripotential, therefore (II.4.1) and (II.4.2) follow from Corollary I.5.4. For (II.4.3) we first observe that for every cog $E \subset U$ every Peirce space

$U_J(E)$ is a w*-closed subsystem of U and thus either vanishes or contains non-zero tripotents by (3.10). Let $\mathcal{O} = (e_i)_{i\in I}$ be a maximal orthogonal system of minimal tripotents. If $U_{00} = \cap_{i\in I} U_0(e_i) \neq 0$ this Peirce space contains a non-zero tripotent e such that $U_2(e) \subset U_{00}$ but no minimal tripotent lies in U_{00} by maximality of $\mathcal{O}$. Hence, by Lemma 3.9, every minimal tripotent $c \in U$ lies in the w*-closed subsystem $U_0(e) \oplus U_1(e) = \ker(Id-2L(e,e))$ contradicting atomicity of U.

We can now apply Lemma II.4.2: The Peirce sum of $\mathcal{O}$ is a direct sum of ideals V_α in $PS_U(\mathcal{O})$. By Corollary 3.12 $PS_U(\mathcal{O})$ is w*-dense in U, whence $\overline{V_\alpha}^{*} =: U_\alpha$ is a w*-closed ideal of U, the w*-closure of the sum of the Peirce spaces of $\mathcal{O}$ which lie in V_α. Therefore $U_\alpha \cap U_\beta = 0$ for $\alpha \neq \beta$, whence $U = \sum^*_{\alpha\in A} U_\alpha \cong \oplus^\infty_{\alpha\in A} U_\alpha$ by Lemma 3.13. By Corollary 3.12 we have $W = \overline{\oplus_{i,j} (W \cap U_{ij})}^{*}$ for every w*-closed ideal W, where $W \cap U_{ij}$ is either zero or contains a non-zero tripotent by (3.10). Hence, Lemma II.4.2 shows that W is the w*-sum of some of the U_α's, in particular every U_α is w*-simple and the decomposition (3.18) is unique.

b) By a) it is enough to consider the case of a w*-simple U. We also know that every $u \in U$ is a w*-sum of elements from the Peirce spaces U_{ij}. By the Main Construction Theorem II.4.7 we may assume that $\mathcal{O}$ is imbedded in a Peirce-dense atomic standard grid E with minimal tripotents, so by Theorem 3.3 all Peirce spaces relative to E are one-dimensional. Moreover, it follows from II.4.7 that in case E is a symplectic, hermitian, Bi-Cayley or Albert grid, E covers $PS_U(\mathcal{O})$, thus (3.19) holds in these cases. In the rectangular resp. quadratic form case we have the density properties of Lemma II.4.5 resp. II.4.4. In particular, (3.19) holds in the rectangular case as soon as we proved that if $C = (c_\ell;\ell\in L) \subset U_{k0}$ is a maximal collinear family, then $u = \sum^*_{\ell\in L} u_\ell$, $u_\ell \in \mathbb{C}c_\ell$, for every $u \in U_{k0}$. This is yet another application of the w*-Summability-Criterion: $P_\ell = P(c_\ell)^2|U_{k0}$ is a family of w*-continuous orthogonal projections which is faithful: Because every c_ℓ is also maximal in U_{k0} we have $\bigcap_{\ell\in L} \ker P_\ell = \bigcap_{\ell\in L} (U_{k0})_1(c_\ell)$ which vanishes by maximality of C. If we know that $(P_\ell(u);\ \ell\in L)$ is w*-summable, then $u = \sum^*_{\ell\in L} P_\ell(u)$ follows from faithfulness, thus, by the w*-Summability- Criterion, it is enough to prove

(*) $\qquad \|\sum_{\ell\in B} P_\ell(u)\| \leq \|u\|$ for every $B \in L_{fin}$

and every $u \in U_{k0}$. To this end we first note that for $u \in U_{k0}$ we have $P(u)u =$ (by (I.1.28)) $\{\{u\,u\,e_k\}e_k\,u\} - \{e_k\,u\,P(u)e_k\} \in \mathbb{R}u$ since $\{u\,u\,e_k\} \in U^+_{kk} = \mathbb{R}e_k$ by (I.1.24) and $P(u)e_k \in U_{00} = 0$. Thus every $u \in U_{k0}$ is a scalar multiple of a tripotent, which is collinear to e_k and therefore minimal. We apply this to the Peirce decomposition of u relative to $(c_\ell)_{\ell\in B}$: $u = u_2 + u_1$ with $u_2 = \sum_{\ell\in B} P_\ell(u)$ and $u_1 \in \bigcap_{\ell\in B} (U_{k0})_1(c_\ell)$. We obtain tripotents e, f where $u_2 = se$, $u_1 = tf$ for some s, $t \in \mathbb{R}$ which we may assume to be non-zero. Since $P(u_2)u_1 = 0$ we have $P(e)f = 0$, so (e,f) are rigidly imbedded by Criterion I.1.12. Then $u^3 = P(u)u = P(se)u + P(se,tf)u + P(tf)u = s^3e + s^2tf + st^2e + t^3f = (s^2 + t^2)u$, thus by (3.5) $\|u\|^2 = s^2+t^2 = \|u_2\|^2 + \|u_1\|^2$ because $\|e\| = 1 = \|f\|$, implying (*).

For the quadratic form case (3.19) follows from the next lemma. The remaining assertions are clear now, except maybe contractiveness in iii) which follows from $4P(e)^2 = (Id - \exp(\frac{\pi}{2}iL(e,e))(Id + \exp(\pi iL(e,e))$ and (3.3). ∎

<u>Lemma 3.15</u> U <u>is a</u> JB*-<u>triple covered by a triangle with minimal tripotents iff</u> U <u>is a complex Hilbert space with respect to a scalar product</u> σ <u>and has a conjugation</u> $^-$ (i.e. $^-$ <u>is a</u> $\mathbb{C}$-<u>antilinear, involutorical map with</u> $\sigma(\bar{x},y) = \sigma(\bar{y},x)$) <u>such that the Jordan triple product of</u> U <u>is given by</u>

(3.21) $$P(x)y = 2\sigma(x,y)x - \sigma(x,\bar{x})\bar{y}$$

<u>and the</u> JB*-<u>norm of</u> U <u>is</u>

(3.22) $$\|x\|^2 = \sigma(x,x) + \sqrt{\sigma(x,x)^2 - |\sigma(x,\bar{x})|^2}$$

<u>In this case we have</u>:

i) <u>the</u> JB*-<u>norm and the Hilbert norm are equivalent, in particular</u> U <u>is a</u> JBW*-<u>triple</u>.

ii) U <u>is (algebraically) simple</u>,

iii) U <u>contains an even- or odd-dimensional quadratic form grid</u> $\mathcal{Q}$ <u>which is a Hilbert space basis of</u> U:

$$U = \oplus^2_{e\in\mathcal{Q}} \mathbb{C}\, e,$$

<u>in particular</u> (3.19) <u>holds</u>.

<u>Proof</u>. Let U be covered by the triangle $(e_0;e_1,e_2)$ with minimal e_i. An application of results from [43] shows that U is a quadratic form triple: By (3.5) U does not contain absolute zero divisors and by Theorem 3.3 $U_2(e_i) = \mathbb{C}\, e_i$, i = 1,2, is a simple Jordan triple system. Thus U is (algebraically) simple by [43]1.15, e_0 is faithful by [43]1.12 and in the notation of [43] $\Delta_1(U_{11};U_{12}) = 0$ by [43]3.8 since clearly $x_{11}\cdot u_{12} = (x_{11})^*\cdot u_{12}$. Hence, by the Clifford Criterion [43]3.6 U is a quadratic form triple, and since $\frac{1}{2} \in \mathbb{C}$ it is of type $[\mathbb{C},^-,U,q,S]$ as defined in Example I.1.6. Note that $^-$ is complex conjugation.

Using these data we define for $x,y \in U$ $\bar{x} := Sx$ and $2\sigma(x,y) := q(x,Sy)$. Then $P(x)y = q(x,Sy)x - q(x)Sy = 2\sigma(x,y)x - \sigma(x,x)\bar{y}$, i.e.

(3.21) holds, and σ is hermitian: $\sigma(x,y) = q(x,Sy) = \overline{q(Sx,y)} = \overline{\sigma(y,x)}$. To show positivity of σ we note

$$U = U_+ \oplus iU_+, \qquad \text{where } U_+ = \{x \in U;\ \bar{x} = x\},$$

so for $0 \neq x = x_+ + iy_+$ we have $2\sigma(x,x) = q(x_+,x_+) + q(y_+,y_+) > 0$ because $L(x_+,x_+) = q(x_+,x_+)\mathrm{Id}$ has non-negative spectrum by (3.4) and $L(x_+,x_+) \neq 0$ for $x_+ \neq 0$ by (3.5).

Next we show (3.22). For $\bar{x} = \pm x$ this formula specializes to $\|x\|^2 = \sigma(x,x)$ which follows from $P(x)x = \sigma(x,x)x$ and (3.5). In the same way the formula follows for $\sigma(x,\bar{x}) = 0$. Hence we may assume $\bar{x} \neq \pm x$, $\sigma(x,\bar{x}) \neq 0$, i.e. x and $\tilde{x} = \sigma(x,\bar{x})\bar{x}$ are linear independent over $\mathbb{C}$. Let q_i, $i = 1,2$, be the two distinct real roots of $0 = q^2 + 2\sigma(x,x)q + |\sigma(x,\bar{x})|^2$ and put $e_i = \beta_i(-q_i x + \tilde{x})$ where β_i is determined by $\sigma(e_i,e_i) = \frac{1}{2}$. Then $(e_1,e_2) \subset \mathbb{R}x \oplus \mathbb{R}\tilde{x}$ satisfy $\sigma(e_i,\bar{e}_i) = 0 = \sigma(e_1,e_2)$ and $\bar{e}_i = \gamma_i e_j$ for $|\gamma_i| = 1$, $i + j = 3$. Therefore, (e_1,e_2) is a pair of orthogonal tripotents and x has a minimal decomposition $x = se_1 + te_2$ with $s,t \in \mathbb{R}$. By (3.13) applied to $U = \mathbb{C}x \oplus \mathbb{C}\tilde{x} = \mathbb{C}e_1 \oplus \mathbb{C}e_2$ we have $\|x\| = \max(|s|,|t|)$ since $\|e_i\| = 1$. On the other hand $\sigma(x,x) +$

$$\sqrt{\sigma(x,x)^2 + |\sigma(x,\bar{x})|^2} = \tfrac{1}{2}(s^2+t^2+|s^2-t^2|) = \max(|s|,|t|)^2.$$

By (3.22) we have $\sigma(x,x) \leq \|x\|^2 \leq 2\sigma(x,x)^2$ showing the equivalence of the Hilbert norm and the JB*-norm. In particular, U is its own pre-dual and hence a JBW*-triple. A maximal system of collinear minimal tripotents in U is nothing else but a maximal system of vectors $(e_i, i \in I)$ satisfying $\sigma(e_i,e_j) = \frac{1}{2}\delta_{ij}$ and $\sigma(e_i,\bar{e}_j) = 0$ for all $i,j \in I$. For a fixed $j \in I$ the family $\{\bar{e}_j\} \cup \{e_i; i \in I\}$ generates an even-dimensional quadratic form grid $\mathcal{Q}_e(I)$ (Theorem II.2.12) which is either Peirce-dense or can be imbedded in a Peirce-dense odd-dimensional quadratic form grid $\mathcal{Q}_o(I)$. In both cases every $x \in U$ is a Hilbert sum $x = \sum^2_{e \in \mathcal{Q}} \alpha_e e$, in particular (3.19) holds.

Finally the converse direction: Given $(U,\sigma,\bar{\ })$, then (3.21) defines a Jordan-*-triple structure on U such that (U,σ) is a positive Hilbert triple, namely a spin factor - see Example 2.13.IV.a). That (3.22) defines a norm is obvious with the exception of the triangle inequality $\|x+y\| \leq \|x\| + \|y\|$ which can be proven by a lengthy estimation or by noting that it suffices to consider the finite-dimensional subsystem $\mathbb{C}x + \mathbb{C}\bar{x} + \mathbb{C}y + \mathbb{C}\bar{y}$ and that $\|x\| = \max(|s|,|t|)$ where $x = se_1 + te_2$, $s, t \in \mathbb{R}$ and (e_1,e_2) is a pair of orthogonal tripotents. Thus $\|\cdot\|$ is the "spectral norm" introduced in [32]3.17. The remaining defining properties of a JB^*-triple are

(3.2): The left multiplication is obviously continuous with respect to the Hilbert space topology, hence also with respect to the equivalent spectral norm.

(3.3): $L(z,z)$ is hermitian with respect to σ and also with respect to $q(x,y) = 2\sigma(x,\bar{y})$, thus $\exp(itL(z,z))$ is an isometry with respect to $\|\cdot\|$.

(3.4): For $z = se_1 + te_2$ as above the spectral values of $L(z,z)$ are $2s^2$, $2t^2$, $s^2 + t^2$.

(3.5): Follows from $P(z)z = s^3e_1 + t^3e_2$ and $\|z\| = \max(|s|,|t|)$. ∎

We want to derive a classification of atomic JBW*-triples. By the Structure Theorem 3.14 every such U contains a grid E with minimal tripotents and a w*-dense cover $C_U(E) = \oplus_{e\in E}\, \mathbb{C}\, e$. The structure of $C_U(E)$ is known, since we know E. To conclude from $C_U(E)$ to U we will use a modified version of an extension theorem due to G. Horn. We will also need

<u>Lemma 3.16</u> ([15]2.11) <u>Every algebraic isomorphism between</u> JB*-<u>triples is an isometry</u>.

<u>Proof</u>. For any such isomorphism ϕ we have $\phi L(x,x)\phi^{-1} = L(\phi x,\phi x)$ whence the spectra of $L(x,x)$ and $L(\phi x,\phi x)$ coincide, hence also the spectral radii $r(L(x,x)) = r(L(\phi x,\phi x))$. But for hermitian operators $r(L(x,x)) = \|L(x,x)\|$ ([10]4.1), thus $\|x\|^2 = \|\phi x\|^2$ by (3.6). ∎

<u>Remark</u>. The converse of 3.16 is proven in [36]5.5. Moreover one can show that any homomorphism between JB*-triples is continuous. However, we will not need this in the following.

It will be helpful to use the notation

$$X^r = \{x \in X;\ \|x\| \leq r\}, \qquad r \geq 0$$

for the closed ball of radius r in a normed space X.

<u>Extension Theorem 3.17</u>. <u>Let</u> U_k, $k = 1,2$, <u>be two atomic</u> JBW*-<u>triples containing grids</u> E_k <u>as described in the Structure Theorem</u> 3.14. <u>Assume that the unit ball of the cover</u> $C_k = C(E_k)$ <u>is</u> w*-<u>dense in</u> U_k^1,

(3.23) $$U_k^1 = \overline{C_k^1}^{\,*}, \quad k = 1,2.$$

<u>Then every triple isomorphism</u> $\phi: C_1 \to C_2$ <u>has a unique extension to an isometric triple isomorphism</u> $\Phi: U_1 \to U_2$.

<u>Proof</u>. (modification of [15]4.1, 4.5). We first show:

i) $\phi: C_1 \to C_2$ is an isometry.
Since every $x \in C_1$ is a finite sum the structure of standard grids shows that x lies in a finite-dimensional subsystem $U(x) \subset C_1$ which is a JB*-triple with respect to the induced norm by (3.9). Hence 3.16 applied to $\phi|U(x)$ shows $\|\phi(x)\| = \|x\|$ for every $x \in C_1$.
ii) $\phi: C_1^1 \to C_2^1$ is a homeomorphism with respect to the induced w*-topologies.
Let V_k be the linear span of all $V_{ke} = \{\xi \in U_{k*} \subset U_k^* \ ; \ \xi = \xi \circ P(e)^2\}$, $e \in E_k$. Since $(P(e)^2; e \in E_k)$ is a faithful family of orthogonal w*-continuous projections, it follows as in the proof of the w*-Summability- Criterion 3.10 that the relative w*-topology on U_k^1 coincides with the topology induced by $\sigma(U_k, V_k)$ on U_k^1. Therefore $\phi|C_1^1$ will be w*-continuous (which implies ii)) if $V_2 \circ \phi \subset V_1|C_1$. But this is easily seen: Let $\xi \in V_{2e}$ for some $e \in E_2$ and put $f = \phi^{-1}(e)$. Then $\zeta := \xi \circ \phi \circ P(f)^2$ is w*- continuous, hence $\zeta \in V_{1f} \subset V_1$, and $\xi \circ \phi = \zeta|C_1$.
iii) Let $r > 0$. We know that C_k^r is w*-dense in U_k^r and $\phi: \ C_1^r \to C_2^r$ is a w*-homeomorphism. Since U_k^r is w*-complete there exists a unique extension of $\phi|C_1^r$ to a homeomorphism $\Phi_r: U_1^r \to U_2^r$. Define Φ by $\Phi|U_1^r = \Phi_r$. Since C_1^r is w*- dense in U_1^r for all $r > 0$, it easily follows that $\Phi: U_1 \to U_2$ is an isometric triple isomorphism. ⊞

Isomorphism Theorem 3.18. Two atomic JBW*-triples U_k = k=1,2, with grids E_k as described in the Structure Theorem 3.14 are isometrically isomorphic if E_1 and E_2 are the same union of the same standard grids.

Proof. We may of course assume that U_1 and U_2 are w*-simple. Then E_1 and E_2 are just different versions of the same standard grid, whence there exists a triple isomorphism $\phi: C_1 \to C_2$ where $C_k = C_k(E_k)$ is the cover of E_k. Obviously we are done if E_k is the Bi-Cayley or Albert grid. For the remaining cases we will use the Extension Theorem 3.17, so we have to verify the assumption (3.23) there. For the quadratic form grids (3.23) easily follows from Lemma 3.14, so we are left with the cases of rectangular, symplectic or hermitian grids.

For easier notation we put $U = U_1$ and $E = E_1$. Let $\mathcal{O} = (e_i)_{i \in I} \subset E$ be a maximal orthogonal system, put $e = \sum^*_{i \in I} e_i$ and denote the Peirce spaces of $\mathcal{O}$ by U_{ij}. We will show:

(*) $\{x \in \oplus_{i,j \in I} U_{ij}; \ \|x\| \le 1\}$ is w*-dense in $U_2(e)^1$.

For any finite subset $F \subset I$ let $P_F = P(\sum_{i \in F} e_i)^2$. Then $(P_F; F \in I_{fin})$ is a directed family of w*-continuous contractive projections which is faithful by Corollary 3.12. It now follows as in the proof of the w*-Summability-Criterion 3.10 (details in [15]4.4) that x is the w*-limit of $(P_F(x); F \in I_{fin})$ which implies (*).

For a rectangular, symplectic or hermitian grid $\oplus_{i,j\in I} U_{ij} = C_U(E) \cap U_2(e)$. Thus (*) implies (3.23) for $U_2(e)$ and we can extend $\phi | C_U(E) \cap U_2(e)$ to $U_2(e)$. If we can also prove that for a symplectic or rectangular grid $\phi | C_U(E) \cap U_1(e)$ can be extended to $U_1(e)$, then the extension will be a global isomorphism since it is w*-continuous and ϕ is an isomorphism. That the extension exists follows by applying the Extension Theorem 3.17 to $U_1(e)$ and $E \cap U_1(e)$. Since $E \cap U_1(e)$ in the symplectic case is a maximal collinear system, the symplectic case is a special case of the rectangular case.

To verify (3.23) in the rectangular case $E = R(I,J)$ we note that every $x \in U_1(e)$ is a w*-sum, $x = \sum^{*}_{i\in I} x_{i0}$ by Corollary 3.12. For any finite subset $F \subset I$ and $e_F = \sum_{i\in F} e_i$ we have $\sum_{i\in F} x_{i0} = L(e_F,e_F)x$. Because $2L(e_F,e_F)|U_1(e) = (Id - \exp(\pi i L(e_F,e_F)))|U_1(e)$ we see that $L(e_F,e_F)|U_1(e)$ is contractive, whence $\|\sum_{i\in F} x_{i0}\| \leq \|x\|$. It is therefore enough to consider $x_{i0} \in U_{i0}$. But the w*-denseness of $[U_{i0} \cap C_U(E)]^1$ in U^1_{i0} was already shown in the proof of the Structure Theorem 3.14. ∎

In view of the Structure Theorem 3.14 and the Isomorphism Theorem 3.16 all what is left to do for a classification of w*-simple atomic JBW*-triples is to give an example for each of the 6 types of standard grids. These examples are well-known - they appear already in [23]§I, see also [15]§6.

Examples 3.19. For the first 3 types we use the fact that B(H), the bounded linear operators on a complex Hilbert space H, is a W*-algebra with respect to the operator norm and the adjoint map as involution ([51]1.15). Hence, by Example 3.1 and (3.9) B(H) and every w*-closed subtriple of B(H) are JBW*-triples.

type I): For complex Hilbert spaces H, K the Banach space B(H,K) of all bounded linear operators from H to K is w*-closed in $B(H \oplus K)$, hence a JBW*-triple. Let $(e_j)_{j\in J}$ resp. $(e_i)_{i\in I}$ be Hilbert space bases for H resp. K. Then the maps

$$E_{ij} = e_i \otimes e_j^* : H \to K: x \to (x,e_j)e_i$$

form a rectangular grid of type $R(I,J)$ which is w*-dense in the sense of (3.19) and consists of minimal tripotents. In particular is B(H,K) atomic and w*-simple.

type II): As in I) let $(e_i)_{i\in I}$ be a Hilbert space basis of K. Then $\sum_{i\in I} s_i e_i \to \sum_{i\in I} \bar{s}_i e_i$ defines a conjugation on K and induces a conjugation $x \to \bar{x}$ on $B(K)$. Then $A(K) = \{x \quad B(K);\ \bar{x} = -x^*\}$ is a w^*-simple atomic JBW*-triple which contains a w^*-dense symplectic grid $S(I) = \{F_{ij};\ i < j\}$ where $F_{ij} = e_i \otimes e_j^* - e_j \otimes e_i^*$ and $(I,<)$ is a total order.

type III): In the notation of type II) $H(K) = \{x \in B(K);\ \bar{x} = x^*\}$ is a w^*-simple atomic JBW*-triple which contains a w^*-dense hermitian grid $H(I) = \{H_{ij};\ i \leq j\}$ where $H_{ii} = e_i \otimes e_i^*$ and $H_{ij} = e_i \otimes e_j^* + e_j \otimes e_i^*$ for $i \neq j$.

type IV): This is the example of Lemma 3.15.

type V): The Bi-Cayley triple $B(\mathbb{C},\kappa)$, κ = complex conjugation, is a simple 16-dimensional positive hermitian Hilbert triple (see Example 2.13.V)), hence a JB*-triple with respect to the spectral norm ([32]3.17). $B(\mathbb{C},\kappa)$ is covered by a Bi-Cayley grid.

type IV): The hermitian Hilbert triple $A(\mathbb{C},\kappa)$ (see Example 2.13.VI) is a simple 27-dimensional JB*-triple with respect to the spectral norm. $A(\mathbb{C},\kappa)$ is covered by an Albert grid.

Taking together the Structure Theorem 3.14 and the Isomorphism Theorem 3.18 we obtain

Classification Theorem 3.20. Every w^*-simple atomic JBW*-triple is isometrically isomorphic to one of the six types in the Examples 3.19.

This result is also obtained in [15]9.12 , however with a different proof - for example the proof there uses the elaborate structure theory of JBW*-algebras which we have completely bypassed. Instead, one can specialize our results to the case of JBW*-algebras and obtain for example a classification of atomic JBW*-algebras. The details are left to the reader.

Remark: Besides the already quoted references [8], [9], [15] and [25], JBW*-triples are also considered in the recent preprints [57] and [59]-[63].

REFERENCES

[1] E. Backes: Geometric applications of Euclidean Jordan triple systems. Manuscripta math. 42, 265-272 (1983)

[2] E. Backes, H. Reckziegel: On symmetric submanifolds of spaces of constant curvature. Math. Ann. 263, 419-433 (1983)

[3] E. Barton, R. Timoney: Weak*-continuity of Jordan triple products and applications. Preprint 1985. To appear in Math. Scand.

[4] H. Braun, M. Koecher: Jordan - Algebren. Die Grundlehren der mathematischen Wissenschaften in Einzeldarstellungen, Band 128, Springer-Verlag 1967

[5] S. Dineen: The second dual of a JB*-triple system. Complex Analysis, Functional Analysis and Approximation Theory, ed. by J. Mujica, North Holland Math. Studies, vol 125, p.67-69 (1986)

[6] J. Faulkner, J. Ferrar: Exceptional Lie algebras and related algebraic and geometric structures. Bull. London Math. Soc. 9, 1-35 (1977)

[7] D. Ferus: Symmetric submanifolds of Euclidean space. Math. Ann. 247, 81-93 (1980)

[8] Y. Friedman, B. Russo: Structure of the predual of a JBW*-triple. J. Reine u. Angew. Math. 356, 67-89 (1985)

[9] Y. Friedman, B. Russo: The Gelfand-Naimark theorem for JB*-triples. Duke Math. J. 53, 139-148 (1986)

[10] K. R. Goodearl: Notes on Real and Complex C*- algebras. Shiva mathematics series, vol.5, Shiva Publishing Limited 1982

[11] H. Hanche-Olsen, E. Størmer: Jordan operator algebras. Monographs and studies in Mathematics 21, Pitman 1984

[12] R. Hartshorne: Algebraic Geometry. Graduate Texts in Mathematics 52, Springer-Verlag 1977

[13] A. Henderson: The Twenty-Seven Lines upon the Cubic Surface. Cambridge Tracts in Mathematics and Mathematical Physics 13, Cambridge University Press 1911

[14] D. Hilbert, S. Cohn-Vossen: Anschauliche Geometrie. Springer-Verlag 1932

[15] G. Horn: Klassifikation der JBW*- Tripel vom Typ I. Dissertation Tübingen 1984. To be published as [59]-[61].

[16] R. Iordanescu: Jordan algebras with applications. Bucarest 1979

[17] B. Iochum: Cônes autopolaires et algèbres de Jordan. Lecture Notes in Mathematics 1049, Springer-Verlag 1984

[18] N. Jacobson: Composition algebras and their automorphisms. Rend. Circ. Mat. Palermo 7, 55-80 (1958)

[19] N. Jacobson: Structure and representations of Jordan algebras. American Mathematical Society Colloquium Publications Vol XXXIX, 1968

[20] N. Jacobson: Lectures on quadratic Jordan algebras. Lecture notes, Tata Institute, Bombay 1969

[21] N. Jacobson: Structure Theory of Jordan Algebras. The University of Arkansas Lecture Notes, vol. 5, Fayetteville 1981

[22] K. Johnson, W. Lichtenstein: Linear spaces on G/P. Preprint
never appeared. No copies exist, according to Ken Johnson, May 9'

[23] W. Kaup: Über die Klassifikation der symmetrischen hermiteschen Mannigfaltigkeiten unendlicher Dimension I. Math. Ann. 257, 463-486 (1981)

[24] W. Kaup: Über die Klassifikation der symmetrischen hermiteschen Mannigfaltigkeiten unendlicher Dimension II. Math. Ann. 262, 57-75 (1983)

[25] W. Kaup: A Riemann mapping theorem for bounded symmetric domains in complex Banach spaces. Math. Z. 183, 503-529 (1983)

[26] W. Kaup, H. Upmeier: Jordan algebras and symmetric Siegel domains in Banach spaces. Math. Z. 157, 179-200 (1977)

[27] E. Kleinfeld: Simple alternative rings. Ann. of Math. 58, 544-547 (1953)

[28] M. Koecher: An Elementary Approach to Bounded Symmetric Domains. Lecture Notes, Rice University, Houston 1969

[29] O. Loos: Jordan triple systems, R-spaces, and bounded symmetric domains. Bull. Amer. Math. Soc. 77, 558-561 (1971)

[30] O. Loos: Lectures on Jordan triples. Lecture notes, University of British Columbia, Vancouver 1971

[31] O. Loos: Jordan Pairs. Lecture Notes in Mathematics 460, Springer-Verlag 1975

[32] O. Loos: Bounded Symmetric Domains and Jordan Pairs. Lecture Notes, University of California, Irvine 1977

[33] O. Loos: Charakterisierung symmetrischer R-Räume durch ihre Einheitsgitter. Math. Z. 189, 211-226 (1985)

[34] O. Loos, K. McCrimmon: Speciality of Jordan triple systems. Comm. in Algebra 5, 1057-1082 (1977)

[35] Y. Manin: Cubic Forms - Algebra, Geometry, Arithmetic. North-Holland Mathematical Library vol. 4, North-Holland 1974

[36] K. McCrimmon: Quadratic Jordan algebras whose elements are all invertible or nilpotent. Proc. Amer. Math. Soc. 35, 309-316 (1972)

[37] K. McCrimmon: Jordan algebras and their applications. Bull. Amer. Math. Soc. 84, 612-627 (1978)

[38] K. McCrimmon: Book review of [31]. Bull. Amer. Math. Soc. 84, 685-690 (1978)

[39] K. McCrimmon: Peirce ideals in Jordan triple systems. Pacific J. Math. 83, 415-439 (1979)

[40] K. McCrimmon: Compatible Peirce decompositions of Jordan triple systems. Pacific J. Math. 103, 57-102 (1982)

[41] K. McCrimmon: The russian revolution in Jordan algebras. Algebras, Groups and Geometries 1, 1-61 (1984)

[42] K. McCrimmon, K. Meyberg: Coordinatization of Jordan triple systems. Comm. in Algebra 9, 1495-1542 (1981)

[43] K. McCrimmon, E. Neher: Coordinatization of triangulated Jordan triple systems. Preprint 1986. To appear in J. of Algebra.

[44] K. Meyberg: Lectures on algebras and triple systems. Lecture Notes, University of Virginia, Charlottesville 1972

[45] D. Mumford: Algebraic Geometry I, Complex Projective Varieties. Grundlehren der mathematischen Wissenschaften 221, Springer-Verlag 1976

[46] E. Neher: Involutive gradings of Jordan structures. Comm. in Algebra 9, 575-589 (1981)

[47] E. Neher: On Jordan triple systems with enough tripotents (A survey). Preprint 1981

[48] E. Neher: Jordan triple forms of Jordan algebras. Proc. Amer. Math. Soc. 87, 386 -388 (1983)

[49] E. Neher: On triagonal elements and centralizers in Jordan triple systems. J. of Algebra 90, 18-36 (1984)

[50] E. Neher: Jordan triple systems with completely reducible derivation or structure algebras. Pacific J. Math. 113, 137-164 (1984)

[51] S. Sakai: C*- algebras and W*- algebras. Ergebnisse der Mathematik und ihrer Grenzgebiete, Band 60, Springer-Verlag 1971

[52] I. Satake: Algebraic structures of symmetric domains. Princeton University Press 1980

[53] R. Schafer: An Introduction to Nonassociative algebras. Pure and Applied Mathematics 22, Academic Press 1966

[54] T. Schwarz: Einfache Jordantripelsysteme. Dissertation, Hagen 1983

[55] H. Upmeier: Symmetric Banach Manifolds and Jordan C*- algebras. North-Holland Mathematics Studies 104, North-Holland 1985

[56] E.I. Zel'manov: Primary Jordan triple systems
I Sibirskii Mat. Zh. 24, 23-37 (1983)
II Sibirskii Mat. Zh. 25, 50-61 (1984)
III Sibirskii Mat. Zh. 26, 71-82 (1985)

After submission of these notes I received resp. finished the following preprints:

[57] T. Barton, T. Dang and G. Horn: Normal representations of Banach Jordan Triple Systems. Preprint 1986.

[58] J.A. Cuenca, A. Rodriguez: Structure theory for noncommutative Jordan H*-algebras. To appear in J. of Algebra.

[59] T. Dang, Y. Friedman: Classification of JBW*-triple factors and applications. Preprint 1986.

[60] G. Horn: Characterization of the predual and ideal structure of a JBW*-triple. To appear in Math. Scand.

[61] G. Horn: Coordinatization theorems for JBW*-triples. To appear in Quart. J. Math. Oxford.

[62] G. Horn: Classification of JBW*-triples of type I. Preprint 1986.

[63] G. Horn and E. Neher: Classification of continuous JBW*-triples. Preprint 1986.

INDEX

LIST OF SYMBOLS

MIX
Papier aus verantwortungsvollen Quellen
Paper from responsible sources
FSC® C105338

If you have any concerns about our products,
you can contact us on
ProductSafety@springernature.com

In case Publisher is established outside the EU,
the EU authorized representative is:
Springer Nature Customer Service Center GmbH
Europaplatz 3, 69115 Heidelberg, Germany

Printed by Libri Plureos GmbH
in Hamburg, Germany